FARM
MANAGEMENT

Fifth Edition

FARM
MANAGEMENT

Ronald D. Kay
Professor Emeritus
Texas A&M University

William M. Edwards
Professor, Iowa State University

Patricia A. Duffy
Professor, Auburn University

 **Higher Education**

Boston   Burr Ridge, IL   Dubuque, IA   Madison, WI   New York   San Francisco   St. Louis
Bangkok   Bogotá   Caracas   Kuala Lumpur   Lisbon   London   Madrid   Mexico City
Milan   Montreal   New Delhi   Santiago   Seoul   Singapore   Sydney   Taipei   Toronto

Higher Education

FARM MANAGEMENT, FIFTH EDITION

Published by McGraw-Hill, a business unit of The McGraw-Hill Companies, Inc., 1221 Avenue of the Americas, New York, NY 10020. Copyright © 2004, 1999, 1994 by The McGraw-Hill Companies, Inc. All rights reserved. Previously published under the title of *Farm Management: Planning, Control, and Implementation.* Copyright © 1986, 1981 by McGraw-Hill, Inc. All rights reserved. No part of this publication may be reproduced or distributed in any form or by any means, or stored in a database or retrieval system, without the prior written consent of The McGraw-Hill Companies, Inc., including, but not limited to, in any network or other electronic storage or transmission, or broadcast for distance learning.

Some ancillaries, including electronic and print components, may not be available to customers outside the United States.

This book is printed on acid-free paper.

1 2 3 4 5 6 7 8 9 0 DOC/DOC 0 9 8 7 6 5 4 3

ISBN 0-07-242868-6

Publishers: *Martin J. Lange/Colin H. Wheatley*
Developmental editor: *Fran Schreiber*
Marketing manager: *Heather K. Wagner*
Senior project manager: *Rose Koos*
Lead production supervisor: *Sandy Ludovissy*
Lead media project manager: *Judi David*
Senior coordinator of freelance design: *Michelle D. Whitaker*
Cover designer: *Rebidas Design & Associates*
Cover image: © *Andy Sacks/Stone*
Senior photo research coordinator: *Lori Hancock*
Compositor: *GAC Indianapolis*
Typeface: *10/12 Times Roman*
Printer: *R. R. Donnelley Crawfordsville, IN*

Opening photos for chapters 2, 3, 4, 5, 6, 7, 10, 11, 13, 14, 15, 16, 17, 20, and 21; parts 1 and 4; and title page courtesy of U.S. Department of Agriculture
Opening photos for chapters 1, 8, 9, 12, and 18 and parts 2, 3, and 5 courtesy of USDA Natural Resource Conservation Service
Opening photo for chapter 19 courtesy of Kirk Withrow
Opening photo for chapter 22 courtesy of USDA Agricultural Research Service

Library of Congress Cataloging-in-Publication Data

Kay, Ronald D.
   Farm management / Ronald D. Kay, William M. Edwards, Patricia A. Duffy. — 5th ed.
     p. cm.
   Includes bibliographical references and index.
   ISBN 0-07-242868-6 (hard copy : alk. paper)
    1. Farm management. I. Edwards, William M. II. Duffy, Patricia A. III. Title.

S561.K36 2004
630'.68—dc21
                                   2003010863
                                   CIP

www.mhhe.com

CONTENTS

PREFACE

Farms and ranches, like other businesses, require sound management to survive and prosper. The continual development of new agricultural technologies means that farm and ranch managers must stay informed of the latest advances and decide whether or not to adopt them. Adopting a risky, unproven technology that fails to meet expectations can cause financial difficulties or even farm financial collapse. On the other hand, failing to adopt a profitable new technology will put the farm business at a competitive disadvantage that could also prove disastrous in the long run. In addition, changing public policies regarding environmental protection, taxes, and price supports can make certain alternatives and strategies more or less profitable than they had been previously. Finally, changes in consumer tastes, the demographic makeup of our population, and world agricultural trade policies affect the demand for agricultural products.

The continual need for farm and ranch managers to keep current and update their skills motivated us to write this fifth edition. It was also a good time to add another author to provide additional farm management experience, skills, and viewpoints to the writing process. Dr. Patricia Duffy from Auburn University joins us as a third author for this edition. She has considerable experience teaching farm management and other agricultural economics courses and is familiar with previous editions of this text. Dr. Duffy has contributed not only to specific chapters in the text, but also did the PowerPoint presentations that are now available for instructors. We look forward to her contributions in future editions.

This book is divided into five parts. Part I begins with the chapter "Farm Management in the Twenty-First Century." It describes some of the forces and technology, which are driving the changes we see in agriculture. By reading this chapter students will find both an incentive to study farm management and an appreciation for the management skills modern farm managers must have or acquire. Part I concludes with an explanation of the concept of management and the decision-making process, with an increased emphasis on the importance of strategic decision-making.

Part II presents the basic information needed to measure management performance, financial progress, and the financial condition of the farm business. It discusses how to collect and organize accounting data and how to construct and analyze balance sheets and income statements. In response to several suggestions, income tax depreciation (MACRS) has been added to Chapter 4, which discusses depreciation methods and asset valuation.

As in the fourth edition, Part III contains three chapters on basic microeconomic principles and four chapters on budgeting and planning tools. The topics in this part provide the basic tools needed to make good management decisions. Students will learn how and when economic principles can be used in management decision-making, along with the importance of the different types of economic costs. The discussion of economies of size has been revised and expanded. Practical use of budgeting is emphasized in the chapters on enterprise, partial, whole farm, and cash flow budgets. An expanded discussion on using sensitivity analysis with partial budgets has been added, and linear programming is covered in more detail than in previous editions.

Topics necessary to increase a manager's decision-making skills are included in Part IV. Farm business organization, analyzing investments, managing risk, income tax management, and whole farm business analysis are discussed. The chapter on income tax management has been updated with the latest changes available. Data from an actual farm is now used to demonstrate the analysis process in the chapter on whole farm business analysis. The chapter on investment analysis includes a new discussion of the concepts of annual equivalent and capital recovery values.

Part V discusses the management alternatives and decisions related to acquiring the resources needed on farms and ranches. This part includes chapters on capital and credit, land, human resources, and machinery. The human resource chapter includes new sections on improving managerial capacity and bridging the cultural barriers that may be encountered in managing agricultural labor.

Last, but certainly not least, this edition's most obvious changes are the new cover design and book style. We hope users will find the new cover to be more attractive and distinguishable. Between the covers is a new style of double-column pages and other related style changes. These changes should increase the readability of the text and improve student comprehension.

Much of the material in the fourth edition's instructor manual has been updated and moved to the text's fifth edition website. An electronic PowerPoint presentation covering each chapter, a test bank, and answers to the end of chapter questions can be found at www.mhhe.com/kay5e

The authors would like to thank the instructors who have adopted the previous edition for their courses and the many students who have used it both in and out of formal classrooms. Your comments and suggestions have been carefully considered and many were incorporated in this edition. Suggestions for future improvements are always welcome. A special thanks goes to the following McGraw-Hill reviewers for their many thoughtful ideas and comments provided during the preparation of this edition.

Gary W. Brester
Montana State University—Bozeman
Stephen B. Davis
University of Minnesota—Crookston
Carl R. Dillon
University of Kentucky
Duane K. Jewell
Northwest Missouri State University
Hezekiah S. Jones
Alabama A & M University
Donald M. Nixon
Texas A & M University—Kingsville
Guido van der Hoeven
North Carolina State University

Ronald D. Kay
William M. Edwards
Patricia A. Duffy

Ronald D. Kay is a professor emeritus in the Department of Agricultural Economics at Texas A&M University. Dr. Kay taught farm management at Texas A&M University for 25 years, retiring at the end of 1996. He was raised on a farm in southwest Iowa and received his B.S. in agriculture and Ph.D. in agricultural economics from Iowa State University. Dr. Kay has experience as both a professional farm manager and a farm management consultant, and he maintains an active interest in a farming operation. He is a member of several professional organizations, including the American Society of Farm Managers and Rural Appraisers, where he is a certified instructor in their management education program. Dr. Kay received the Society's Excellence in Education award for 2002.

William M. Edwards is a Professor of Economics at Iowa State University, from which he received his B.S., M.S., and Ph.D. degrees in agricultural economics. He grew up on a family farm in south-central Iowa, and worked as an agricultural economist with the Farmer's Home Administration and the Colombian Agrarian Reform Institute. Since 1974, he has taught on-campus and distance education courses, and carried out extension programs in farm management at Iowa State University. In 1997, he received the Iowa State University Extension Achievement Award. Dr. Edwards has also collaborated in farm management educational programs in Latin America and Eastern Europe.

Patricia A. Duffy is a professor in the Department of Agricultural Economics and Rural Sociology at Auburn University, where she has taught farm management since 1985. She grew up in Massachusetts and received her B.A. from Boston College. After finishing this degree, she served as a Peace Corps volunteer for two years, teaching basic agriculture sciences in a vocational secondary school. She received her Ph.D. in agricultural economics from Texas A&M University. Her research papers in farm management and policy have been published in a variety of professional journals. In 1994, she received an award from the Southern Agricultural Economics Association for distinguished professional contribution in teaching programs. In 2001, she received Auburn University's College of Agriculture teaching award.

MANAGEMENT

Good management is a crucial factor in the success of any business. Farms and ranches are no exception. To be successful, farm and ranch managers today need to spend more time making management decisions and developing management skills than their parents and grandparents did.

This is because production agriculture in the United States and other countries is changing along the following lines: more mechanization, increasing farm size, continued adoption of new production technologies, growing capital investment per worker, more borrowed or leased capital, new marketing alternatives, and increased business risk. These factors create new management problems, but also present new opportunities for managers with the right skills.

These trends will likely continue throughout the twenty-first century. Farmers will make the same type of management decisions as in the past, but will be able to make them faster and more accurately. Advances in the ability to collect, transfer, and store data about growing conditions, pest and disease problems, and product quality will give managers more signals to which to respond. Moreover, future farm and ranch operators will have to balance their personal goals for an independent lifestyle, financial security, and rural living against societal concerns about food safety, environmental quality, and agrarian values.

The long-term direction of a ranch or farm is determined through a process called strategic planning. Farm families establish goals for themselves and their businesses based on their personal values, individual skills and interests, financial and physical resources, and the economic and social conditions facing agriculture in the next generation. They can choose to emphasize wider profit margins or higher volumes of production, or to produce special services and products. After identifying and selecting strategies to help achieve their goals, farm and ranch operators employ tactical management to carry them out. There are many decisions to be made, and many alternatives to analyze. Finally, the results of those decisions must be monitored and evaluated.

Chapter 1 discusses some factors that will affect the management of farms and ranches in the twenty-first century. These and other factors will require a new type of manager. This manager must be able to absorb, organize, and use large amounts of information, particularly information related to new technologies. Resources will be a mix of owned, rented, and borrowed assets. Products will need to be more differentiated to match consumer tastes and safety standards. The profitability of a new technology must be determined quickly and accurately before it is or is not adopted. A twenty-first century manager also will need new human resource skills as the number of employees per farm increases.

Chapter 2 explains the concept of management, including strategic planning and tactical decision making. What is management? What functions do managers perform? How should managers make decisions? What knowledge and skills are needed to be a successful manager? The answers to the first three questions are discussed in Chapter 2. Answers to the last question will require studying the remainder of the book.

FARM MANAGEMENT IN THE TWENTY-FIRST CENTURY

CHAPTER OBJECTIVES

1. To discuss how changes in the structure and technology of agriculture in the twenty-first century will affect the next generation of farm and ranch managers

2. To identify the skills that future farm and ranch managers will need in order to respond to these changes

What will future farm managers be doing as we enter the twenty-first century? They will be doing what they are doing now, making decisions. They will still be using economic principles, budgets, record summaries, investment analyses, financial statements, and other management techniques to help make those decisions. What kinds of decisions will managers be making in future decades? They will still be deciding input and output levels and combinations, and when and how to acquire additional resources. They will continue to analyze the risks and returns from adopting new technology, making new capital investments, adjusting farm size, and changing enterprises.

Will anything about management decisions in the future be different? Yes. While the broad types of decisions being made will be the same, the details and information used will change. Technology will continue to provide new inputs to employ and new, more specialized products for production and marketing. Management

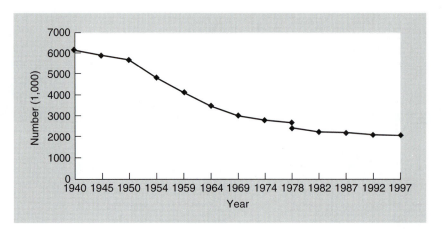

Figure 1-1   **Number of farms in the United States.**
Source: Census of Agriculture, USDA.

information systems, aided by electronic innovations, will provide more accurate and timely information for use in making management decisions. Farmers and ranchers will have to compete more aggressively with nonagricultural businesses for the use of land, labor, and capital resources. As in the past, the better managers will adapt to these changes and efficiently produce commodities that consumers want.

STRUCTURE OF FARMS AND RANCHES

The number of farms in the United States has been decreasing since 1940, as shown in Figure 1-1. Because the amount of land in farms and ranches has been relatively constant, this means the average production per farm has increased considerably, as shown in Figure 1-2. Several factors have contributed to this change.

First, labor-saving technology in the form of larger agricultural machinery, automated equipment, and specialized livestock buildings has made it possible for fewer farm workers to produce more. Second, employment opportunities outside agriculture have become more attractive and plentiful, encouraging labor to move out of agriculture. Also during this period of change, the cost of labor has increased faster

than the cost of capital, making it profitable for farm managers to substitute capital for labor in many areas of production.

Third, farm and ranch operators have aspired to earn higher levels of income and to enjoy a standard of living comparable to that of nonfarm families. One way to achieve a higher income has been for each farm family to control more resources and produce more output while holding costs per unit level or even decreasing them. The desire for an improved standard of living has provided much of the motivation for increasing farm size, and new technology has provided the means for growth.

Fourth, some new technology is available only in a minimum size or scale, which encourages farmers to expand production and spread the fixed costs of the technology over enough units to be economically efficient. Examples include grain drying and handling systems, four-wheel drive tractors, large harvesting machines, confinement livestock buildings, and automated cattle feedlots. Perhaps even more important are the time and effort required for a manager to learn new skills in production, marketing, and finance. These skills also represent a fixed investment and thus generate a larger return to the operator when they are applied to more units of production.

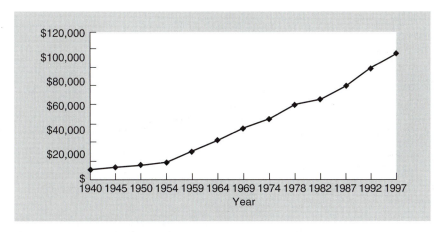

Figure 1-2   **Total sales per farm in 1997 dollars.**
Source: Census of Agriculture, USDA.

These forces are likely to continue throughout the twenty-first century, meaning that farm sizes will continue to increase. Future managers will find themselves making more use of hired labor or machines to perform tasks requiring less skill. They will spend their time applying more sophisticated management skills to producing, financing, and marketing more and more units of production.

As illustrated in Figure 1-3, farmers and ranchers in the twenty-first century will choose from among four general business strategies: low volume, high value producers; high volume, low margin producers; specialty product and service providers; and part-time operators.

Low Volume, High Value Producers

Lack of access to additional land and capital effectively limits the potential of many growers for expanding their businesses. For them, the key to higher profits is producing higher valued commodities. Some look for alternative enterprises such as emus, bison, asparagus, or pumpkins. Promotion and marketing become critical to their success. Others try variations of traditional commodities, such as organically grown produce, tofu soybeans, free-range poultry, or seed crops. Margins may be increased even more through added processing and direct marketing.

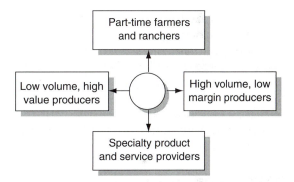

Figure 1-3   **Alternative paths for farm and ranch businesses.**

Such enterprises often involve high production risks and uncertain markets, but can be quite profitable even on a small scale.

High Volume, Low Margin Producers

There will always be a demand for generic feed grains, oil seeds, cotton, and livestock products. Many producers choose to stick with familiar enterprises and simply expand production as a means of increasing their income. For them, squeezing every nickel out of production costs is critical. Growing the business usually involves leveraging it with borrowed or rented assets. Because profit margins are thin, it is critical to set a floor under market prices or total revenue through insurance products and marketing contracts.

Specialty Product and Service Providers

A third strategy is to specialize in just one or two skills and become one of the best at performing them. Examples are custom harvesting, custom feedlots, raising seed stock or replacement breeding stock, repairing and refurbishing equipment, and applying pesticides and fertilizers. Often a key component of this strategy is making maximum use of expensive, highly specialized equipment and facilities. Marketing the services of the business and interacting with customers are also important ingredients for success.

Part-time Operators

Part-time farms and ranches account for nearly one-half of all U.S. farms. However, they produce only 15 percent of total agricultural sales. Many of these small-scale operations are "lifestyle" farms run by people who enjoy producing crops and livestock even when the potential profits are low. Their primary management concerns are to limit their financial risk and balance farm labor needs with off-farm employment. A combination of farming and nonfarm employment may provide the most acceptable level of financial security and job satisfaction for many families.

Farms of all sizes will continue to find their niche in U.S. agriculture. Naturally, the largest farms contribute the highest proportion of total sales of farm products, as shown in Table 1-1. The consolidation of small and medium sized farms and ranches into larger units will continue, as older operators retire.

Management and operation of farms by family units will continue to be the norm. This is especially true for agricultural enterprises that cannot concentrate production into a small geographic area, such as crop production or extensive grazing of cattle or sheep. Enterprises that can centralize production, such as poultry and hog production or cattle feeding, will be more

TABLE 1-1	Distribution of Farm Sales	
Sales Class	Percent of Farms	Percent of Sales
Less than $50,000	73.6	6.8
$50,000–$99,999	8.3	5.8
$100,000–$249,999	9.9	15.3
$250,000–$999,999	6.8	30.4
$1,000,000 or more	1.4	41.7

Source: 1997 Census of Agriculture, USDA.

easily organized into large-scale business entities. Management of these farms will be segregated into several layers, and areas of responsibility will be more specialized. Most managers of centralized production enterprises will be salaried employees rather than owner-operators.

Some family farm businesses will find that by cooperating with their neighbors and relatives they can achieve many of the same advantages that larger-scale operations enjoy. Decades ago, farmers formed grain threshing or haying crews to take advantage of new harvesting technology. Today, several farmers join together to guarantee a constant, uniform supply of livestock or crops in a quantity that can be transported and processed efficiently. As the number of input suppliers and processing firms diminishes, producers must collaborate in order to maintain their bargaining position. This is one example of how a cooperative effort or *strategic alliance* can provide economic benefits. Another example is several operators forming an input purchasing group to achieve quantity discounts. A small amount of managerial independence must be sacrificed to conform to the needs of the group. However, personal ownership and operation of each individual business is preserved.

THE INFORMATION AGE

Many decision-making principles and budgeting tools have been under-utilized in the past. Individual farm data needed to use them were

> ### Box 1-1 A Twenty-First Century Example: Berilli Farms
>
> Berilli Farms, Inc. consists of only a few hundred acres. These acres have been transformed from growing common field grains to producing high-value specialty crops. Fresh vegetables are sold to a local wholesale grocer. High-protein alfalfa has been contracted to a dairy in the next county. High-quality turf grass seed goes to a chain of nurseries.
>
> Keeping a stable work crew of 25 machinery operators, truck drivers, sorters, and crop scouts is a real test of the Berilli family's human relations skills. All of their employees are trained to gather data on crop growth and yields from monitors mounted on machinery or in fields, and to download it into their handheld computers. Each morning before chemicals or fertilizers are applied, a variable-rate application plan is read into the control units of the applicators.
>
> The Berillis use sophisticated crop simulation computer models to formulate these recommendations, taking into account current input prices and the selling prices for their products that they have contracted or hedged. Each week they review their cash-flow position and transfer operating funds into their business account electronically. All their crops are protected by multiple peril crop insurance, and are committed to delivery according to a detailed production contract.
>
> The grocers, dairies, and nurseries they supply send them real-time data about the results of quality tests performed on their products, and the varieties that are selling the fastest. At the end of the year, the Berillis analyze the costs and returns from each crop, field, and buyer, and replace the least profitable ventures with more promising ones.

not available, or the process for analyzing the data was too complex. The coming years will see rapid changes in methods of data collection, analysis, and interpretation. Electronic sensors and processors used in large-scale industries are now accessible and affordable to farms and ranches.

Not only will more whole-farm data be available, but data specific to very small land areas or to individual animals will become more common. These very specific data will help managers customize the treatment of each acre of land or each head of livestock. Yields can be monitored and recorded as harvesting machines move across the field. Global positioning systems (GPS) can use satellite signals to identify the exact position of harvesting units when the data are collected. Automated machines may be able to take a soil sample every few yards, analyze it instantly, and record the results by field location. Satellite photographs and other techniques may provide information on the specific location of weed and insect infestations or

moisture, permitting a limited, pinpoint application of pesticide or irrigation water.

Miniature electronic sensors will be able to collect and record information from livestock by continuously monitoring individual animal performance levels, feed intake, and health status. When undesirable changes are detected, there could be automatic adjustment in environmental conditions and feed rations. This information could also be related back to genetic background, physical facilities, feed rations, health programs, and other management factors to improve and fine-tune animal performance. Electronic implants can provide *identity preservation* of both crops and livestock from the original producer to the final consumer.

Financial transactions may be recorded and automatically transferred to accounts through the use of debit cards and bar-code symbols whenever purchases and sales occur. Smaller purchases may be made with preloaded cash cards. These transactions can also be posted automatically to the accounting system for an individual farm, and classified by enterprise, vendor, or business unit. These technological advances mean that the information in a farmer's accounting system can be accurate and up to date at the end of each day.

Personal computers have greatly enhanced capacities to receive, process, and store information, and to communicate with outside data sources. Portable computers and personal data recorders allow precise decisions to be made in the pickup or on the tractor, as well as in the office. The first computers were used primarily to sort data and do calculations, but increasingly computers are being designed and used as communication tools. Fiber optic systems, wireless transmission technology, and global computer networks are increasing the availability, speed, and accuracy of information sharing.

Managers in the last century often found the lack of accurate, timely, and complete information to be frustrating. Managers of the twenty-first century may also be frustrated by information, only the cause of their frustration will be the large quantity and continual flow of information available to them. A vital task of twenty-first century managers will be to determine which information is critical to their decision making, which is useful, and which is irrelevant. Even when this is done, the critical and useful information must be analyzed and stored in an easily accessible manner for future reference.

FINANCIAL MANAGEMENT

Outside capital will continue to be needed to finance large-scale operations. Management of traditional sources of farm credit, such as rural banks, is becoming more vertically integrated, and funds will come from national money markets. Credit will be available from nontraditional sources such as input suppliers and processors. Farm managers will increasingly have to compete with nonfarm businesses for access to capital, as the rural and urban financial markets become more closely tied together. This competition will necessitate more detailed documentation of financial performance and credit needs,

and more conformity to generally accepted accounting principles and performance measures. Farmers will need to use standard accounting methods and principles and perhaps even have audited financial statements to gain access to commercial capital markets.

Standardized records and on-line data bases will help make comparative analysis with similar farms more meaningful. The farm manager will have to decide whether to train an employee to carry out the required accounting and analysis, or hire this expertise from outside the business. Even if outside help is used, the manager must have the skills and knowledge to read, interpret, and use this accounting information.

Controlling assets is becoming more important than owning them. Farmers have long gained access to land by renting it. Leasing machinery, buildings, and livestock has been less common, but will likely increase in use. Custom farming and contract livestock production are other means by which a good manager can apply his or her expertise without taking the financial risks of ownership. When other parties supply much of the capital, the operator can produce a larger volume at less risk, although the profit margin may be smaller.

HUMAN RESOURCES

Farm managers of the twenty-first century are depending more and more on a team of employees or partners to carry out specific duties in the operation. Working with other people will become a more important factor in the success of the operation. Motivation, communication, evaluation, and training of personnel will become essential skills.

Farm businesses will have to offer wages, benefits, and working conditions that are competitive with nonfarm employment opportunities. They will likely have to follow more regulations regarding worker safety in handling farm chemicals and equipment, and see that employees are properly trained in the use of new technologies. Many of the most efficient farms and ranches in the twenty-first century will be those with a small number of operators or employees who have specialized responsibilities. They will have mastered the communication and decision-making techniques needed in such operations.

Modern managers will need to take advantage of the expertise of paid consultants and advisors. For some very technical decisions, such as diagnosing animal and plant diseases, developing legal contracts, or executing commodity pricing strategies, the manager may simply pay a consultant to make recommendations. In other cases, the farm manager will obtain information from outside sources but do the analysis and decision making. Examples include formulating livestock rations or crop fertility programs based on the results of laboratory tests. In either case, the successful manager must learn to communicate clearly and efficiently with the consultant. This means understanding the terminology and principles involved, and summarizing information in a concise form before submitting it.

PRODUCING TO MEET CONSUMER DEMANDS

Agriculture has long been characterized by the production of "undifferentiated" commodities. Historically, grain and livestock products from different farms have been treated alike by buyers if these products met basic quality standards and grades. Because the trend is to offer more highly specialized and processed food products to the consumer, buyers are beginning to implement stricter product standards for producers.

For example, livestock processors want uniform animals with specific size and leanness characteristics to fit their processing equipment, packaging standards, and quality levels. Improved measuring devices and data processing will make it easier to pay differential prices to producers based on product characteristics, and to trace each lot to its point of origin. As processors invest in larger-scale plants, they must operate them at full capacity to remain competitive. Producers who can assure the packer of a continuous supply of high-quality, uniform animals will receive a premium price. Those who cannot may find themselves shut out of many markets or forced to accept a lower price.

In crop production, the protein and oil content of grain and forages is becoming easier to measure, making differential pricing possible. Biotechnology research will allow plant characteristics to be altered and genetically engineered varieties to be produced for specific uses, regions, and production technologies.

So-called niche markets will also become more important. Organic produce, extra-lean meat, specialty fruits and vegetables, and custom-grown products for restaurants and food services will be in greater demand. As international trade

---

Box 1-2   Custom Pork Production: Producing for the Market

Howard Berkmann continues to produce traditional cross-bred, uniform lean hogs for the local packing plant. One morning each week he delivers a load of hogs, and by evening, he receives by electronic mail a summary of the carcass data and pricing formula from the packer. He downloads the information into his swine production computer software and prints out a current summary for the facility from which the hogs came and the genetic group they represented.

A few years ago, Howard started a specialty group of Berkshire hogs designed for the Japanese market. The particular coloring and marbling of the meat earns him a premium price. He negotiated an agreement with a Berkshire breeder in a neighboring state to supply him with a regular stream of replacement gilts. Several times a week, he checks the Japanese livestock markets for forward pricing opportunities, and he has visited his marketing contact in Tokyo.

barriers continue to fall, foreign markets also will be more important. These markets may require products with special characteristics. Farm managers who seek out these markets and learn the production techniques necessary to meet their specifications can realize a higher return from their resources. Of course, the manager will have to evaluate the additional costs and increased risks associated with specialty markets and then compare them with the potentially higher returns.

ENVIRONMENTAL AND HEALTH CONCERNS

As the availability of an adequate quantity of food becomes ever more taken for granted, concerns about food quality, food safety, and the present and future condition of our soil, water, and air will continue to receive high priority from the nonfarm population. Farmers and ranchers have always had a strong interest in maintaining the productivity of natural resources under their control. However, the off-farm and long-term effects that new production technologies have on the environment have not always been well quantified or understood. As more people decide to live in rural areas, the contact between farm and nonfarm residents will increase. This will lead to increased concern about agricultural wastes and their effects on air and water quality. Pressure from nonfarm rural residents may even cause some production systems such as concentrated livestock feeding to shift to less populated regions. Farm managers will have to choose between discontinuing those enterprises or moving their businesses.

As research and experience improve understanding of the interactions among various biological systems, education and regulation will be used to increase the margin of safety for preserving resources for future generations. Top agricultural managers of today recognize the need to keep abreast of the environmental implications of their production practices and are often leaders in developing sustainable production systems. All farm managers in the twenty-first century must be aware of the effects their production practices have on the environment, both on and off the farm, and take the steps necessary to keep our agricultural resources productive and environmentally safe.

The value of agricultural assets, particularly farmland, will be affected by environmental conditions and regulations. When farms are sold or appraised, environmental audits are becoming routine to warn potential buyers of any costs that might be incurred to clean up environmental hazards. The crop production combinations and practices allowed by a farm's conservation plan also affect its value. Farm managers will have to evaluate every decision for both profitability and how it affects the environment. The successful managers are those who can generate a profit while sustaining resources on the farm and minimizing environmental problems off the farm.

NEW TECHNOLOGY

Agricultural technology has been evolving for many decades and will continue to do so. The field of biotechnology offers possible gains in production efficiency, which may include crop varieties that are engineered to fit growing conditions at particular locations, are resistant to herbicide damage or to certain insects and diseases, or have a more highly valued chemical composition such as higher protein or oil content. Livestock performance may be improved by introducing new genetic characteristics or by improving nutrient utilization. New nonfood uses for agricultural products will open new markets but may also cause changes in the desired characteristics or composition of products grown specifically for these uses.

One example of a recent technology is the use of global positioning systems (GPS) to pinpoint the exact location of equipment in a field. Combined with other technology, this may have widespread application in the twenty-first century. For example, by combining satellite reception with a yield monitor on harvesting equipment, the crop yield can be measured and recorded continuously for every point in the field. Variations in yield due to soil type, previous crops, different tillage methods, and fertilizer rates can be identified quickly and recommendations made to correct problems. This technology is now being used to automatically adjust the application rates of fertilizer and chemicals as the applicator moves across the field. With this equipment, fertilizer and chemicals are applied only at the rates and locations needed.

This technology and others yet to be developed will provide the farm manager of the twenty-first century with a continual challenge. Should this or any new technology be adopted? The cost of any new technology must be weighed against its benefits, which may come in several forms. There may be increased yields, an improvement in product quality, less variation in yield, a reduced impact on the environment, or most likely, some combination of these benefits. Decisions about if and when to adopt a new technology will affect the profitability and long-term viability of a farm or ranch business.

SUMMARY

Farmers and ranchers in the twenty-first century will be making most of the same basic decisions that they have made in the past century. The difference is they will be making them faster and with more accurate information. Farm businesses will continue to become larger, and their operators will have to acquire specialized skills in managing personnel, interpreting data, competing for resources with nonfarm businesses, and customizing products to meet the demands of new markets. All this must be done while balancing the need to earn a profit in the short run with the need to preserve agricultural resources and environmental quality into the future. While some farm managers will look at these trends as threats to the way they have traditionally operated their businesses, others will see them as new opportunities to gain a competitive advantage and to prosper.

QUESTIONS FOR REVIEW AND FURTHER THOUGHT

1. What forces have caused farms and ranches to become larger? Is this trend likely to continue? Why or why not?

2. How will quick access to more information help farm managers in the twenty-first century make better decisions?

3. List two examples of specialty agricultural markets and the changes a conventional producer might have to make to fill them.

4. List other new challenges not discussed in this chapter that you think farm and ranch managers may have to face in the future.

MANAGEMENT AND DECISION MAKING

CHAPTER OBJECTIVES

1. To understand the functions of management

2. To present the steps in developing a strategic management plan for a farm or ranch

3. To identify common goals of farm and ranch managers and show how they affect decision making

4. To explain the steps in the decision-making process

5. To describe some unique characteristics of the decision-making environment for agriculture

Successful managers cannot simply memorize answers to problems, nor can they do exactly as their parents did. Some managers make decisions by habit. What worked last year also will work this year, and maybe again next year. But good managers learn to continually rethink their decisions as economic, technological, and environmental conditions change. Farmers and ranchers are continually bombarded by new information about prices, weather, technology, public regulations, and consumers' tastes. This information affects the organization of their businesses; what commodities to produce; how to produce them; what inputs to use; how much of each input to use; how to finance their businesses; and how, where, and when to market their products. New information is vital for making new decisions, and will often cause old management strategies to be reconsidered.

Important changes can occur in climate, weather, government programs and policies, imports and exports, international events, and many other factors that affect the supply and demand situation for agricultural commodities. Long-term trends must be recognized and taken into account.

Technology is also a constant source of new information. Examples include the development of new seed varieties; new methods for weed and insect control; new animal health products and feed ingredients; and new designs, controls, and monitors for machinery. Other changes occur in income tax rules, environmental regulations, and farm labor laws. These factors are all sources of new information that the manager must take into account when formulating strategies and making decisions.

Some managers achieve better results than others. Table 2-1 contains some evidence of this from a group of farms in a farm business association. Farms in the top one-third of the group had an average net farm income and return on assets many times higher than those of farms in the lowest one-third. However, the high-profit farms actually had less land, machinery, and labor than

the low-profit farms. Therefore, the wide range in net farm income and return on assets cannot be explained by the different quantities of resources available. The explanation must lie in the management ability of the farm operators.

FUNCTIONS OF MANAGEMENT

Farm and ranch managers perform many functions. Much of their time is spent doing routine jobs and chores. However, the functions that distinguish a manager from a worker are those that involve a considerable amount of thought and judgment. They can be summarized under the general categories of *planning, implementation, and control.*

Planning

The most fundamental and important of the functions is planning. It means choosing a course of action, policy, or procedure. Not much will happen without a plan. To formulate a plan, managers must first establish *goals,* or be sure they clearly understand the business owner's goals. Second, they must identify the quantity and quality of *resources* available to meet the goals. In agriculture, such resources include land, water, machinery, livestock, capital, and labor. Third, the resources must be allocated among several competing uses. The manager must identify all possible *alternatives,* analyze them, and select those that will come closest to meeting the goals of the business. All these steps require the manager to make both long-run and short-run decisions.

Implementation

Once a plan is developed, it must be implemented. This includes acquiring the resources and materials necessary to put the plan into effect, plus overseeing the entire process. Coordinating, staffing, purchasing, and supervising are steps that fit under the implementation function.

TABLE 2-1	Comparison of Low- and High-Profit Farms in Iowa	

Item	Highest Third (Average)	Lowest Third (Average)
Value of farm production	$334,981	$241,861
Net farm income	$90,800	$2,358
Return on assets	10.6%	−1.6%
Crop acres farmed	729	802
Machinery value	$166,212	$196,490
Months of labor utilized	16.6	17.6

Source: 2001 Iowa Farm Costs and Returns, Iowa State University Extension, Iowa State University, Ames, Iowa.

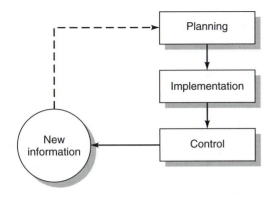

Figure 2-1   Management flow chart based on three functions of management.

Control

The control function includes monitoring results, recording information, comparing results to a standard, and taking corrective action when needed. It ensures that the plan is being followed and producing the desired results, or provides an early warning so adjustments can be made if it is not. Outcomes and other related data become a source of new and improved information to use for adjusting a current plan or improving future plans.

Figure 2-1 illustrates the flow of action from planning through implementation to control. It also shows that information obtained from the control function can be used for revising future plans. This circular process of constant improvement and refinement of decisions can continue through many cycles. But first, some very basic decisions must be made about exactly why the business even exists and where it is headed.

STRATEGIC FARM MANAGEMENT

Management of a farm or ranch can be divided into two broad categories: strategic and tactical. *Strategic management* consists of charting the overall long-term course of the business. *Tactical management* consists of taking short-run

actions that keep the business moving along that course until the destination is reached.

It has been said many times that doing things right in farming is not enough to ensure success. Farmers and ranchers must also do the right things. Strategic management seeks to discover what the right things are for a particular business at a particular time. Simply doing what the previous generation did will not keep the farm competitive in the long run.

Strategic management is an ongoing process. However, this process can be broken down into a series of logical steps:

1. Define the mission of the business.
2. Formulate the goals of the business.
3. Assess the resources of the business (internal scanning).
4. Survey the business environment (external scanning).
5. Identify and select strategies.
6. Implement and refine the selected strategies.

Defining the Mission of the Business

A mission statement is a short description of why a business exists. For some farms and ranches, the mission statement may include strictly business accomplishments. For a family-owned and operated business, the mission of the farm may be only one component of the overall family mission, which may reflect social, religious, and cultural values as well as economic considerations. Mission statements should emphasize the special talents and concerns of each farm business and its managers.

Formulating the Goals of the Business

Goals provide a reference point for making decisions and measuring progress. For a family-owned and operated farm, the goals of the business may be a subset of the overall family goals. For larger farms where hired managers are employed, the owners may define the goals while the manager simply strives to achieve them.

Box 2-1   A Mission Statement

George and Connie Altman have been milking cows and growing crops since their early twenties. At age 35, they decided to assess where their farming operation was and where they wanted it to go. They chose the following mission statement for their business: "Our mission is to produce safe and nutritious milk at a reasonable cost, to maintain and enhance the quality of the natural resources under our control, and to contribute toward making our community a satisfying place to live."

Not all farm managers will have the same goals, even when their resources are similar. This is because people have different *values*. Values influence the goals people set and the priorities they put on them. Table 2-2 lists some typical values held by farmers and ranchers. How strongly they feel about each of them will affect their business and family goals. When more than one person is involved in setting goals, it is important to recognize differences in values and to be willing to compromise if necessary to arrive at a mutually acceptable set of goals.

TABLE 2-2	Common Values Among Farmers and Ranchers

Do you agree or disagree?
1. A farm or ranch is a good place to raise a family.
2. A farm or ranch should be run as a business.
3. It is acceptable for farmers to borrow money.
4. A farmer should have at least two weeks of vacation each year.
5. It is better to be self-employed than to work for someone else.
6. It is acceptable for a farmer to also work off the farm.
7. It is more enjoyable to work alone than with other people.
8. Farmers should strive to conserve soil and keep water and air resources clean.
9. A family farm should be passed on to the next generation.
10. All family members should be involved in the operation.

When goals are being established, it is important to keep the following points in mind:

1. Goals should be written. This allows everyone involved to see and agree on them and provides a record for review at later dates.
2. Goals should be specific. "To own 240 acres of class I farm land in Washington County" is a more useful goal than simply "to own land." It helps the manager determine whether a goal has been reached, and provides a sense of accomplishment and an opportunity to think about defining new goals.
3. Goals should be measurable. The goal of owning 240 acres is measurable, and each year the manager can gauge progress toward the goal.
4. Goals should have a timetable. "To own 240 acres within five years" is more useful than a goal with an open-ended or vague completion date. The deadline helps keep the manager focused on the goal.

Because of the close and direct involvement of family members with the farm business, a farm operated by a family unit often has more than one set of goals. There can be personal goals as well as business goals, and each individual within the family unit may have different goals. In these situations, it is important to use a family conference or discussion to agree on at least the business goals. Without an agreement, everyone may go in different directions, and none of the business goals will be reached. This also applies to farms and ranches with multiple partners or shareholders.

Because individuals and the businesses they manage differ, many potential goals exist. Surveys of farm operators have identified the following common farm and ranch goals:

- Survive, stay in business, do not go broke, avoid foreclosure.
- Maximize profits, get the best return on investment.
- Maintain or increase standard of living, attain a desirable family income.
- Increase equity, accumulate assets.
- Reduce debt, become free of debt.
- Avoid years of low profit, maintain a stable income.
- Pass the entire farm on to the next generation.
- Increase leisure and free time.
- Increase farm size, expand, add acres.
- Maintain or improve the quality of soil, water, and air resources.

These goals are stated in a general manner, and would need to be made more specific before they would be useful to an individual business. Rarely does a single goal exist; farm operators usually have multiple goals. When this occurs, the manager must decide which goals are most important. Some combinations of goals may be impossible to achieve simultaneously, which makes the ranking process even more important. Another job for the manager is to balance the tradeoffs among conflicting goals.

Any of the goals listed may rank first for a certain individual, depending on the time and circumstances. Goals can and do change with changes in age, financial condition, family status, and experience. Also, long-run goals may differ from short-run goals. Profit maximization is often assumed to be a major goal of all business owners, particularly in the study of economics. However, farm operators often rank survival or staying in business above profit maximization.

It is important to note that achieving a profit plays a direct, or at least an indirect, role in meeting many other possible goals, including business survival.

Profit is needed to pay family living expenses and taxes, increase owner equity, decrease debt, and expand production. However, several possible goals on the list imply minimization or avoidance of risk, which may conflict with profit maximization. The most profitable long-run production plans and strategies are often the most risky as well. Highly variable profits from year to year may greatly reduce the chances for survival, conflicting with the desire for a stable income. For these and other reasons, profit maximization is not always the most important goal for all farm operators. Profit may be maximized subject to achieving minimum acceptable levels of other goals, such as security, leisure, and environmental stewardship. Nevertheless, profit maximization has the advantage of being easily measured, quantified, and compared across different businesses.

Assessing the Resources of the Business

Farms and ranches vary widely in the quantity and quality of physical, human, and financial resources available to them. An honest and thorough assessment of these resources will help the manager choose realistic strategies for achieving the goals of the business. This process is often called *internal scanning.*

Physical resources The land base is probably the most critical physical resource. Productivity, topography, drainage, and fertility are just a few of the qualities that determine the potential of land for agricultural use. The number of acres available and their location are also important. In many states, detailed databases exist that describe the important characteristics of a particular tract of land.

Other physical resources that should be evaluated include breeding livestock, buildings and fences, machinery and equipment, irrigation installations, and established perennial crops such as orchards, vineyards, and pasture.

Human resources The skills of the operator(s) and other employees often determine the success or failure of certain enterprises. Some workers are very talented with machinery, while others do better with livestock. Still others excel at marketing or accounting. Equally important is the degree to which each person in the operation likes or dislikes doing certain jobs. It is a good idea to conduct a thorough audit of personal skills and preferences before identifying competitive strategies for a farm business.

Financial resources Even when the physical and human resources are present to carry out certain enterprises, capital may be a limiting factor. Financial resources can be evaluated by completing a set of financial statements, and by exploring the possibility of obtaining additional capital from lenders or outside investors. These tools and strategies will be discussed in detail in later chapters.

An honest and thorough appraisal of the farm's physical, human, and financial strengths and weaknesses will steer it toward realistic strategies. Particular attention should be given to identifying resources that will give the farm or ranch a competitive advantage over other firms.

Surveying the Business Environment

Critically analyzing the business environment surrounding a ranch or farm is called *external scanning*. Although the major types of livestock and crops grown in many parts of the world have not changed recently, many of their characteristics have. Changing consumer tastes and expanded international markets have led some customers to pay premiums for lean meat or high-protein grains, for example.

Other trends affect the availability of new resources and the choices of technology. Changes in government regulations may create new constraints, or even remove some. The prudent manager must be aware of all these changes in the external environment and react to them early. If new production practices that lower costs per unit are adopted by most producers,

then the operation that does not change will soon be at a competitive disadvantage.

Some trends may represent threats to the farm or ranch, which could decrease profits if no corrective action is taken. For example, decreasing consumption of a crop such as tobacco may require alternative crops to be considered. Other trends, such as a desire for diets lower in fat, may actually present opportunities for a farm that could help it reach its goals faster. Whether a trend represents an opportunity or a threat will sometimes depend on the particular nature and location of the farm.

Identifying and Selecting Strategies

Everyone connected with the farm should brainstorm about possible plans for the future. By matching up the most promising opportunities with the strong points of the particular farm or ranch, an overall business strategy with a high chance of success can be formulated. Changes may have to be made, but they will be part of a deliberate, integrated plan, not just haphazard reactions.

Several general strategies were identified in Chapter 1: a low volume, high value producer; a high volume, low margin producer; and a special service provider. Some businesses can expand their options by forming *strategic alliances* with other farms or ranches that have complementary skills, such as a feeder pig producer and a custom hog finisher. Alliances can also be formed with processors and wholesalers.

Some businesses have more possible strategies for reaching their goals than others. In the arid regions of the western United States, for example, the land resource is such that the only alternative may be to use it as pasture for livestock production. But even in this situation, the manager must still decide whether to use the pasture for cow/calf production, for grazing stocker steers during the summer, or for sheep and goat production. Other regions have land suitable for both crop and livestock production, so more alternatives exist. As the number of alternatives for

```
┌─────────────────────────────────────────────────────────────────────────┐
│   Box 2-2          Internal and External Scanning                         │
```

June and Carl Washington have raised corn, tobacco and hay on their rolling 460-acre farm for nearly 18 years. They have also run 50 stock cows on their rough pastureland, and farrowed and sold feeder pigs from 35 sows each year. Through hard work and careful budgeting they have managed to pay down their mortgage and send the children off to school.

This year they lost their tobacco quota, and it is getting harder and harder to sell the feeder pigs through local sale barns. They would like to sell pigs directly to one of the finishing operations in the area, but all of them want a larger volume of pigs, delivered at regular intervals. Without their children around to help, June and Carl don't think they can handle increased hog chores. Besides, they would have to buy extra corn.

Their county beef producers association is negotiating a contract to sell high quality feeder calves to an out-of-state feedlot, by pooling calves from all their members. Carl has always enjoyed working with cattle. After comparing a series of whole farm budgets developed with the help of their farm business association consultant, the Washingtons decide to liquidate their swine operation and purchase 30 first-calf heifers. They also plan to gradually renovate their pastures and subdivide them, and increase their hay acres. By selecting their best heifer calves for replacements, they hope to build their herd up to 100 females in five years. They will keep Supervised Performance Analysis (SPA) records to measure both the production and financial success of their venture.

---

the farm's limited resources increases, so does the complexity of the manager's decisions.

Implementing and Refining the Selected Strategies

Even the best strategy does not happen by itself. The manager must formulate action steps, place them in a timetable, and execute them promptly. In some cases a formal *business plan* will be developed and presented to potential lenders or partners. Concrete, short-term objectives need to be set so that progress toward long-term goals can be measured. The manager then needs to decide what information will be needed to evaluate the success or failure of the strategy, and how to collect and analyze the data.

Above all, strategic management should not be considered a one-time, limited process. It is an ongoing activity in which the manager is constantly alert for new threats or opportunities, ready to take advantage of new resources, and willing to adapt the farm's strategies to

changes in the values and goals of the individuals involved.

TACTICAL DECISION MAKING

Once an overall strategy for the farm or ranch has been developed, the manager must make *tactical decisions* about how to implement it. This is why most definitions of management include the process of decision making. Without decisions, nothing will happen. Even allowing things to drift along as they are implies a decision, perhaps not a good decision, but a decision nevertheless.

The decision-making process can be formalized into several logical and orderly steps:

1. Identify and define the problem or opportunity.
2. Identify alternative solutions.
3. Collect data and information.
4. Analyze the alternatives and make a decision.
5. Implement the decision.

6. Monitor and evaluate the results.
7. Accept responsibility.

Following these steps will not make every decision perfect. It will, however, help a manager act in a logical and organized manner when confronted with choices.

Identifying and Defining the Problem or Opportunity

Many problems confront a farm or ranch manager. Major management problems include choosing what seed to use, selecting a livestock ration, deciding how to market production, and deciding how to obtain access to land.

Problems may be identified by comparing actual results from the business to the levels that could be attained, or that similar farms are achieving. For example, a farm may have a cotton yield that is 100 pounds per acre lower than the average for other farms in the same county on the same soil type. This difference between what is, the farm yield, and what should be, the county average yield or better, identifies a condition that needs attention. What appears to be a problem is often just a symptom of a deeper problem, however. The low cotton yields could be caused by low fertility or inadequate pest control. These, in turn, could be caused by even more fundamental production problems.

A manager must constantly be on the alert to identify problems as quickly as possible. Most problems will not go away by themselves. Once a problem area is identified, it should be defined as specifically as possible. Good problem definition will minimize the time required to complete the remainder of the decision-making steps.

Identifying Alternative Solutions

Step two is to begin listing potential solutions to the problem. Some may be obvious once the problem is defined, while others may require time and research. Still others may become apparent during the process of collecting data and information. This is the time to brainstorm and list any idea that comes to mind. Custom, tradition, or habit should not restrict the number or types of alternatives considered. The most feasible ones can be sorted out later.

Collecting Data and Information

The next step is to gather data, information, and facts about the alternatives. Data may be obtained from many sources, including university extension services, bulletins and pamphlets from agricultural experiment stations, electronic data services, farm input dealers, salespersons of agricultural inputs, radio and television, computer networks, farm magazines and newsletters, and neighbors. Perhaps the most useful source of data and information is an accurate and complete set of past records for the manager's own farm or ranch. New technology for collecting and analyzing data has made it much easier to have current and complete information available. Whatever the source, the accuracy, usefulness, and cost of the information obtained should be considered carefully.

Decision making typically requires information about future events, because plans for producing crops and livestock must be made long before the final products are ready to market. The decision maker may have to formulate some estimates or expectations about future prices and yields. Past observations provide a starting point, but will often need to be adjusted for current and projected conditions.

It is important to distinguish between data and information. Data can be thought of as an unorganized collection of facts and numbers obtained from various sources. In this form, they may have very little use. These data and facts need to be organized, sorted, and analyzed. Information can be thought of as the final product obtained from analyzing data in such a way that useful conclusions and results are obtained. Not all data need to be organized and summarized into useful information, but most are not useful until some type of analysis is performed.

Gathering data and facts and transforming them into useful information can be a never-ending task. A manager may never be satisfied with the accuracy and reliability of the data and the resulting information. However, this step must end at some point. Gathering data has a cost in terms of both time and money. Too much time spent gathering and analyzing data may result in a higher cost than can be justified by the extra benefit received.

Analyzing the Alternatives and Making a Decision

Each alternative should be analyzed in a logical and organized manner. The principles and procedures discussed in Part Three provide the basis for sound analytical methods. Some alternatives may require additional information to complete the analysis. Good judgment and practical experience may have to substitute for information that is unavailable or available only at a cost that is greater than the additional return from its use.

Choosing the best solution to a problem is not always easy, nor is the best solution always obvious. Sometimes the best solution is to change nothing, or to go back, redefine the problem, and go through the decision-making steps again. These are legitimate actions, but they should not be used to avoid making a decision when a promising alternative is available.

After carefully analyzing each alternative, the one that will best meet the established goals is normally selected. Sometimes none of the alternatives appears to be definitely better than any other. If profit maximization is the primary goal, the alternative resulting in the largest profit or increase in profit would be chosen. However, the selection is often complicated by uncertainty about the future, particularly future prices. If several alternatives have nearly the same profit potential, the manager must then assess the probability that each will achieve the expected outcome, and the potential problems that could arise if it doesn't.

Making decisions is never easy, but it is what people must do when they become managers. Most decisions will be made with less than the desired level of information. An alternative must be selected from a set of possible actions, all of which have some disadvantages and carry some risk. Just because a decision is difficult is no reason to postpone making it, though. Many opportunities have been lost by delay and hesitation.

Implementing the Decision

Nothing will happen and no goals will be met by simply making a decision. That decision must be correctly and promptly implemented, which means taking some action. Resources need to be acquired, detailed budgets made, a timetable constructed, and expectations communicated to partners and employees. This takes organizational skills. Remember that *not* implementing a decision is the same as not making the decision at all.

Monitoring and Evaluating the Results

Managers must know the results of their decisions. The longer it takes for the results of a decision to become known, the more likely it is that the results will be different than expected. Sometimes even a good decision will have bad results. Good managers will monitor the results of a decision with an eye toward modifying or changing it.

The more frequently a decision is repeated, the more useful it is to evaluate it. Deciding where to market crops or what genetic lines to choose for livestock is done over and over. Monitoring prices received or performance traits achieved allows better alternatives to be identified over time. Decisions that can be easily reversed also deserve to be evaluated more closely than those that cannot be changed.

Managers must set up a system to assess the results of any decision, so that any deviation from the expected outcome can be identified quickly. This is part of the control function of management. Profit and loss statements summarize the economic impact of a decision, yield

---

Box 2-3   The Decision-Making Process: an Example

*S*tep 1. Identify the problem
Soil erosion rates on the more sloping parts of the farm are above acceptable rates.

Step 2. Identify alternatives
Some of the neighbors use terraces or strip cropping on similar slopes. Many farmers are experimenting with reduced tillage or no-till practices.

Step 3. Collect information
Study research results from similar soils, comparing reduced tillage, terraces, and strip cropping. Obtain prices for different equipment and for building terraces. Visit with neighbors about their results. Consult extension experts about what changes in crop production practices and fertility would be needed.

Step 4. Analyze the alternatives and select one
Taking into account long-term costs and effects on yield, reduced tillage seems to be the most profitable way to bring soil erosion down to an acceptable level.

Step 5. Implement the decision
Purchase a new planter and modify the tillage implements.

Step 6. Monitor the results
Compare yields, calculate machinery and chemical costs, and measure erosion rates for several years.

Step 7. Accept responsibility for results
Yields and erosion are acceptable, but costs have increased. Fine tune fertilizer and pesticide applications.

---

records measure the impact on crop production, and daily milk or feed efficiency logs monitor livestock performance. Careful observation and good records will provide new data to be analyzed. The results of this analysis provide new information to use in modifying or correcting the original decision and making future decisions. Evaluating decisions is a way to "learn from your past mistakes."

Accepting Responsibility

Responsibility for the outcome of a decision rests with the decision maker. A reluctance to bear responsibility may explain why some individuals find it so difficult to make decisions. Sometimes even good decisions bring bad results, due to uncertainties of markets and production. Blaming the government, the weather, or suppliers and processors when a decision turns out badly will not improve the results of the next decision. The manager must simply try to control the damage and then turn attention to the future.

CHARACTERISTICS OF DECISIONS

The amount of time and effort a manager devotes to making a decision will not be the same in every case. Some decisions can be made almost instantly, while others may take months or years of investigation and thought. Some of the characteristics that affect how the steps in the decision-making process are applied to a specific problem include:

1. Importance
2. Frequency
3. Imminence
4. Revocability
5. Number of available alternatives

Importance

Some farm and ranch decisions are more important than others. Importance can be measured in several ways, such as the number of dollars involved in the decision or the size of the potential gain or loss. Decisions risking only a few dollars, such as sorting hogs into pens or buying small tools, may be made rather routinely. For these, little time needs to be spent gathering data and proceeding through the steps in the decision-making process.

On the other hand, decisions involving a large amount of capital or potential profits and losses need to be analyzed more carefully. Decisions about purchasing land, installing an irrigation system, and constructing a new total-confinement hog building easily justify more time spent on gathering data and analyzing the alternatives.

Frequency

Some decisions may be made only once in a lifetime, such as choosing farming or ranching as a vocation, or buying a farm. Other decisions are made almost daily, such as scheduling field work activities, balancing livestock rations, and setting breeding schedules. Frequent decisions are often made based on some rule of thumb or the operator's intuitive judgment. Nevertheless, small errors in frequent decisions can accumulate into a substantial problem over a period of time.

Imminence

A manager is often faced with making a decision very quickly or before a certain deadline to avoid a potential loss. Grain prices that are moving up or down rapidly call for quick action. Other decisions have no deadline, and there is little or no penalty for delaying the decision until more information is available and more time can be spent analyzing the alternatives. The thoroughness with which any decision is made will depend on the time available to make it.

Revocability

Some decisions can be easily reversed or changed if later observations indicate that the first decision was not the best. Examples are calibrating a seeder or adjusting a feeder, which can be changed rather quickly and easily. Managers may spend less time making the initial decision in these situations, since corrections can be made quickly and at very little cost.

Other decisions may not be reversible or can be changed only at a very high cost. Examples would be the decision to drill a new irrigation well or to construct a new livestock building. Once the decision is made to go ahead with these projects, and they are completed, the choice is to either use them or abandon them. It may be very difficult or impossible to recover the money invested. These nonreversible decisions justify much more of the manager's time.

Number of Available Alternatives

Some decisions have only two possible alternatives. They are of the "yes or no" or "either, or" type. The manager may find these decisions to be easier and less time-consuming than others that have many alternative solutions or courses of action. Where many alternatives exist, such as selecting seed varieties, the manager may be forced to spend considerable time identifying the alternatives and analyzing each one.

THE DECISION-MAKING ENVIRONMENT IN AGRICULTURE

Is managing a farm or ranch greatly different from managing other types of businesses? The basic functions, principles, and techniques of management are the same everywhere, but a typical farm or ranch business has some unique characteristics that affect the way decisions are made.

Biological Processes and Weather

One distinguishing characteristic of agriculture is the limitation placed on a manager's decisions by the biological and physical laws of nature. Managers soon find there are some things they cannot control. Nothing can be done to shorten the gestation period in livestock production, there is a limit on how much feed a pig can consume in a day, and crops require some minimum time to reach maturity. Even attempts to control the effects of climate with irrigation equipment and confinement buildings can be thwarted by sudden rainstorms and blizzards.

The unpredictability of the production process is unique to agriculture. Not even the best manager can forecast with certainty the effects of variations in rainfall, temperature, disease, or genetic combinations. This introduces an element of risk that most nonfarm businesses do not face.

Fixed Supply of Land

In most industries, a business can buy more raw materials or replicate production facilities when demand for their products increases. However, the supply of the most valuable resource in agricultural production, land, is essentially fixed. Farm managers can only try to increase the productivity of the existing land base or outbid other farms for the limited amount of land available for sale or rent. This makes decisions about buying, selling, and renting land particularly crucial. It also causes the sale and rental prices of farmland to be especially sensitive to changes in the prices of agricultural commodities and inputs.

Small Size

Eighty-six percent of the farms in the United States are sole proprietorships. Even many farm partnerships and corporations are small, family-owned businesses. No other sector of the economy is so dominated by small-scale operations. By contrast, in a large corporation the stockholders own the business, while the board of directors sets goals, defines policies, and hires managers to follow them. It is generally easy to identify three distinct groups: owners, management, and labor. These distinct groups do not exist on the typical family farm or ranch, where one individual or a small group owns the business, provides the

management, and contributes most or all of the labor. This makes it difficult to separate the management activity from labor, because the same individuals are involved. It also causes the constant need for labor "to get a job done," placing management in a secondary role, with decisions constantly being delayed or ignored. Moreover, when the manager's residence and home life are located on the farm, personal and business goals and activities become more intertwined. Except on large, multi-person farms, there is little opportunity for task specialization among personnel.

Perfect Competition

Production agriculture is often used as an example of a perfectly competitive industry. This means that each individual farm or ranch is only one of many and represents a very small part of the total industry. Therefore, the individual manager usually cannot affect either the prices paid for resources or the prices received for products sold. Prices are determined by national and international supply and demand factors, over which an individual manager has very little control except possibly through some type of collective action. Some managers attempt to overcome this by finding niche markets or very localized markets where they are the only suppliers. Examples include growers of organic produce or exotic animal species.

All these factors taken together create a unique environment for making management decisions in agriculture. The remaining chapters will explain in more detail the principles and tools used by farm and ranch managers, and the particular types of problems to which they apply them.

SUMMARY

Good management usually means the difference between earning a profit or suffering a loss in a commercial farm or ranch business. Managers must make plans for the farm business, implement the plans, and monitor their success.

The overall direction in which the farm is headed is defined through the process of strategic planning. A strategic plan begins with a vision statement of why the business exists. Goals provide the direction and focus for the process, and reflect the values of the managers. After assessing both the internal resources and the external environment of the business, strategies can be identified and selected. Finally, the strategies must be implemented and the results monitored. The strategic plan is carried out by making a large number of short-run tactical decisions. Tactical decisions are made by defining the problem, gathering information about and analyzing alternative solutions, choosing and implementing an alternative, and evaluating the results.

Farm and ranch managers operate in a different environment than managers of other businesses. Agriculture relies heavily on biological processes; the supply of farmland is essentially fixed; many of the same people combine ownership, labor, and management; and the firms usually operate in a perfectly competitive economic environment.

QUESTIONS FOR REVIEW AND FURTHER THOUGHT

1. What is your own definition of management? Of a farm manager?

2. Do farm and ranch managers need different skills than managers of other businesses? If so, which skills are different? Which are the same?

3. How do strategic management and tactical management differ?

4. Would you classify the following decisions as strategic or tactical?
   a. deciding whether a field is too wet to till
   b. deciding whether to specialize in beef or dairy production
   c. deciding whether to take a partner into the business
   d. deciding whether to sell barley today or wait until later

5. Why are goals important? List some examples of long-term goals for a farm or ranch business. Make them specific and measurable, and include a time line.

6. What are some common goals of farm and ranch families that might be in conflict with each other?

7. What are your personal goals for the next week? For next year? For the next 5 years?

8. What internal characteristics of the farm should a manager consider when developing a strategic plan?

9. Identify several trends in technology or consumer tastes that a farm manager should consider when developing a strategic plan.

10. List the steps in the decision-making process. Which steps are part of the planning function of management? Implementation? Control?

11. What characteristics of a decision affect how much time and effort a manager devotes to making it?

12. What are some characteristics of agriculture that make managing a farm or ranch different from managing other businesses?

MEASURING MANAGEMENT PERFORMANCE

T wo points made in Chapter 2 were: (1) setting goals is important, and (2) control is one of the functions of management. Part Two will assume a goal of profit maximization. It will describe how to measure and analyze profit and other financial characteristics of a farm or ranch business. The results allow the manager to determine how well and to what degree this goal is being attained.

This discussion is also related to the control function of management. Control is a monitoring system to see whether the farm's business plan is being followed and how close the farm is toward meeting the goal(s) of the plan. Many of the same records needed to measure profit and the financial status of the business are also needed to perform the control function of management. They provide a method of measuring not only how well the business is doing but also how well the manager or management is doing.

This discussion introduces the need for a record-keeping or accounting system for the farm or ranch business. There are many choices and options in the design and implementation of a farm record system, and they range from the very simple to the very complex. The right system for any business will depend on many factors, including the size of the business, the form of business organization, the amount of capital borrowed, lender requirements, and what specific financial reports are needed and in what detail.

Chapter 3 discusses some of these choices and options in a farm record system. It also covers the purposes and parts of a farm record system. Chapter 4 contains a discussion of depreciation, depreciation methods, and valuation of assets. The next two chapters cover the two most common financial reports: the balance sheet and the income statement. Chapter 5 emphasizes that a balance sheet is not intended to measure profit but to record the financial condition of the business at a point in time. It provides the data needed to evaluate the overall financial health of the business.

Chapter 6 introduces the income statement, which provides an estimate of profit or net farm income for an accounting period. The accuracy of the reported profit depends on many factors, including the type of record system employed and the effort put forth to keep good records. Many of the choices and options discussed in Chapter 3 will be shown, along with their effect on the estimated profit. The emphasis will be on understanding what it takes to accurately measure profit or net farm income. Without an accurate measurement, the effects of past management decisions will be distorted, and poor information will be used to make future decisions.

ACQUIRING AND ORGANIZING MANAGEMENT INFORMATION

CHAPTER OUTLINE

CHAPTER OBJECTIVES

1. To appreciate the importance and value of establishing a good farm or ranch accounting system

2. To discuss some choices that must be made when selecting an accounting system

3. To outline the basic concepts of the cash accounting method

4. To present the basic concepts of the accrual accounting method and to compare them with cash accounting

5. To review some recommendations of the Farm Financial Standards Council related to choice of accounting method

6. To introduce some financial records that can be obtained from a good accounting system

A business with poor or no records can be likened to a ship in the middle of the ocean that has lost the use of its rudder and navigational aids. It does not know where it has been, where it is going, or how long it will take to get there. Records tell the manager where the business has been and whether it is now on the path to making profits and creating financial stability. Records are, in one respect, the manager's "report card," because they show the results of management decisions over past time periods. Records may not directly show where a business is going, but they can provide considerable information to correct or amend past decisions and to improve future decision making. In that way, they at least influence the future direction of the business.

For a number of reasons, farm and ranch records have traditionally been poorly kept. Even the better record systems have not been totally consistent with the standards that the accounting profession follows for other types of businesses. Financial problems on farms and ranches during the 1980s focused attention on the poor records kept by many farmers and ranchers, the many different styles and formats of financial reports being used, differences in terminology, and inconsistent treatment of some accounting transactions that are unique to agriculture.

PURPOSE AND USE OF RECORDS

Several uses for records have already been mentioned. A more detailed and expanded list of the purpose and use of farm records is possible. One such list is:

1. Measure profit and assess financial condition.
2. Provide data for business analysis.
3. Assist in obtaining loans.
4. Measure the profitability of individual enterprises.
5. Assist in the analysis of new investments.
6. Prepare income tax returns.

This is not a complete list of all possible reasons for keeping and using farm records. Other possible uses are for establishing insurance needs, planning and valuing estates, monitoring inventories, dividing landlord/tenant expenses, reporting to partners and shareholders, and developing marketing plans. However, the six uses in the list are the more common and will be discussed in more detail.

Measure Profit and Assess Financial Condition

These two reasons for keeping and using farm records are among the more important. Profit is estimated by developing an income statement, which will be the topic for Chapter 6. The financial condition of the business as shown on a balance sheet will be covered in detail in Chapter 5.

Provide Data for Business Analysis

After the income statement and balance sheet are prepared, the next logical step is to use this

---

Box 3-1   Farm Financial Standards Council

A Farm Financial Standards Task Force (FFSTF) was formed in 1989 to address accounting and record problems on farms and ranches. Their 1991 report included recommendations to bring farm records into closer compliance with basic accounting principles. They also recommended more standardized and consistent financial reports, uniform financial terminology, procedures to handle the more difficult and unusual accounting transactions unique to farms and ranches, and specific measures to analyze farm financial reports. The group's name was changed to the Farm Financial Standards Council (FFSC), which issued revised reports in 1995 and 1997. This and later chapters of this text follow the recommendations of the FFSC.

information to do an in-depth business analysis. There is a difference between just making "a profit" and having "a profitable" business. Is the business profitable? How profitable? Just *how* sound is the financial condition of the business? The answers to these and related questions require more than just preparing an income statement and balance sheet. A financial analysis of the business can provide information on the results of past decisions, and this information can be very useful when making current and future decisions.

Assist in Obtaining Loans

Lenders need and require financial information about the farm business to assist them in their lending decisions. After the financial difficulties of the 1980s, many agricultural lenders and bank examiners are requiring more and better farm records. Good records can greatly increase the odds of getting a loan approved and receiving the full amount requested.

Measure the Profitability of Individual Enterprises

A farm or ranch that is showing a profit may include several different enterprises. It is possible that one or two of the enterprises are producing all or most of the profit, and one or more of the other enterprises are actually losing money. A record system can be designed that will show revenue and expense not only for the entire business but for each individual enterprise. With this information, the unprofitable or least profitable enterprises can be eliminated, and resources can be redirected for use in the more profitable ones.

Assist in the Analysis of New Investments

A decision to commit a large amount of capital to a new investment can be difficult and may require a large amount of information to do a proper analysis. The records from the past operation of the business can be an excellent source of information to assist in analyzing the potential investment. For example, records on the

same or similar investments can provide data on expected profitability, expected life, and typical repairs over its life.

Prepare Income Tax Returns

Internal Revenue Service (IRS) regulations *require* keeping records that permit the proper reporting of taxable income and expenses. This can often be done with a minimal set of records that is not adequate for management purposes. A more detailed management accounting system can actually produce income tax benefits. It may identify additional deductions and exemptions, for example, and allow better management of taxable income from year to year, reducing income taxes paid over time. In case of an IRS audit, good records are invaluable for proving and documenting all income and expenses.

FARM BUSINESS ACTIVITIES

In designing a farm accounting system, it is useful to think of the three types of business activities that must be incorporated into the system. Figure 3-1 indicates that an accounting system must be able to handle transactions relating not only to the *production* activities of the business, but also to the *investment* and *financing* activities.

Production Activities

Accounting transactions for production activities are those related to the production of crops and livestock. Revenue from their sale, and

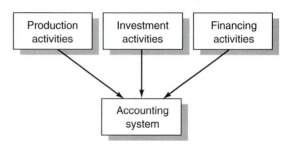

Figure 3-1   Farm business activities to be included in an accounting system.

other farm revenue such as government program payments and custom work done for others, would be included here. Expenses incurred in producing that revenue—such as feed, fertilizer, chemicals, fuel, interest, and depreciation—are also part of the production activities that need to be recorded in the accounting system.

Investment Activities

These activities are those related to the purchase, depreciation, and sale of long-lived assets. Examples would be land, buildings, machinery, orchards, vineyards, and breeding livestock. Records kept on each asset should include purchase date, purchase price, annual depreciation amount, book value, current market value, sale date, sale price, and gain or loss when sold.

Financing Activities

Financing activities are all transactions related to borrowing money and paying interest and principal on debt of all kinds. This includes money borrowed to finance new investments, operating money borrowed to finance production activities for the year, and accounts payable at farm supply stores.

Dividing the farm business activities into these three types illustrates the broad range of transactions that should be recorded in any accounting system. It also shows some interrelations among these activities. Interest expense comes from financing activities, but it is a production or operating expense. Depreciation results from investment in a depreciable asset, but it is also a production or operating expense. Therefore, a good accounting system must be able to not only record all the various types of transactions but also to assign them to the appropriate activity and enterprise of the operation.

BASIC ACCOUNTING TERMS

A person does not need an accounting degree to keep and analyze a set of farm or ranch records. However, some knowledge of basic accounting and accounting terminology is extremely useful. It is needed to fully understand and use any accounting system and to accurately communicate accounting information to others. The following terms and definitions provide the foundation to understand the material in the remainder of this chapter and those that follow. Other terms will be defined as they are introduced.

Account payable—an expense that has been incurred but not yet paid. Typical accounts payable are for items charged at farm supply stores where the purchaser is given 30 to 90 days to pay the amount due.

Account receivable—revenue for a product that has been sold or a service provided but for which no payment has yet been received. Examples would be custom work done for a neighbor who has agreed to make payment by the end of next month, or grain sold on a deferred payment contract.

Accrued expense—an expense that accrues or accumulates daily but has not yet been paid. An accrued expense has typically not been paid because the due date or payment date is still some time in the future. Examples are interest on loans and property taxes.

Asset—any item of value, tangible or financial. On a farm or ranch, examples would be machinery, land, bank accounts, buildings, grain, and livestock.

Credit—in accounting, a credit is simply an entry in the right-hand side of a double-entry ledger. A credit entry is used to record a decrease in the value of an asset or an increase in a liability, an owner equity, or an income account.

Debit—an entry in the left-hand side of a double-entry ledger. A debit entry is used to record an increase in an asset or expense account, and a decrease in a liability or owner equity account.

Expense—a cost or expenditure incurred in the production of revenue.

Inventory—the physical quantity and financial value of products produced for sale that have not yet been sold. Farm or ranch examples would be grain in storage, or livestock that are ready for sale or could be sold at the time the inventory is taken.

Liability—a debt or other financial obligation that must be paid at some time in the future. Examples would include loans from a bank or other lending institution, accounts payable, and accrued expenses.

Net farm income—revenue minus expenses. The same as profit.

Owner equity—the difference between business assets and business liabilities. It represents the net value of the business to the owner(s) of the business.

Prepaid expense—a payment made for a product or service in an accounting period before the one in which it will be used to produce revenue.

Profit—revenue minus expenses. The same as net farm income.

Revenue—the value of products and services produced by a business during an accounting period. Revenue may be either cash or noncash.

OPTIONS IN CHOOSING AN ACCOUNTING SYSTEM

Before anyone can begin entering transactions into an accounting system, several decisions must be made about the type of system to be used. A number of options are available and generally fall into the following areas:

1. What accounting period should be used?
2. Should it be a cash or an accrual system?
3. Should it be a single- or double-entry system?
4. Should it be a basic or complete accounting system?

It is often difficult to make certain types of changes in an accounting system once one is established and users are familiar with it. Therefore, considerable thought should be given and advice obtained when making the initial selection of an accounting system.

Box 3-2   Debits and Credits

It is easy to fall into the trap of thinking all debits are "bad" or decreases and all credits are "good" or increases. Accounting procedures make no such distinction. A debit is simply an entry in the left column of a two-column ledger account and a credit is an entry in the right column. Whether a debit or credit is "good" or "bad" or an increase or a decrease depends on the type of account. For example, a checking account is an asset account where a debit is an increase and a credit a decrease in the account balance. The receipt of cash for a sale would be recorded in the debit column of the checking account and the same amount in the credit column of a sales account, a revenue account. In this example, both the debit and credit are "good" representing an increase in both sales and cash. When a check is written to pay a bill the appropriate expense account has a debit entry and the checking account a credit entry. A debit entry to an expense account increases it and the credit entry to the checking account decreases it. Both the debit and credit entries have "bad" results, an increase in expenses and a decrease in cash.

REMEMBER, debit means left column and credit means right column. Nothing more can be inferred until something is known about which accounts are affected by the transaction.

Accounting Period

An accounting period is that period of time used to summarize revenue and expenses and estimate profit. It can be either a *calendar year* or a *fiscal year.* With a calendar-year accounting period, all transactions occurring between January 1 and December 31 of each year are organized and summarized into financial reports. Reporting can be done for shorter periods, such as quarterly, but everything would still be consolidated into annual reports. A fiscal year is any 12-month period that begins on some date other than January 1. Accounting can be done on a fiscal-year basis both for management purposes and for income tax purposes.

It is generally recommended that a firm's accounting period follow the production cycle of the major enterprises and end at a time when business activities are slow. For most crop production activities and some livestock, a December 31 ending date fits this recommendation, so most farmers and ranchers use a calendar-year accounting period. However, winter wheat, citrus crops, and winter vegetables are examples of crops where intensive production or harvesting activities may be underway around December 31. These producers may want to consider a fiscal-year accounting period that ends after harvesting is completed. Large dairies and commercial feedlots with continuous feeding activities would have a difficult time finding a month when business transactions are much slower than any other month. They could just as well use a calendar-year or a fiscal-year accounting period.

Cash Versus Accrual Accounting

This topic will be covered again when discussing income taxes in Chapter 16. However, the discussion here will be restricted to accounting for management purposes and not for income taxes. While the concepts are the same in either case, the advantages and disadvantages of each accounting method may be different, depending on the use for management or for income tax purposes. The basics of cash and accrual accounting will be discussed in later sections of this chapter.

Single Versus Double Entry

With a single-entry cash system, only one entry is made in the books to record a receipt or expenditure. A sale of wheat would simply have the dollar amount recorded under the "Grain Sales" column in the ledger. A check written to pay for feed would have the amount entered under the "Feed Expense" column. The other side of the transaction is always assumed to be cash, which changes the balance in the checking account. In practice, the checkbook register might be thought of as the other entry, but one that is not included in the ledger.

A double-entry system records changes in the values of assets and liabilities as well as revenue and expenses. There must be equal and offsetting entries for each transaction. This system will result in more transactions being recorded during an accounting period, but it has two important advantages:

1. improved accuracy, because the accounts can be kept in balance more easily
2. the ability to produce complete financial statements, including a balance sheet, at any time, directly from data already recorded in the system

The improved accuracy of double-entry accounting comes from the two offsetting entries, which means that *debits* must equal *credits* for each transaction recorded. It also means that the basic accounting equation of

$$\text{Assets} = \text{Liabilities} + \text{Owner Equity}$$

will be maintained. The double-entry system maintains the current values of assets and liabilities within the accounting system, allowing financial statements to be generated directly from the accounting system without any need for outside information.

Basic Versus Complete System

The most basic and simple accounting system would be one which is manual and uses cash accounting only. A complete system would be computerized with capabilities for both cash accounting for tax purposes and accrual accounting for management purposes. It would also be able to track inventories, compute depreciation, track loans, perform enterprise analysis, and handle all employee payroll accounting.

Between these two extremes are many possibilities. For example, a simple basic system can be maintained on a computer with any one of a number of personal finance software programs available. The next step up would be any of several small business accounting programs which can be used for farm accounting by simply changing the account names. These programs can do either cash or basic accrual accounting. Most are inexpensive but relatively powerful programs which often do a good job of "disguising" the fact that the user is dealing with debits and credits. Very little accounting knowledge is needed to use these programs and to get useful and accurate output.

Most farmers and ranchers desiring a complete system use computer software accounting programs written specifically for agricultural use. These programs use agricultural terminology and are designed to handle many of the situations unique to agriculture such as accounting for raised breeding livestock, government farm program payments and commodity loans, and quantities of product sold or in inventory. Many also allow the user to maintain both cash and accrual records for each year within the same program. This makes it easy to pay income taxes on a cash basis while still having complete accrual records for making management decisions. Some software companies sell a basic program along with a number of optional add-on modules. This allows the user to become familiar with the program and its basic output without being overwhelmed by its complexity. Later, modules for inventory, depreciation, payroll, and production records can be added as needed or desired.

How complete the accounting system should be for any given farm business will depend primarily on the answers to three questions:

1. How much accounting knowledge does the user have?
2. How large and complex is the farm business and its financial activities?
3. How much and what kind of information is needed or desired for management decision making?

The lack of accounting training should not deter farmers and ranchers from using one of the more complete software accounting programs. Many of them require only a limited knowledge of accounting. Additional accounting information is available in the form of self-help manuals, adult education courses, and community college courses. The price of the more complete and therefore more expensive farm accounting programs may include some training on using that specific software program as well as free technical assistance for some period of time.

The larger the farm business, the more enterprises involved, the more employees hired, the more depreciable assets owned, and the more money borrowed, the more need there is for a complete accounting system. Software programs that compute depreciation, track inventories, and complete employee payrolls as part of the accounting system become increasingly necessary and useful. The type and quantity of information provided by the accounting system will depend on the knowledge and interest of the user and the size of the business. However, users who begin with a basic computer accounting system often find additional output would be useful and perhaps needed. This is one advantage of beginning with a complete system or one that can be easily upgraded.

BASICS OF CASH ACCOUNTING

The term "cash" in the name is perhaps the best description of this accounting method. With only a few exceptions, no transaction is recorded unless cash is spent or received.

Revenue

Revenue is recorded only when cash is received for the sale of products produced or services provided. The accounting period during which the products were produced or the service provided is not considered when recording revenue. Revenue is recorded in the accounting period when cash is received regardless of when the product was produced or the service provided.

Note that cash accounting can and often does result in revenue being recorded in an accounting period other than the one in which the product was produced. A common example is a crop produced in one year, placed in storage, and sold the following year. Any accounts receivable at the end of an accounting period also result in cash being received in an accounting period after the product was produced or the service provided.

Expenses

Cash accounting records expenses in the accounting period during which they are paid; that is, when the cash is expended. The accounting period in which the product or service was purchased is not considered. Expenses can therefore be recorded in an entirely different year or accounting period than the one in which the purchased product or service generated a product and related revenue. Items may be purchased and paid for late in one year but not used until the next, and items used in one year may not be paid for until the following year. The first case is an example of a prepaid expense, and the latter of an account payable. An accrued expense such as interest is another example. Here, an item (borrowed money) is being used in one year, but the cost of that item (interest) will not be paid until the next year when the annual payment is due.

There is one major exception to the rule that only cash expenses are recorded in a cash accounting system. Although depreciation is a noncash expense, it is generally considered an expense when using cash accounting, if for no other reason than that it is allowed for income tax purposes.

Advantages and Disadvantages

Cash accounting is a relatively simple, easy-to-use system that requires very little knowledge of accounting. It also has some definite advantages for many farmers and ranchers when computing taxable income for income tax purposes.[1]

However, these advantages are offset by one major disadvantage. As noted above, it is very common to have revenue and expenses recorded in a year other than the year the product was produced or the expense was used to produce a product. Therefore, neither the revenue nor the expenses may have any direct relation to the actual production activities for a given year. The result is an estimated profit that may not truly represent the profit from the year's production activities.

This inability of cash accounting to properly match revenue and expenses within the same year that the related production took place is a major disadvantage of this method. Compared to the true profit, the profit shown by cash accounting can be very distorted. It may be overestimated in some years and underestimated in others. If this estimate of profit is then used to make management decisions for the future, the result is often poor decisions.

BASICS OF ACCRUAL ACCOUNTING

Accrual accounting is the standard of the accounting profession. It requires more entries and accounting knowledge than cash accounting. However, it provides a much more accurate estimate of annual profit than does cash accounting.

Revenue

Accrual accounting records as revenue the value of all products produced and all services provided during a year. Whenever the products are produced, sold, and the cash received all in the same year, this is no different than cash accounting. The difference occurs when the product is produced in one year and sold in the next, or

[1]There can be differences between accounting cash revenue and cash revenue for income tax purposes. See *Farmer's Tax Guide,* Publication 225, Department of Treasury, Ch. 3.

when cash for a service is not received until a later year. Accrual accounting emphasizes that the value of a product or service should be counted as revenue in the year it was produced, *no matter* when the cash is received.

The handling of inventories is a major difference between cash and accrual accounting on farms and ranches. Accrual accounting records inventories as revenue. A simple example will show why this is important. Assume a farmer produces a crop but places it all in storage for sale the following year, when prices are expected to be higher. In the year of production, there would be no cash sales but presumably some cash expenses. Under cash accounting, there would be a negative profit (loss) for the year. This result is a poor indicator of the results of the production activities for the year and completely ignores the value of the crop in storage.

Accrual accounting includes an estimate of the value of the crop in storage as revenue in the year it was produced. This is done by adding an inventory *increase* from the beginning to the end of the year to other revenue. An inventory *decrease* is deducted from other revenue. The result is an estimated profit that more accurately describes the financial results of the production activities for the year. In the same manner and for the same reason, any uncollected amounts for services provided (accounts receivable) are also recorded as revenue.

Expenses

One principle of accounting is "matching." This principle states that once revenue for a year is determined, all expenses incurred in the production of that revenue should be recorded in the same year. The results for items purchased and paid for in the same year are the same for cash and accrual accounting. Differences arise when items are purchased in the year before the one in which they produce revenue (prepaid expenses), or payment is not made until the year after the items are used (accounts payable and accrued expenses).

To match expenses with revenue in the proper year, an accounting entry must cause:

1) prepaid expenses to show up as expenses in the year after the item or service was purchased and paid for; 2) accounts payable to be entered as expenses, although no cash has yet been expended to pay for the items; and 3) accrued expenses at year-end to be entered as expenses, although no cash has been expended. A typical example of the latter is interest that has accrued from the last interest payment to the end of the year. This accrued expense recognizes that the borrowed capital was used to produce revenue in one year, but the next cash interest payment may not be due for several months into the following year.

Advantages and Disadvantages

A major advantage of accrual accounting is that it produces a more accurate estimate of profit than can be obtained with cash accounting. Related to this is the accurate information it provides for financial analysis and management decision making. These advantages make accrual accounting the standard of the accounting profession and generally required for all corporations selling stock to the public. The latter requirement ensures that potential investors can base their investment decision on the best financial information possible.

The disadvantages of accrual accounting are primarily the additional time and knowledge required to properly use this method. Also, accrual accounting may not be the best choice for all farmers and ranchers to use when calculating their taxable income.

A CASH VERSUS ACCRUAL EXAMPLE

The differences between cash and accrual accounting and the resulting effect on annual profit may be best explained with an example. Assume that the information at the top of page 44 contains most of the relevant transactions related to producing a crop in the year 20X6. Note, however, that transactions related to the 20X6 crop year occur in three years. How would these transactions be handled with both cash and

Month/Year	Transaction
November 20X5	Purchased, paid for, and applied fertilizer to be used by the 20X6 grain crop, $8,000
May 20X6	Purchased and paid for seed, chemicals, fuel, etc., $25,000
October 20X6	Purchased and charged to account fuel for drying, $3,000
November 20X6	One-half of grain sold and payment received, $50,000. The remaining one-half is placed in storage and has an estimated value of $50,000
January 20X7	Paid bill for fuel used to dry grain, $3,000
May 20X7	Remaining 20X6 grain sold, $60,000

accrual accounting, and what would the estimated profit be with each method? Each transaction will be looked at individually, and then the 20X6 profit for each method will be calculated.

(a) November 20X5

Cash: Increase fertilizer expense by $8,000. The result is a 20X5 fertilizer expense $8,000 higher than it should be, because this fertilizer will not be used to produce a crop or revenue until 20X6.

Accrual: Decrease cash by $8,000, and increase prepaid expense by the same amount. The result is an exchange of one asset (cash) for another (prepaid expense), with no effect on fertilizer expense for 20X5.

(b) May 20X6

Cash: Seed, chemicals, and fuel expenses each increased by their appropriate share of the $25,000.

Accrual: Cash is decreased by $25,000, and seed, chemical, and fuel expenses are each increased by their appropriate share of the $25,000.

(c) May 20X6

Sometime during 20X6, the prepaid expense must be converted to fertilizer expense.

Cash: No entry. It has already been counted as an expense.

Accrual: Decrease prepaid expense, and increase fertilizer expense each by $8,000. The prepaid expense is eliminated, and the $8,000 fertilizer expense now shows up as a 20X6 expense, as it should.

(d) October 20X6

Cash: No entry for the drying fuel, because no cash has been expended.

Accrual: Increase fuel expense and increase account payable each by $3,000. The result places the fuel expense in the proper year (it was used to dry 20X6 grain) and establishes a liability, an account payable in the amount of $3,000.

(e) November 20X6

Cash: Increase grain sold by $50,000.

Accrual: Increase cash by $50,000 and grain revenue by $50,000. As a part of this entry, or as a separate entry, grain revenue should be increased by another $50,000, and inventory (a new asset) also should be increased by $50,000. These entries result in all $100,000 of the 20X6 grain being included in 20X6 revenue, even though cash has been received for only one-half of it.

(f) January 20X7

Cash: Increase fuel expense by $3,000.

Accrual: Decrease cash by $3,000, and decrease account payable by $3,000. This eliminates the account payable but does

	20X6 Profit			
	Cash Accounting		**Accrual Accounting**	
Cash grain sales	50,000 (e)		50,000 (e)	
Grain inventory *increase*	N/A		50,000 (e)	
Total Revenue		$50,000		$100,000
Fertilizer	0		8,000 (c)	
Seed, chemicals, fuel	25,000 (b)		25,000 (b)	
Drying fuel	0		3,000 (d)	
Total Expenses		25,000		36,000
Net Farm Income (Profit)		$25,000		$64,000

not increase fuel expense, because that was done in October 20X6.

(g) May 20X7

Cash: Increase grain revenue by $60,000.

Accrual: Increase cash by $60,000, increase grain revenue by $10,000, and decrease inventory by $50,000. This sale indicates there was more grain in the bin than was estimated in November 20X6, or that the price increased since then. To adjust for either or both of these pleasant outcomes, grain revenue must be increased by $10,000. Inventory is decreased by the original amount to give it a $0 balance. It must have a $0 balance, because all grain has now been sold.

It can be seen from this example that accrual accounting requires more entries and more knowledge of accounting than does cash accounting. However, there are several benefits from this extra work and knowledge. The most important is a more accurate estimate of profit.

Review the transactions again. This business produced grain with a value of $100,000 during 20X6 but only sold one-half of it. The other one-half was in storage at the end of the year. Expenses to produce this grain totaled $36,000. However, $8,000 of this amount was paid in 20X5, and $3,000 was not paid until

20X7. A comparison of the 20X6 profit under both cash and accrual accounting will show how this distribution of cash revenue and cash expenditures affects profit.

Because cash accounting includes only cash receipts and cash expenditures during 20X6, there are only two entries. The calculated profit is $25,000. Accrual accounting, by the use of an inventory change, includes the value of all grain produced in 20X6, even though not all was sold. Similarly, adjustments through the use of a prepaid expense and an account payable record all expenses incurred in producing this grain as 20X6 expenses. The result is a profit of $64,000. This is a much more accurate estimate of what the 20X6 production activities contributed to the financial condition of this business than the $25,000 estimated by cash accounting.

While cash accounting shows a lower profit than accrual accounting in 20X6, note that the opposite may occur in 20X7. Assume all grain produced in 20X7 is sold at harvest. The result is cash receipts for both one-half the 20X6 grain and all the 20X7 crop being received in the same year. Cash accounting would show large cash receipts, only one year's cash expenses, and therefore, a very large profit. Accrual accounting would show the same cash receipts, but this would be offset by an inventory *decrease* of $50,000, because no grain is in storage at the end of 20X7. Net total revenue then would

---

Box 3-3   **Annual Versus Lifetime Profit**

The different results from using cash or accrual accounting only show up in *annual* estimates of profit. Over the full lifetime of the farm business, *total lifetime* before-tax profit would be the same using either method. The two methods only serve to divide the total lifetime profit in a different manner among individual years in the life of the business. Total lifetime, *after-tax* profit could be different because of differences in timing and amounts of annual income tax.

---

be only the value of the 20X7 grain plus the $10,000 increase in value from the 20X6 grain. Again, accrual accounting would result in a more accurate estimate of what the 20X7 production activities contributed to the financial position of the business than would cash accounting.

FARM FINANCIAL STANDARDS COUNCIL RECOMMENDATIONS

The Farm Financial Standards Council (FFSC) report assumes an accrual-based system throughout its discussion and recommendations on financial analysis measures. But it also recognizes that a large majority of farmers and ranchers currently use cash accounting and will continue to do so for some time. Simplicity, ease of use, and very often income tax advantages account for the popularity of cash accounting. Therefore, the FFSC accepts the use of cash accounting during the year but strongly recommends that end-of-year adjustments be made to convert the cash accounting profit to an "accrual adjusted" profit. The latter should then be used for analysis and management decision making.

The type and nature of some of these accrual adjustments may be evident in the cash versus accrual example above. A full discussion of these adjustments and how they should be done will be delayed until Chapter 6, The Income Statement and Its Analysis.

OUTPUT FROM AN ACCOUNTING SYSTEM

Any accounting system should be able to produce some basic financial reports. Computerized accrual systems can generate many different reports. Figure 3-2 expands on Figure 3-1 to show the possible products from an accounting system. The balance sheet and income statement are shown first, for two reasons. First, they are the two most common reports to come out of an accounting system, and second, they are the subjects of the next two chapters. Some other possible reports are often necessary and useful but may not be available from all systems nor widely used.[2]

> **Balance Sheet** The balance sheet is the report that shows the financial condition of the business at a point in time. A detailed discussion of this report and its use will be covered in Chapter 5.
>
> **Income Statement** An income statement is a report of revenue and expenses ending with an estimate of net farm income. This report will be discussed in detail in Chapter 6.

---

[2]The 12 possible reports discussed here were adapted from J. F. Guenthner and R. L. Wittman: *Selection and Implementation of a Farm Record System,* Western Regional Extension Publication WREP 99, Cooperative Extension Service, University of Idaho, 1986.

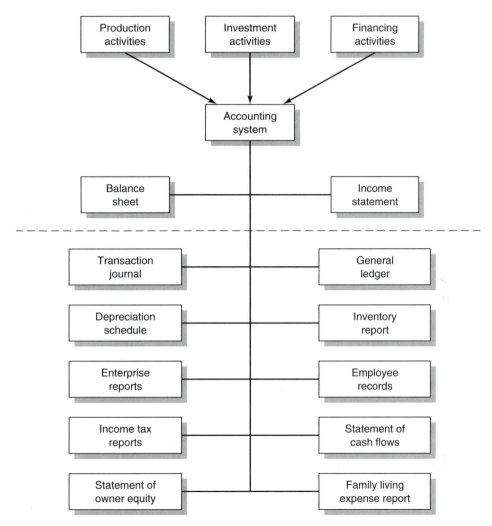

Figure 3-2   **Twelve possible reports from an accounting system.**

Transaction Journal This is a record of all financial transactions, including check and deposit numbers, dates, payees and payers, amounts, and descriptions. A check register is a form of a transaction journal but does not contain all of the above information. This journal is used to make entries into the general ledger and to provide an audit trail.

General Ledger The general ledger contains the different financial accounts for the business and the balances in these accounts. Balances in the revenue and expense accounts are used to prepare an income statement, and balances in the asset, liability, and owner equity accounts are used to prepare a balance sheet.

Depreciation Schedule A depreciation schedule is a necessary part of any accounting system. Annual depreciation on all depreciable assets must be computed and recorded as an expense before an income statement can be produced. This is true whether a cash or accrual system is used. Depreciation and the information contained in a depreciation schedule are discussed in detail in the next chapter.

Inventory Report This is a useful report, particularly for large crop farms and livestock operations. It tracks the quantity and value of crops and livestock on hand by recording purchases, sales, births, deaths, amounts harvested, and amounts fed. This report is useful for monitoring feed availability and usage, for developing a marketing program, and for monitoring any inventory pledged as collateral for a loan.

Enterprise Reports These look like income statements for each individual enterprise. They are very useful for determining which enterprises are contributing the most profit to the business and are therefore candidates for expansion. Enterprises identified as unprofitable become candidates for elimination.

Employee Records Any business with employees must keep considerable data related to each employee. This includes not only information such as hours worked but financial data on gross pay, deductions for income taxes and social security, and so forth. Several payroll reports must be filed in a timely manner with both state and federal agencies. All of this payroll-related work can be done by hand, but there are many computer programs designed especially to compute and record gross pay, deductions, and net pay. If the payroll program is part of a general accounting program, all of this information can be automatically entered into the general ledger.

Income Tax Reports The information from any farm accounting system must be sufficient to prepare the farm tax return. In some systems, it will be necessary to take values from the accounting reports and enter them on IRS Schedule F, Form 1040. It is helpful if the general ledger accounts are named and organized in the same way as the categories on this schedule. Some computer accounting systems can compile and print the tax information in the same format, making it easy to transfer the data. Other programs can actually duplicate a Schedule F and print a completed return.

Statement of Cash Flows This statement summarizes all sources and uses of cash during the accounting period and is useful when analyzing the business activities during that period. When prepared monthly, it allows comparison of actual cash flows with budgeted cash flows. It is also important as a source of data when completing a cash flow budget for the next accounting period.

Statement of Owner Equity Financial transactions during the accounting period will affect the owner equity or net worth of the business. This statement identifies and summarizes the sources of these changes.

Family Living Expense Report While not really a part of the business financial activities, it is desirable to keep detailed records of family living expenses. This is particularly so for any expenditures that may be deductible on income tax returns. Again, this can be done manually or included as part of the farm accounting system, provided care is taken to be sure business and personal records are not mixed. There are also rather inexpensive computer programs designed solely for recording, summarizing, and analyzing personal expenses and investments.

SUMMARY

This chapter discussed the importance, purpose, and use of records as a management tool. Records provide the information needed to measure how well the business is doing in terms of meeting its goals. They also provide feedback so the results of past decisions can be evaluated, as can be the decision-making ability of the manager. Finally, individual farm records are perhaps the single best source of information needed to make current and future decisions.

Any accounting or record system must be able to handle transactions from the production, investment, and financing activities of the farm business. The choices of accounting period, cash or accrual accounting, single- or double-entry accounting, and a basic or complete system are important. They affect the quantity, quality, and accuracy of the information provided by the accounting system and the time required to maintain the records. The output required or desired from an accounting system also must be considered when making these choices.

QUESTIONS FOR REVIEW AND FURTHER THOUGHT

1. Can farmers and ranchers use a fiscal-year accounting period? If so, should they? Under what conditions?

2. How would one construct a balance sheet if the accounting was done using a single-entry, cash system?

3. Is it possible to use double entry with a cash accounting system? If so, what are the advantages and disadvantages?

4. Is it possible to use single entry with an accrual system? Why or why not?

5. Check advertising material for several farm accounting software programs. Are they cash or accrual systems? Single- or double-entry? How many of the 12 reports from an accounting system discussed in this chapter are available from each program? Are any additional reports available?

6. Place an "X" under the column(s) to indicate whether each business event is an operating, investment, or financing activity.

Event	Operating	Investment	Financing
Pay cash for tractor repairs	X		
Borrow $10,000			X
Pay interest on the loan			X
Charge $1,000 of feed	X		
Equipment depreciates	X	X	
Sell $12,000 of corn	X		
Purchase pickup with a $15,000 loan			
Pay principal on a loan			X

7. Explain the difference between an account payable and an account receivable.

8. What products might a typical farm or ranch have in inventory at the end of a year?

9. What is the basic accounting equation? What part of it would a business owner be most interested in seeing increase over time?

10. Why are the results from an accrual accounting system recommended for use when making management decisions?

DEPRECIATION AND ASSET VALUATION

CHAPTER OBJECTIVES

1. To define depreciation and related terms

2. To illustrate the different methods of computing depreciation

3. To compare economic and income tax depreciation

4. To outline the different methods that can be used to value farm and ranch assets

Establishing and maintaining a set of records often requires computing depreciation for depreciable assets and determining the value of all types of assets. These functions are related, because depreciation is a factor that determines value for depreciable assets. This chapter will examine depreciation, methods for computing depreciation, and alternatives for valuing assets.

DEPRECIATION

Machinery, buildings, and similar assets are purchased because they are required or helpful in the production of farm products, which in turn produce revenue. Their use in the production process over time causes them to grow old, wear out, and become less valuable. This loss in value is considered a business expense, because it is a direct result of the asset's use in producing revenue and profit.

Depreciation is often defined as the annual loss in value due to use, wear, tear, age, and technical obsolescence. It is both a business expense that reduces annual profit and a reduction in the value of the asset. What types of assets would be depreciated? To be depreciable, an asset must have the following characteristics:

1. a useful life of more than one year
2. a determinable useful life but not an unlimited life
3. a use in a business in order for the depreciation to be a *business* expense (loss in value on a personal automobile or personal residence is not a business expense)

Examples of depreciable assets on a farm or ranch would be vehicles, machinery, equipment, buildings, fences, livestock and irrigation wells, and purchased breeding livestock. Land is not a depreciable asset, because it has an unlimited life. However, some improvements to land, such as drainage tile, can be depreciated.

How much does an asset depreciate each year? Several methods or mathematical equations can be used to compute annual depreciation. However, it should always be remembered that they are only estimates of the actual loss in value. The true depreciation can be determined only by finding the asset's current market value and comparing it to the original cost. This would require an appraisal or the actual sale of the asset.

Several terms used with depreciation must be defined before reviewing the depreciation methods. *Cost* is the price paid for the asset, including taxes, delivery fees, installation, and any other expenses directly related to placing the asset into use. *Useful life* is the number of years the asset is expected to be used in the business. This may be something less than the total potential life of the asset if it will be traded or sold before it is completely worn out. *Salvage value* is the expected market value of the asset at the end of its assigned useful life. Therefore, the difference between cost and salvage value is the total depreciation or loss in value expected over the useful life.

A salvage value will generally be some positive value. However, it may be zero if the asset will be used until it is completely worn out and will have no scrap or junk value at that time. Note that there should be a relation between useful life and salvage value. The shorter the useful life, the higher the salvage value, and vice versa.

Book value is another term related to depreciation. It is equal to the asset's cost less accumulated depreciation. The latter is all depreciation from purchase date to the current date. Note that book value will always be somewhere between cost and salvage value. It will never be less than salvage value, and it will equal salvage value at the end of the asset's useful life. While book value is one way to determine the value of an asset, it should not be confused with market value. Both useful life and salvage value are only estimates made at the time the asset is purchased. It is unlikely that either or both will turn out to be the actual values. Therefore, book value and market value will be equal only by chance.

DEPRECIATION METHODS

Several methods or equations can be used to compute annual depreciation. There is no single correct choice for every business or every asset. The correct choice will depend on the type of asset, its pattern of use over time, how rapidly or slowly its market value declines, and other factors. The three most common depreciation methods are illustrated in this section.

Straight Line

The straight-line method of calculating depreciation is widely used. This easy-to-use method gives the same annual depreciation for each full year of an item's life.

Annual depreciation can be computed from the equation

$$\text{Annual depreciation} = \frac{\text{cost} - \text{salvage value}}{\text{useful life}}$$

Straight-line depreciation can also be computed by an alternative method using the equation

$$\text{Annual depreciation} = (\text{cost} - \text{salvage value}) \times R$$

where R is the annual straight-line percentage rate found by dividing 100 percent by the useful life (100 percent ÷ useful life).

Box 4-1 **The Accounting Explanation of Depreciation**

There is another way to explain the concept of depreciation as an annual expense. Consider the purchase of a depreciable asset that costs $100,000 and produces revenue for 10 years. What if the entire $100,000 was considered a business expense in the year of purchase? Expenses relative to revenue would be too high in the purchase year and too low (no expense) for the next nine years. Profit would be distorted each year of the asset's life. There needs to be some way to divide the cost of the asset over the years it will produce revenue, rather than calling the entire purchase price an expense in the purchase year. In other words, the asset's cost should somehow be allocated to or matched with the time periods during which it produces revenue. This is done with an annual depreciation expense. Accountants think of depreciation as a way to distribute a long-lived asset's cost over its productive life. Accounting principles do not permit the full purchase price to be called an expense in the year of purchase.

For example, assume the purchase of a machine for $10,000 that is assigned a $2,000 salvage value and a 10-year useful life. The annual depreciation using the first equation would be

$$\frac{\$10,000 - \$2,000}{10 \text{ years}} = \$800$$

Using the second equation, the percentage rate would be 100 percent $\div$ 10, or 10 percent, and the annual depreciation is

$$(\$10,000 - \$2,000) \times 10\% = \$800$$

The result is the same for either procedure, and the total depreciation over 10 years would be $800 $\times$ 10 years = $8,000, reducing the machine's book value to its salvage value of $2,000.

Sum-of-the-Year's Digits (SOYD)

The annual depreciation using the sum-of-the-year's digits method is computed from the equation

$$\frac{\text{Annual}}{\text{depreciation}} = (\text{cost} - \text{salvage value}) \times \frac{\text{RL}}{\text{SOYD}}$$

where RL = remaining years of useful life as of the beginning of the year for which depreciation is being computed

SOYD = sum of all the numbers from 1 through the estimated useful life. For example, for a 5-year useful life, SOYD would be $1 + 2 + 3 + 4 + 5 = 15$, and it would be 55 for a 10-year useful life.[1]

Continuing with the same example used in the previous section, the SOYD would be 55 (the sum of the numbers 1 through 10). The annual depreciation would be computed in the following manner:

Year 1: $(\$10,000 - \$2,000) \times \dfrac{10}{55} = \$1,454.55$

Year 2: $(\$10,000 - \$2,000) \times \dfrac{9}{55} = \$1,309.09$

Year 3: $(\$10,000 - \$2,000) \times \dfrac{8}{55} = \$1,163.64$

$\cdot \qquad \cdot \qquad \cdot \qquad \cdot \qquad \cdot$

$\cdot \qquad \cdot \qquad \cdot \qquad \cdot \qquad \cdot$

$\cdot \qquad \cdot \qquad \cdot \qquad \cdot \qquad \cdot$

Year 10: $(\$10,000 - \$2,000) \times \dfrac{1}{55} = \$145.45$

---

[1] A quick way to find the sum-of-the-year's digits (SOYD) is from the equation $[(n)(n + 1)] \div 2$ where n is the useful life. For example, SOYD for a 10-year life is $[(10)(11)] \div 2 = 55$.

Notice that the annual depreciation is highest in the first year and declines by a constant amount each year thereafter.

Declining Balance

There are a number of variations or types of declining-balance depreciation. The basic equation for all types is

$$\begin{array}{c} \text{Annual} \\ \text{depreciation} \end{array} = \left(\begin{array}{c} \text{book value at} \\ \text{beginning of year} \end{array}\right) \times R$$

where R is a constant percentage value or rate. The same R value is used for each year of the item's life and is multiplied by the book value, which declines each year by an amount equal to the previous year's depreciation. Therefore, annual depreciation declines each year with this method. Notice that the percentage rate is multiplied by each year's book value, *not* cost minus salvage value as was done with the straight-line and SOYD methods.

The various types of declining balance come from the determination of the R value. For all types, the first step is to compute the straight-line percentage rate, which was 10 percent in our previous example. The declining balance method then uses a multiple of the straight-line rate, such as 200 (or double), 175, 150, or 125 percent, as the R value. If double declining balance is chosen, R would be 200 percent, or two times the straight-line rate. The R value would be determined in a similar manner for the other variations of declining-balance depreciation.

Using the previous example, the double declining balance rate would be 2 times 10 percent, or 20 percent, and the annual depreciation would be computed in the following manner:

Year 1: $10,000 $\times$ 20% = $2,000
Year 2: $ 8,000 $\times$ 20% = $1,600
Year 3: $ 6,400 $\times$ 20% = $1,280
. . . .
. . . .
. . . .
Year 7: $ 2,622 $\times$ 20% = $524

Year 8: $ 2,098 $\times$ 20% = $420 (but this amount would reduce the book value below the $2,000 salvage value, so only $98 of depreciation can be taken)

Years 9 and 10: No remaining depreciation

This example is not unusual, as double declining balance will often result in the total allowable depreciation being taken before the end of the useful life, and depreciation must stop when the book value equals salvage value. Notice also that the declining-balance method will never reduce the book value to zero. With a zero salvage value, it is necessary to switch to straight-line depreciation at some point to get all the allowable depreciation, or to take all remaining depreciation in the last year.

If a 150 percent declining balance is used, R is one and one-half times the straight-line rate, or 15 percent for this example. The annual depreciation for the first 3 years would be

Year 1: $10,000 $\times$ 15% = $1,500
Year 2: $ 8,500 $\times$ 15% = $1,275
Year 3: $ 7,225 $\times$ 15% = $1,084

Notice that each year's depreciation is smaller than if the double declining balance is used, and therefore, the book value declines at a slower rate.

Partial-Year Depreciation

The example used here assumed that the asset was purchased at the beginning of the year, with a full year's depreciation that year. An asset purchased during the year should have the first year's depreciation prorated according to the length of time it was actually owned. For example, a tractor purchased on April 1 would be eligible for 9/12 of a full year's depreciation, and a pickup purchased on October 1 would get only 3/12 of a year's depreciation in the year of purchase. Any time there is a partial-year depreciation the first year there will be less than a full year of depreciation in the final year of depreciation.

Comparing Depreciation Methods

Figure 4-1 graphs the annual depreciation for each depreciation method illustrated, based on a $10,000 asset with a $2,000 salvage value and a 10-year life. The annual depreciation is very different for each method. Double declining balance and sum-of-the-year's digits have a higher annual depreciation in the early years than straight line, with the reverse being true in the later years.

It is important to note that the choice of depreciation method does not change the *total* depreciation over the useful life. That is determined by the cost and salvage value ($10,000 − $2,000 = $8,000 in this example) and is the same regardless of depreciation method. The different methods only spread or allocate the $8,000 in a different pattern over the ten-year life. The most appropriate depreciation method will depend on the type of asset and the use to be made of the resulting book value. For example, the ac-

tual market value of vehicles, tractors, and other motorized machinery tends to decline most rapidly during the first few years of life and more slowly in the later years. If it is important for depreciation on these items to approximate their actual decline in value as closely as possible, double declining balance or sum-of-the-year's digits should be used. Assets, such as fences and buildings, have little or no market value without the land they are attached to, and they provide a rather uniform flow of productive services over time. Straight-line depreciation might be the more appropriate depreciation method to use.

Since the amount of annual depreciation depends on the depreciation method used, so does the book value at the end of each year. With straight line, book value declines by the constant annual depreciation each year, i.e. from $10,000 to $2,000 over the useful life. Using SOYD or double declining balance causes book value to

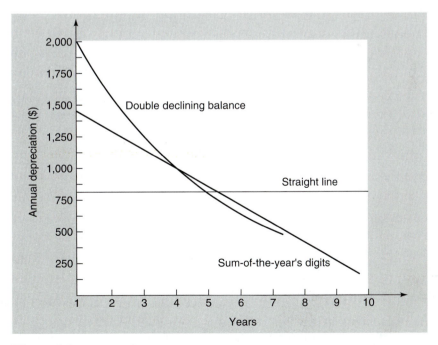

Figure 4-1 **Comparison of annual depreciation for three depreciation methods.**

decrease faster in the early years because annual depreciation is higher. Book value will decline more slowly in the later years of useful life than for straight line. Regardless of the method used though, book value will be equal to salvage value at the end of the useful life.

INCOME TAX DEPRECIATION

Computing depreciation for income tax purposes must be done using the methods and rules allowed by the Internal Revenue Service (IRS). These methods, or systems, differ in a number of ways from those discussed so far. Even though any method can be used for management accounting purposes, most farmers and ranchers use only the tax system in their records. For that reason, a brief description of the income tax depreciation system is included in this chapter (See Chapter 16 for more details.)

The Modified Accelerated Cost Recovery System

The current tax depreciation system is called the Modified Accelerated Cost Recovery System (MACRS). While several alternatives are allowed under this system, most farmers and ranchers use the General Depreciation System (GDS) alternative because it allows for the most rapid depreciation. This will be the only alternative discussed here and will be referred to as regular MACRS.

Regular MACRS has the same basic assumptions for all depreciable property:

- an implied salvage value of $0 for all items
- one-half year of depreciation allowed in the year of purchase regardless of the purchase date (with certain exceptions)
- a system of property classes which fixes the useful life for each type of property

Depreciable assets on farms and ranches generally fall into the 3-, 5-, 7-, 10-, 15-, or 20-year classes. Some examples for each class are:

3-year: breeding hogs

5-year: cars, pickups, breeding cattle and sheep, dairy cattle, computers, trucks

7-year: most farm machinery and equipment, fences, grain bins, silos, office furniture

10-year: single-purpose agricultural and horticultural structures such as confinement swine facilities and greenhouses as well as trees bearing fruit or nuts

15-year: water wells, paved lots, drainage tile

20-year: general-purpose buildings such as machine sheds and hay barns

After the proper class is found for a particular property, the annual depreciation is computed using fixed percentage recovery rates. For agricultural property, the regular MACRS rates are based on the 150% declining balance method. Even though the rates shown in Table 4-1 vary from year to year, they will result in the same annual depreciation as using 150% declining

TABLE 4-1	Regular MACRS Recovery Rates for Farm and Ranch Assets in the 3-, 5-, and 7-Year Classes (Half-Year Convention)

	Recovery percentages		
Recovery year	3-year class	5-year class	7-year class
1	25.00	15.00	10.714
2	37.50	25.50	19.133
3	25.00	17.85	15.033
4	12.50	16.66	12.249
5		16.66	12.249
6		8.33	12.249
7			12.249
8			6.124

Source: IRS Publication 225, Farmer's Tax Guide.

Box 4-2   **"Fast" Depreciation**

Depreciation methods with high depreciation in the early years are often referred to as "fast" depreciation methods or "accelerated" methods. Double declining balance is obviously a "fast" method, and sum-of-the-year's digits is generally considered a "fast" method. Relative to other methods, straight line would have to be considered a "slow" method.

---

balance. However, since the declining balance method will never decrease book value to zero (the assumed salvage value here), the system changes to straight line depreciation in the year when this results in greater depreciation than continuing with 150% declining balance.

As an example of how these rates are used, assume the purchase of a business pickup for $25,000 It would be a 5-year class property and annual depreciation would be computed as follows:

Year 1: $25,000 × 15.00% = $ 3,750
Year 2: $25,000 × 25.50% = $ 6,375
Year 3: $25,000 × 17.85% = $ 4,463
Year 4: $25,000 × 16.66% = $ 4,165
Year 5: $25,000 × 16.66% = $ 4,165
Year 6: $25,000 × 8.33% = $ 2,082
                    Total $25,000

Depreciation for other classes would be computed in the same manner using the appropriate percentages for that class.

Since the percentages sum to 100 percent and are multiplied by the cost of the property, there is no salvage value in this system. Total depreciation will be 100 percent of the cost and book value will be $0 at the end of the class life. Note also that the first year percentage for 5-year class assets should be 30 percent based on the usual computation for 150% declining balance. The 15 percent shown is an adjustment to allow only a half year of depreciation in year 1. The depreciation in year 6 is also computed only for a half year to complete a total of five years of depreciation spread over six tax years.

Economic Versus Tax Depreciation

Economic depreciation can be thought of as the decline in value due to an asset's reduced ability to produce revenue now and in the future. This happens due to a reduction in annual productive capacity and fewer years of life remaining to produce revenue. The depreciation methods discussed earlier are used in accounting to estimate this decline in value and the amount of depreciation to include as an expense for the accounting period.

Like other business expenses, depreciation is a deductible expense for income tax purposes. MACRS is a "fast" depreciation system given the relatively short useful lives allowed and the use of 150% declining balance. With a zero salvage value, 100 percent of the cost is depreciated over the class life regardless of any actual salvage value. As discussed in Chapter 16, some types of property are eligible for an optional depreciation deduction called "expensing." This can be taken only in the first year of ownership but is in addition to regular depreciation. The result of all these factors is an annual depreciation expense for tax purposes which is generally higher to much higher in the early years than the true economic depreciation. In later years this result may be reversed with tax depreciation becoming less than economic depreciation. However, at the end of the asset's useful life, it will have a tax book value of $0 but may still be useable and have a positive market value.

Depreciation must be computed for tax purposes anyway, so it is easy and convenient to use the tax values throughout the accounting system. However, if the tax depreciation is higher (or faster) than the true economic depreciation, there are several results. First, there is a higher depreciation expense during the first few years an asset is owned, and therefore a lower net farm income or profit. The opposite may be true in later years. Second, any item valued by using its book value will have a lower value, in turn causing a lower asset value and owner equity for the business.

The effects and implications of these results will become more evident after reading Chapters 5 and 6. However, it is clear that a misstatement of values can lead to poor conclusions about the financial operations and condition of the business, and hence, poor management decisions. To prevent this, it may be necessary or desirable to complete two depreciation schedules. One would estimate economic depreciation and be part of the records used for business analysis and management decision making. The other would contain tax depreciation and be used only for preparing the income tax return. Table 4-2 is an example of a depreciation schedule that can be used for either purpose.

There are other places to use economic depreciation besides the income statement. It should be used when calculating the annual costs of owning a long-life asset such as machinery (see Chapter 22). Economic depreciation should also be used for many types of budgeting. Best estimates of actual physical life and salvage values should be used, *not* tax lives and salvage values. Using these values in the straight-line equation gives the average annual decline in economic value of the item. This is the proper value to use in cost and budgeting analyses, not tax depreciation. The latter will often be too high based on the relatively short tax lives and the zero salvage value. Using tax values can distort the cost or budget analysis and result in poor management decisions.

VALUATION OF ASSETS

While beginning, maintaining, or improving an accounting system, it is often necessary to estimate the value of business assets. These values may be needed for existing or newly acquired

TABLE 4-2	Depreciation Schedule										
						20___		20___		20___	
Item	Date purchased	Cost or basis	Salvage value	Life	Depreciation method	Depreciation	Book value	Depreciation	Book value	Depreciation	Book value

assets, or for those that have been produced in the normal course of business.

There are several valuation methods that can be used, and the choice will depend on the type of asset and the purpose of the valuation. Whichever method is used, the accounting concepts of "conservatism" and "consistency" should be kept in mind. Conservatism cautions against placing too high a value on any asset, while consistency stresses using the same valuation method or methods over time. Use of these concepts makes financial statements directly comparable from year to year and prevents an overly optimistic portrayal of the firm's financial condition. The following are commonly used valuation methods.

Market Value

This method values an asset using its current market price. It may sometimes be called the fair market value or net market price method. Any normal marketing charges, such as transportation, selling commissions, and other fees, are subtracted to find the *net* market value. This method can be used for many types of property, but it works particularly well for items that could or will be sold in a relatively short period of time as a normal part of the business activities, and for which current market prices are available. Examples would be hay, grain, feeder livestock, stocks, and bonds.

Cost

Items that have been purchased can be valued at their original cost. This method works well for items that have been purchased recently and for which cost records are still available. Land is generally valued at cost for a conservative valuation. Feed, fertilizer, supplies, and purchased feeder livestock are often valued at cost. Items such as buildings and machinery, which normally lose value or depreciate over time, should *not* be valued with this method. Livestock and crops that have been raised cannot be valued at cost, because there is no purchase price to use.

Lower of Cost or Market

This valuation method requires valuing an item at both its cost and its market value and then using whichever value is lower. This is a conservative method, because it minimizes the chance of placing an overly high value on any item. Using this method, property that is increasing in value because of inflation will have a value equal to its original cost. Valuing such property at cost eliminates any increase in value over time caused solely by inflation or a general increase in prices. When prices have decreased since the item was purchased, this method results in valuation at market value.

Farm Production Cost

Items produced on the farm can be valued at their farm production cost. This cost is equal to the accumulated costs of producing the item but should not include profit nor any opportunity costs associated with the production. Grain, hay, silage, and raised livestock can be valued by this method if good cost-of-production or enterprise records are available. Established but immature crops growing in a field are generally valued this way, with the value equal to the actual, direct production expenses incurred to date. It is not appropriate to value a growing crop using expected yield and price, because poor weather, a hailstorm, or lower prices could drastically change the value before harvest and final sale. This procedure is another example of conservatism in valuation. However, a detailed set of enterprise accounts would be needed to use this method.

Cost Less Accumulated Depreciation

Obviously, this valuation method can be used only for depreciable assets such as machinery, buildings, and purchased breeding livestock. Book value was defined earlier as cost less accumulated depreciation. Therefore, this method results in an asset's estimated value being the same as its current book value.

SUMMARY

A depreciation schedule containing the annual depreciation for each depreciable asset is a necessary part of any accounting system. The total annual depreciation expense for the business is necessary for computing annual profit, and depreciation on each individual depreciable asset has an effect on its value. This chapter discussed the concept of depreciation and defined related terms. The three most common accounting methods for computing depreciation were discussed and compared. Next, the Modified Accelerated Cost Recovery System used to compute depreciation for income tax purposes was introduced.

Any accounting system from time to time will require values for business assets. Several possible valuation methods were discussed, along with the usual methods for computing economic depreciation for accounting purposes. Depreciation for tax purposes uses different methods, and these values may not be the best choice to use in an accounting system whose results are used to make management decisions.

Questions for Review and Further Thought

1. Assume that a new tractor is purchased on January 1 for $62,000 and given a salvage value of $10,000 and a useful life of 8 years. What would the annual depreciation be for the first 2 years under each depreciation method?

	Year 1	Year 2
Straight line	_____	_____
Sum-of-the year's digits	_____	_____
Double declining balance	_____	_____

2. For the problem above, what would the tractor's book value be at the end of year 2 under each depreciation method?

3. Rework the answers to question one assuming the purchase date was May 1.

4. Compute the MACRS depreciation for the tractor in question one using Table 4-1. A tractor is a 7-year class life asset.

5. What makes a depreciation method "fast" or "slow"? Give an example of each.

6. What are the advantages and disadvantages of each valuation method? Give examples of property that might be valued by each method. What types of property could or could not be valued by each method?

C H A P T E R

5

THE BALANCE SHEET AND ITS ANALYSIS

CHAPTER OBJECTIVES

1. To discuss the purpose of a balance sheet

2. To illustrate the format and structure of a balance sheet

3. To outline some problems when valuing assets, and the recommended valuation methods for different types of assets

4. To show the differences between a cost-basis and a market-basis balance sheet

5. To define owner equity or net worth and show its importance

6. To analyze a firm's solvency and liquidity through the use of a number of financial ratios derived from the balance sheet

7. To introduce the statement of owner equity and explain its construction

---

> ### *Box 5-1*   **Farm Financial Standards Council**
>
> As a result of the farm financial crisis of the mid-1980s, the Farm Financial Standards Task Force (FFSTF) was formed. Its mission was to develop standards for the reporting and analysis of financial data from agricultural producers. Its 1993 report recommended: a) standards for format and content of financial reports from agricultural producers and, b) financial measures and ratios for analyzing these financial reports along with a standardized method for calculating each measure. Several additions and clarifications have been made to the original report. The FFSTF has also changed its name to Farm Financial Standards Council (FFSC).
>
> The current recommendations of the FFSC form the basis for accuracy, uniformity, and consistency in the reporting and analysis of financial data from farms and ranches. This and following chapters use the standards, definitions, terminology, and analytical measures recommended by the FFSC unless stated otherwise.

Chapter 3 introduced the balance sheet and the income statement as two of the products or outputs from an accounting system. They are both part of a complete set of financial statements but are meant to serve different purposes. A balance sheet summarizes the financial condition of the business *at a point in time,* while an income statement summarizes those financial transactions that affected revenue and expenses *over a period of time.* The purpose of an income statement is to provide an estimate of net farm income or profit, while the balance sheet concentrates on estimating net worth or owner equity by valuing and organizing assets and liabilities.

Most transactions affect both the balance sheet and the income statement. Because transactions can occur daily, the balance sheet can change daily. That is why the point-in-time concept is emphasized when discussing the balance sheet. While intended for different purposes, there is a key relation, or connection, between these two financial statements, which will be discussed in Chapter 6.

PURPOSE AND USE OF A BALANCE SHEET

A balance sheet is a systematic organization of everything "owned" and "owed" by a business or individual at a given point in time. Anything of value owned by a business or individual is called an asset, and any debt or other financial obligation owed to someone else is referred to as a liability. Therefore, a balance sheet is a listing of assets and liabilities that concludes with an estimate of net worth or owner equity. These two terms mean the same thing: the difference between total assets and total liabilities. It is the amount the owner(s) have invested in the business. The "balance" in balance sheet comes from the requirement that the ledger be in balance through the basic accounting equation of

$$\text{Assets} = \text{liabilities} + \text{owner equity.}$$

Rearranging this equation allows finding owner equity once assets and liabilities are known:

$$\text{Owner equity} = \text{assets} - \text{liabilities.}$$

A balance sheet can be completed any time during an accounting period. One of the advantages of a computerized accounting system is the ease with which an interim balance sheet can be prepared, such as might be needed for a loan application. However, most balance sheets are prepared at the end of the accounting period, which is December 31 on most farms and ranches. This procedure allows a single balance sheet to be both an end-of-the-year statement for one accounting period and a beginning-of-the-year statement for the next accounting period. For purposes of comparison and analysis, it is necessary to have a balance sheet available for both the beginning and end of each year.

Measuring the financial position of a business at a point in time is done primarily through the use of two concepts.

1. *Solvency,* which measures the liabilities of the business relative to the amount of owner equity invested in the business. It also provides an indication of the ability to pay off all financial obligations or liabilities if all assets were sold; that is, it measures the degree to which assets are greater than liabilities. If assets are not greater than liabilities, the business is insolvent and a possible candidate for bankruptcy proceedings.
2. *Liquidity,* which measures the ability of the business to meet financial obligations as they come due without disrupting the normal operations of the business. Liquidity measures the ability to generate cash in the amounts needed and at the time it is needed. These cash requirements and possible sources of cash are generally measured only over the next accounting period, making liquidity a short-run concept.

BALANCE SHEET FORMAT

A condensed and general format for a balance sheet is shown in Table 5-1. Assets are shown on the left side or top part of a balance sheet, and liabilities are placed to the right of or below assets.

Assets

An asset has value for one or both of two reasons. First, it can be sold to generate cash, or

TABLE 5-1	General Format of a Balance Sheet		
Assets		**Liabilities**	
Current assets	$100	Current liabilities	$ 60
Noncurrent assets	400	Noncurrent liabilities	200
		Total liabilities	$260
		Owner's equity	240
		Total liabilities and	
Total assets	$500	owner's equity	$500

second, it can be used to produce other goods that in turn can be sold for cash at some future time. Goods that have already been produced, such as grain and feeder livestock, can be sold quickly and easily without disrupting future production activities. They are called *liquid assets.* Assets such as machinery, breeding livestock, and land are owned primarily to produce agricultural commodities that can then be sold to produce cash income. Selling income-producing assets to generate cash would affect the firm's ability to produce future income, so they are less liquid, or illiquid. These assets are also more difficult to sell quickly and easily at their full market value.

Current Assets

Accounting principles require current assets to be separated from other assets on a balance sheet. Current assets are the more liquid assets, which will either be used up or sold within the next year as part of normal business activities. Cash on hand and checking and savings account balances are current assets and are the most liquid of all assets. Other current assets include readily marketable stocks and bonds, accounts and notes receivable (which represent money owed to the business because of loans granted, products sold, or services rendered), and inventories of feed, grain, supplies, and feeder livestock. The latter are livestock held primarily for sale and not for breeding purposes.

Noncurrent Assets

Any asset that is not classified as a current asset is, by default, a noncurrent asset. On a farm or ranch, these would include primarily machinery and equipment, breeding livestock, buildings, and land.

Liabilities

A liability is an obligation or debt owed to someone else. It represents an outsider's claim against one or more of the business assets.

Current Liabilities

Current liabilities must be separated from all other liabilities for the balance sheet to follow basic accounting principles. Current liabilities are financial obligations that will become due and payable within one year from the date of the balance sheet, and will therefore require that cash be available in these amounts within the next year. Examples would be accounts payable at farm supply stores for goods and services received but not yet paid for, and the full amount of principal and accumulated interest on any short-term loans or notes payable. Short-term loans are those requiring complete payment of the principal in one year or less. These would typically be loans used to purchase crop production inputs, feeder livestock, and feed for the feeder livestock.

Loans obtained for the purchase of machinery, breeding livestock, and land are typically for a period longer than 1 year. Principal payments may extend for three to five years for machinery and for 20 years or more for land. However, a principal payment is typically due annually or semiannually, and these payments will require cash within the next year. Therefore, all principal payments due within the next year, whether they are for short-term loans or for noncurrent loans, are included as current liabilities.

The point-in-time concept requires identifying all liabilities that exist as of the date of the balance sheet. In other words, what obligations would have to be met if the business sold out and

ceased to exist on this date? Some expenses tend to accrue daily but are only paid once or twice a year. Interest and property taxes are examples. To properly account for these, *accrued expenses* are included as current liabilities. Included in this category would be interest that has accumulated from the last interest payment on each loan to the date of the balance sheet. Accrued property taxes would be handled in a similar manner, and there may be other accrued expenses, such as wages and employee tax withholdings, that have been incurred but not yet paid. Income taxes on farm income are typically paid several months after the close of an accounting period. Therefore, a balance sheet for the end of an accounting period should also show accrued income taxes, or income taxes payable, as a current liability.

Noncurrent Liabilities

These include all obligations that do not have to be paid in full within the next year. As discussed, any principal due within the next year would be shown as a current liability, and the remaining balance on the debt would be listed as a noncurrent liability. Care must be taken to be sure the current portion of these liabilities has been deducted, and only the amount remaining to be paid after the next year's principal payment is recorded as a noncurrent liability.

Owner Equity

Owner equity represents the amount of money left for the owner of the business should the assets be sold and all liabilities paid as of the date of the balance sheet. It is also called net worth. Equity can be found by subtracting total liabilities from total assets and is therefore the "balancing" amount, which causes total assets to exactly equal total liabilities plus owner equity. Owner equity is the owner's current investment or equity in the business.

Owner equity can and does change for a number of reasons. One common and periodic change comes from using assets to produce crops and livestock, with the profit from these production activities then used to purchase additional assets or to reduce liabilities. This production process takes time, and one of the reasons for comparing a balance sheet for the beginning of the year with one for the end of the year is to study the effects of the year's production on owner equity and the composition of assets and liabilities. Owner equity will also change if there is a change in an asset's value, a gift or inheritance is received, cash or property is contributed or withdrawn from the business, or an asset is sold for more or less than its balance sheet value.

However, changes in the composition of assets and liabilities may not cause a change in owner equity. For example, if $10,000 cash is used to purchase a new machine, owner equity does not change. There is now $10,000 less current assets (cash) but an additional $10,000 of noncurrent assets (the machine). Total assets remain the same, and therefore, so does owner equity. If $10,000 is borrowed to purchase this machine, both assets and liabilities will increase by the same amount, leaving the difference or owner equity the same as before. This purchase will not affect owner equity until its depreciation and resulting loss in value are recognized. Using the $10,000 to make a principal payment on a loan will also have no effect on equity. Assets have been reduced by $10,000, but so have liabilities. Equity will remain the same.

These examples illustrate an important point. The owner equity in a business changes only when the owner puts additional personal capital into the business, withdraws capital from the business, or when the business shows a profit or loss. Changes in asset values due to changes in market prices also affect equity if assets are valued at market value. However, many business transactions only change the mix or composition of assets and liabilities and do not affect owner equity.

Alternative Format

Farm and ranch balance sheets have traditionally included three categories of both assets and liabilities, while the accounting profession uses two, as shown in Table 5-1. The FFSC restates the accounting principle of separating current from noncurrent items on a balance sheet. However, it does not restrict all noncurrent items to a single category. A three-category balance sheet is permitted if the preparer believes such a format would add information and the definition used for intermediate assets and liabilities is disclosed.

The usual categories for the traditional farm or ranch balance sheet are shown in Table 5-2. Both current assets and current liabilities are separated from other assets and liabilities as required by accounting principles. These categories would each contain exactly the same items as mentioned in the discussion of Table 5-1. The difference in the two formats comes in the division of noncurrent assets and liabilities into two categories, intermediate and fixed or long-term.

Intermediate assets are generally defined as less liquid than current assets and with a life greater than 1 year but less than about 10 years. Machinery, equipment, perennial crops, and breeding livestock would be the usual intermediate assets found on farm and ranch balance sheets. Fixed assets are the least liquid and have a life greater than 10 years. Land and buildings are the usual fixed assets.

TABLE 5-2	**Format of a Three-Category Balance Sheet**		
Assets		**Liabilities**	
Current assets	$100	Current liabilities	$60
Intermediate assets	120	Intermediate liabilities	75
Fixed assets	280	Long-term liabilities	125
		Total liabilities	$260
		Owner's equity	$240
		Total liabilities and	
Total assets	$500	owner's equity	$500

Intermediate liabilities are debt obligations where repayment of principal occurs over a period of more than 1 year and up to as long as 10 years. Some principal and interest would typically be due each year, and the current year's principal payment would be shown as a current liability, as discussed earlier. Most intermediate liabilities would be loans where the money was used to purchase machinery, breeding livestock, and other intermediate assets. Long-term liabilities are debt obligations where the repayment period is for a length of time exceeding 10 years. Farm mortgages and contracts for the purchase of land are the usual long-term liabilities, and the repayment period may be 20 years or longer.

Intermediate assets and liabilities are often a substantial and important part of both total assets and total liabilities on farms and ranches. This may explain why the three-category balance sheet came into use in agriculture. However, the FFSC encourages the use of the two-category format and predicts a movement away from the use of a three-category balance sheet.

ASSET VALUATION AND RELATED PROBLEMS

The general approach to asset valuation, and the proper method to use for specific assets, have long been debated. Much of the discussion is whether or not agriculture should use a cost-basis or a market-basis balance sheet. A cost-basis balance sheet is required when following basic accounting principles. It values all assets using the cost, cost less depreciation, or farm production cost methods discussed in Chapter 4. The one general exception would be inventories of grain and market livestock. Stored grain can be valued at market value less selling costs, provided it meets several conditions, which it normally would. While market livestock are not specifically mentioned in accounting principles, the FFSC suggests that there is little difference between grain and market livestock, particularly

when the livestock are nearly ready for market. They recommend market valuation for *raised* market livestock even on a cost-based balance sheet. For *purchased* market livestock, market valuation is acceptable but lower of cost or market is preferred.

A market-basis balance sheet would have all assets valued at market value less estimated selling costs. Long-run inflation could cause land owned for a number of years to have a market value higher than its cost. Inflation and fast depreciation methods also would result in machinery and breeding livestock market values being higher than their book values. Therefore, a market-basis balance sheet typically would show a higher total asset value, and consequently a higher equity, than one using cost valuation. An exception would be during periods of falling prices and corresponding decreases in asset values.

Advantages of cost-basis balance sheets include conformity to generally accepted accounting principles, conservatism, and their direct comparability with balance sheets from other types of businesses using cost basis. Also, changes in equity come only from net income that has been earned and retained in the business, not from asset price changes. The main advantage of market-basis balance sheets is the more accurate indication of the current financial condition of the business and the value of collateral available to secure loans. Because lenders have been the primary users of farm and ranch balance sheets, market-basis valuation, which shows the current value of available collateral, has been in general use.

The FFSC considered both types of balance sheets and concluded that information on both cost and market values is needed to properly analyze the financial condition of a farm or ranch business. They state that acceptable formats would be: (1) a market-basis balance sheet with cost information included as footnotes or shown in supporting schedules, or (2) a double-column balance sheet, with one column containing cost values and the other market values.

> ### Box 5-2   Valuing Farm Inventories
>
> Accounting principles allow grain and certain other commodities to be valued at market value on a balance sheet, provided they meet the following conditions: (1) have a reliable, determinable, and realizable market price; (2) have relatively small and known selling expenses; and (3) are ready for immediate delivery for sale. As the FFSC suggests, some market livestock would also meet these conditions.

Table 5-3 contains FFSC recommended or acceptable methods for valuing assets on both types of balance sheets. Cost-basis valuation, as used in this table, represents one of three valuation methods: cost, cost less depreciation, or farm production cost. Cost less depreciation (or book value) would be used for all depreciable assets, such as machinery, purchased breeding livestock, and buildings. Farm production cost, or the accumulated direct expenses incurred to date, would be the value used for the investment in growing crops.

Raised breeding livestock are a difficult valuation problem. The FFSC recommends the farm production cost method. However, this method requires segregating and accumulating all costs (that is, not showing them on the income statement) associated with raising each animal from birth to productive age. When the animal becomes productive, it is added to the depreciation schedule, and the total accumulated expense is depreciated just as if it were a purchase price. The FFSC recognized that, while recommended, this is a procedure most farmers and ranchers would not use because of the difficulty of separating individual animal costs from total livestock costs. As an alternative, they recommend using a fixed base value for each age and type of raised breeding livestock. This value should approximate the cost of raising the animal but would remain fixed over time. Changes in the total value of raised breeding livestock would then occur only when there was a change in the number of animals. A change in market value would have no effect.

Every asset is not valued at market, even on a market-basis balance sheet. Accounts receivable and prepaid expenses are valued at cost, which is simply their actual dollar amount. Investment in growing crops is valued under both valuation methods at an amount equal to the

TABLE 5-3	**Valuation Methods for Cost-Basis and Market-Basis Balance Sheets**

Asset	Cost Basis	Market Basis
Marketable securities	Cost	Market
Inventories of grain and market livestock	Market*	Market
Accounts receivable	Cost	Cost
Prepaid expenses	Cost	Cost
Investment in growing crops	Cost	Cost
Purchased breeding livestock	Cost	Market
Raised breeding livestock	Cost or a base value	Market
Machinery and equipment	Cost	Market
Buildings and improvements	Cost	Market
Land	Cost	Market

*Market is an acceptable method for *raised* grain and market livestock. Lower of cost or market is the preferred method for *purchased* grain and market livestock.

direct expenses incurred for that crop to date. The crop is not yet harvested, not ready for market, and still subject to production risks. Valuing it at "expected" market value would be a very optimistic approach and not in compliance with conservative accounting principles.

BALANCE SHEET EXAMPLE

Table 5-4 is an example of a balance sheet with a format and headings that follow the FFSC rec-

ommendations. It uses the double-column format to present both cost and market values on a single balance sheet. For simplicity, only farm business assets and liabilities are included. Many farm and ranch balance sheets include personal as well as business assets and liabilities. It is often difficult to separate them, and lenders may prefer they be combined for their analysis. This presentation of a complete balance sheet with both cost and market values includes some new headings and concepts that

TABLE 5-4	**Balance Sheet for I. M. Farmer, December 31, 20XX**					

Assets				Liabilities		
Current Assets:	**Cost**	**Market**		**Current Liabilities:**	**Cost**	**Market**
Cash/checking acct.	$ 5,000	$ 5,000		Accounts payable	6,000	6,000
Marketable securities	1,000	2,200		Notes payable within 1 year	15,000	15,000
Inventories				Current portion of term debt	28,000	28,000
Crops	40,000	40,000		Accrued interest	15,700	15,700
Livestock	52,000	52,000		Income taxes payable	8,000	8,000
Supplies	4,000	4,000		Current portion—deferred taxes	15,020	15,260
Accounts receivable	1,200	1,200		Other accrued expenses	900	900
Prepaid expenses	500	500		Total Current Liabilities	$88,620	$88,860
Investment in growing crops	7,600	7,600		**Noncurrent Liabilities:**		
Other current assets	0	0				
Total Current Assets	$111,300	$112,500		Notes payable		
				Machinery	20,000	20,000
Noncurrent Assets:				Breeding livestock	40,000	40,000
				Real estate debt	175,000	175,000
Machines and equipment	67,500	95,000		Noncurrent portion		
Breeding livestock (purchased)	48,000	60,000		—deferred taxes	—	45,000
Breeding livestock (raised)	12,000	24,000				
Buildings & improvements	27,000	50,000		Total Noncurrent Liabilities	$235,000	$280,000
Land	288,000	400,000		Total Liabilities	$323,620	$368,860
Other noncurrent assets	0	0		**Owner Equity:**		
Total Noncurrent Assets	$442,500	$629,000		Contributed capital	50,000	50,000
Total Assets	$553,800	$741,500		Retained earnings	180,180	180,180
				Valuation adjustment	—	142,460
				Total Equity	$230,180	$372,640
				Total liabilities and owner equity	$553,800	$741,500

need to be discussed along with a review of the asset valuation process.

Asset Section

With inventories valued at market in both cases, there is often little or no difference between the total value of current assets under the two valuation methods. However, marketable securities, such as stocks and bonds that can be easily sold, can have market values that differ substantially from their original cost. In this example, the $1,200 difference between the two valuations of current assets is due entirely to the marketable securities.

Most of the difference in asset values between cost- and market-based balance sheets will show up in the noncurrent asset section. A combination of inflation and rapid depreciation can result in the book values for machinery, equipment, purchased breeding livestock, and buildings being much less than their market values. The market value of raised breeding livestock can be, and hopefully is, higher than the cost of raising them. Land that has been owned for a number of years during which there was only moderate inflation can still have a market value considerably higher than its original cost. All of these factors can combine to make the noncurrent asset value much higher with market valuation than with cost valuation. Of course, it is also possible for market values to be less than cost during periods when asset values are declining.

Liability Section

There is little difference in the valuation of the usual liabilities on a cost- or market-based balance sheet. However, there are several entries related to income taxes in the liability section of this balance sheet that have not been discussed before. *Income taxes payable* under current liabilities represent taxes due on any taxable net farm income for the past year. Taxes on farm income are generally paid several months after the year ends, but they must be paid. Because they

are caused by the past year's activity but are not yet paid, they are like an account payable, which would still have to be paid even if the business ceased to operate as of December 31. This is true regardless of the valuation method being used.

The current portion of *deferred* income taxes represents taxes that would be paid on revenue from the sale of current assets, less any current liabilities that would be tax-deductible expenses. They are called deferred taxes, because the assets have not been sold nor the expenses paid. Therefore, no taxes are payable at this time. If cash accounting is used for tax purposes, taxes are deferred into a future accounting period when the assets are converted into cash and the expenses actually paid. However, because the assets exist and the expenses have been incurred, there is no question that there will be taxable events in the future.

The noncurrent portion of deferred taxes arises from the difference between cost and market values for noncurrent assets. Market values are generally higher than cost and result in a balance sheet presenting a stronger financial position than one done on a cost basis. If the assets were sold for this market value, the business would have to pay income taxes on the difference between market value and the asset's cost or tax basis. Ignoring these taxes on a market balance sheet results in owner equity that is higher than what would actually result from a complete liquidation of the business. Therefore, noncurrent deferred taxes are included on a market-based balance sheet and are an estimate of the income taxes that would result from a liquidation of the assets at their market value. They are called deferred taxes because they have not been incurred at this time and are deferred until the time the assets are sold.

Owner Equity Section

Owner equity has three basic sources: (1) capital contributed to the business by its owner(s), (2) earnings or business profit that has been left in the business rather than withdrawn, and (3) any

---

Box 5-3   Notes to a Balance Sheet

The numbers shown on a balance sheet provide little information about the nature of the business, its accounting procedures, and the computations done to arrive at a balance sheet value. Accounting principles require this and other information be disclosed in notes included with the balance sheet.

The FFSC recommends that, at a minimum, three types of information be included in notes to the balance sheet.

1. *Basis of Accounting*—a brief description of the accounting methods, procedures, and policies used and whether the balance sheet is based on market or cost values
2. *Nature of the Operation*—a brief description of the operation including acres owned and

rented, types and number of crops and livestock raised, and the form of business organization used
3. *Depreciation Methods*—information on the years of depreciable life and depreciation methods used for each type of depreciable asset

Additional notes may be desirable or necessary in some cases to help the user (often a lender) to understand and trust the values shown. Examples would be details on the valuation methods used, details about inventory assets (bushels, head, weights, etc.), a listing of machinery, equipment, and buildings, the calculation of deferred income taxes, and details about debt and debt payments including names of creditors.

---

change caused by fluctuating market values when market valuation, rather than cost, is used. The FFSC recommends showing all three sources of equity separately, rather than simply combining them into one value. This provides additional information for anyone analyzing the balance sheet.

The example in Table 5-4 shows $50,000 of contributed capital. This is the value of any personal cash or property the owner used to start the business and any that might have been contributed since that time. In the absence of any further contributions or withdrawals charged against contributed capital, this value will remain the same on all future balance sheets.

Any before-tax net farm income that is not used for family living expenses, income taxes, or withdrawals for other purposes remains in the business. These retained earnings end up in the form of additional assets (not necessarily cash), decreased liabilities, or some combination of the two. Retained earnings, and therefore owner equity, will increase during any year in which net farm income is greater than the combined total

of income taxes paid and net withdrawals. If these latter two items are greater than net farm income, retained earnings and cost-basis owner equity will decrease. In the example, a total of $180,180 has been retained since the business began. Retained earnings will be discussed again in further detail in Chapter 6.

There is no valuation adjustment on a cost-basis balance sheet. Owner equity will change only when there are changes in contributed capital or retained earnings. Whenever a market-based balance sheet includes an asset valued at more than its cost, it creates equity that is neither contributed capital nor retained earnings. For example, during inflationary periods, the market value of land may increase substantially. A balance sheet with market values would include an increase in the land value each year, which would cause an equal increase in equity. However, this increase is due only to the ownership of the land and not from its direct use in producing agricultural products. Any differences between cost and market values, which can be either positive or negative, should be shown as a valuation adjustment in the equity section of the balance sheet. This makes it easy for an analyst to determine what part of total equity on a market-based balance sheet is a result of valuation differences.

BALANCE SHEET ANALYSIS

A balance sheet is used to measure the financial condition of a business and, more specifically, its liquidity and solvency. Analysts often want to compare the relative financial condition of different businesses and of the same business over time. Differences in business size cause potentially large differences in the dollar values on a balance sheet, and therefore cause problems comparing their *relative* financial condition. A large business can have serious liquidity and solvency problems, as can a small business, but the difficulty is measuring the size of the problem relative to the size of the business.

To get around this problem, ratios are often used in balance sheet analysis. They provide a standard procedure for analysis and permit comparison over time and between businesses of different sizes. A large business and a small one would have substantial differences in the dollar values on their balance sheets, but the same ratio value would indicate the same relative degree of financial strength or weakness. Ratio values can be used as goals and easily compared against the same values for other businesses. In addition, many lending institutions use ratio analysis from balance sheet information to make lending decisions and to monitor the financial progress of their customers.

Most farm and ranch balance sheets are used for loan purposes, where market values are of the most interest to the lender. Therefore, the analysis of the balance sheet in Table 5-4 will use values from the "market" column.

Analyzing Liquidity

An analysis of liquidity concentrates on current assets and current liabilities. The latter represent the need for cash over the next 12 months, and the former, the sources of cash. Liquidity is a relative rather than an absolute concept, because it is difficult to state that a business is or is not liquid. Based on an analysis, however, it is possible to say that one business is more or less liquid than another.

Current Ratio
This is one of the more common measures of liquidity and is computed from the equation

$$\text{Current ratio} = \frac{\text{Current asset value}}{\text{Current liability value}}$$

The current ratio for the balance sheet example in Table 5-4 would be

$$\text{Current ratio} = \frac{\$112,500}{\$88,860} = 1.27$$

This ratio measures the amount of current assets relative to current liabilities. A value of 1.0 means current liabilities are just equal to current assets, and while there are sufficient current assets to cover current liabilities, there is no safety margin. Asset values can and do change with changes in market prices, so values larger than 1.0 are preferred to provide a safety margin for price changes and other factors. The larger the ratio value, the more liquid the business, and vice versa.

Working Capital
Working capital is the difference between current assets and current liabilities:

$$\text{Working capital} = \text{current assets} - \text{current liabilities}$$

This equation computes the dollars that would remain after selling all current assets and paying all current liabilities. It is an indication of the margin of safety for liquidity measured in dollars. I. M. Farmer's balance sheet in Table 5-4 shows working capital of $112,500 minus $88,860, or $23,640.

Working capital is a dollar value, not a ratio. This makes it difficult to use working capital to compare the liquidity of businesses of different sizes. A larger business would be expected to have larger current assets and liabilities, and would need more working capital to have the same relative liquidity as a smaller business. Therefore, it is important to relate the amount of working capital to the size of the business.

Analyzing Solvency

Solvency measures the relative relations among assets, liabilities, and equity. It is a way to analyze the business debt and to see if all liabilities could be paid off by the sale of all assets. The latter requires assets to be greater than liabilities, indicating a solvent business. However, solvency is generally discussed in relative terms by measuring the degree to which assets exceed liabilities.

Three ratios are commonly used to measure solvency and are recommended by the FFSC. Each of them uses two of the three items mentioned earlier, making them all related to one another. Any one of these ratios will, when properly computed and analyzed, provide full information about solvency. However, all three have been in common use, and some individuals prefer one over another.

Debt/Asset Ratio

The debt/asset ratio is computed from the equation

$$\text{Debt/asset ratio} = \frac{\text{Total liabilities}}{\text{Total assets}}$$

and measures what part of total assets is owed to lenders. This ratio should have a value less than 1.0, and even smaller values are preferred. Notice that a debt/asset ratio of 1.0 means debt or liabilities equal assets, and therefore, equity is zero. Ratios greater than 1.0 would be obtained for an insolvent business. I. M. Farmer has a debt/asset ratio of $368,860 ÷ $741,500, or 0.50 using market values.

Equity/Asset Ratio

This ratio is computed from the equation

$$\text{Equity/asset ratio} = \frac{\text{Total equity}}{\text{Total assets}}$$

and measures what part of total assets is financed by the owner's equity capital. Higher values are preferred, but the equity/asset ratio cannot exceed 1.0. A value of 1.0 is obtained when equity equals assets, which means liabilities are zero. An insolvent business would have a negative equity/asset ratio, because equity would be negative. For the example in Table 5-4, the equity/asset ratio is $372,640 ÷ $741,500, or 0.50, using the market values.

Debt/Equity Ratio

The debt/equity ratio is also called the leverage ratio by some analysts and is computed from the equation

$$\text{Debt/equity ratio} = \frac{\text{Total liabilities}}{\text{Owner equity}}$$

This ratio compares the proportion of financing provided by lenders with that provided by the business owner. When the debt/equity ratio is equal to 1.0, lenders and the owner are providing an equal portion of the financing. Smaller values are preferred, and the debt/equity ratio will approach zero as liabilities approach zero. Very large values result from very small equity, which means an increasing chance of insolvency. I. M. Farmer has a debt/equity ratio of $368,860 ÷ $372,640, or 0.99.

Summary of Analysis

Table 5-5 summarizes the values obtained in the analysis of the I. M. Farmer balance sheet. These values show current assets to be only 27 percent higher than current liabilities. Using market values, the business financing is about equally divided between borrowed and equity capital.

Performing an analysis of this balance sheet using the cost values would show two general results. First, there would be little difference in the liquidity measures, because there is little difference in the values of current assets and current

TABLE 5-5	Summary of I. M. Farmer's Financial Condition

Measure	Market Basis
Liquidity:	
Current ratio	1.27
Working capital	$23,640
Solvency:	
Debt/asset ratio	0.50
Equity/asset ratio	0.50
Debt/equity ratio	0.99

liabilities under the two valuation methods. Second, cost values, being lower, would show a weaker solvency position for each measure used. These results would be typical. Market valuation will have little effect on liquidity measures, but it has the potential for a large impact on solvency measures. This illustrates the importance of knowing how assets were valued on any balance sheet being analyzed, and shows why the FFSC recommends including information on both cost and market values. Providing complete information allows a more thorough analysis and eliminates any possible confusion about which valuation method was used.

Other Measures

The FFSC recommends the use of the five analysis measures discussed above. However, there are others which have been widely used and continue to be used. One is the net capital ratio which is another measure of solvency. It is computed from the equation

$$\text{Net capital ratio} = \frac{\text{Total assets}}{\text{Total liabilities}}$$

A net capital ratio of 1.0 results from zero equity and higher values indicate a greater degree of solvency. For example, a value of 2.0 means liabilities are one-half of assets and equal to

equity. This is often considered the minimum safe value.

Another popular ratio is the debt structure ratio computed as

$$\text{Debt structure ratio} = \frac{\text{Current liabilities}}{\text{Total liabilities}}$$

This ratio shows what proportion current liabilities are to total liabilities and can be converted to a percentage by multiplying by 100. It cannot be greater than one or 100 percent which would result when all liabilities are current liabilities. Since all current liabilities must be paid within the next year, smaller numbers are preferred. Higher values mean a large proportion of total liabilities must be paid within the next year which may require more cash than will be available. This could indicate a need to convert some current liabilities into noncurrent which would reduce current payments by spreading payments over more years. However, a relatively high debt structure ratio may not indicate a problem if both current and total liabilities are small amounts.

STATEMENT OF OWNER EQUITY

The balance sheet shows the amount of owner equity at a *point in time,* but not what caused changes in this value *over time.* Because of this, the FFSC recommends that a statement of owner equity be part of a complete set of financial records. This statement shows the sources of changes in owner equity and the amount that came from each source. It also serves to reconcile the beginning and ending owner equity.

An Example

The balance sheet for I. M. Farmer in Table 5-4 shows an owner equity of $372,640 using market-basis valuation. However, it does not show how much this total changed over the past year. With a balance sheet from a year earlier, the change could be calculated by simple subtraction, but the sources or causes of any change would still not be evident. To show the sources and amounts of any changes, a statement of owner equity is needed.

Table 5-6 is an example of a statement of owner equity for I. M. Farmer. It shows an owner equity of $344,490 at the beginning of Year 20XX, which increased to $372,640 at the end of the year. Where did this increase of $28,150 come from? An item that always affects equity is net farm income for the year, which in this case was $47,900. However, there were cash income taxes paid during the year and a change in taxes payable amounting to $8,150. Therefore, the net effect of the year's *after-tax* profit is a $39,750 increase in equity.

Any money I. M. takes from the business for personal or other use reduces farm equity. This statement shows that $36,000 was withdrawn from the business during the year. However, nonfarm income of $9,500 was contributed to the business, leaving a net withdrawal of $26,500. This is shown as a negative amount because it reduces equity. Possible changes in equity can also come from other contributions to or from the business. Contributions to the business

may be in the form of gifts or inheritances of either cash or property. In a similar manner, property may be removed from the business and gifted to someone else, or converted to personal use rather than business use. Contributions to the business would increase equity, and contributions from the business would decrease equity.

A statement of owner equity for someone using cost-basis balance sheets would end at this point, showing an increase in equity of $11,650. The only things affecting equity on a cost-basis balance sheet are net farm income, owner withdrawals, and other contributions to and from the business. However, changes in the market value of assets will affect equity when market values are used.

This balance sheet assumes that the increase in market values for the year totaled $22,500. An increase in land values, an increase in the difference between market value and book value for depreciable assets, and other factors could contribute to this increase. However, should these assets be sold, income taxes would be due on this increase in value, reducing the net effect on equity. Additional income taxes due were es-

TABLE 5-6	Statement of Owner Equity for I. M. Farmer for December 31, 20XX (Market Valuation)		
Owner equity, January 1, 20XX			$344,490
Net farm income for 20XX	$47,900		
Less adjustment for income taxes paid and payable	(8,150)		
Net after-tax farm income			39,750
Less increase in current portion-deferred income taxes			(1,600)
Owner withdrawals from farm business	(36,000)		
Nonfarm income contributed to farm business	9,500		
Net owner withdrawals from farm business			(26,500)
Other capital contributions to farm business			0
Other capital distributions from farm business			0
Increase in market value of farm assets	22,500		
Less increase in noncurrent portion of deferred income taxes	(6,000)		
Net increase in valuation equity			16,500
Owner equity, December 31, 20XX			$372,640

timated to be $6,000. This should be the increase from January 1 to December 31 in the noncurrent portion—deferred taxes item shown in noncurrent liabilities. Therefore, the net after-tax effect of the increase in market values for the year is an increase of $16,500. This amount, *plus* the $39,750 after-tax net farm income, *less* the $26,500 net owner withdrawals *less* the $1,600 increase in current portion-deferred taxes accounts for and reconciles the $28,150 increase in market-value owner equity during 20XX.

A statement of owner equity pulls together accounting information from a number of sources to document and reconcile the reasons behind any change in owner equity. Any failure to completely explain the changes indicates a less than adequate accounting system. It takes an accounting system that is complete, accurate, detailed, and consistent to provide the information needed.

SUMMARY

A balance sheet shows the financial position of a business at a point in time. It does this by presenting an organized listing of all assets and liabilities belonging to the business. The difference between assets and liabilities is owner equity, which represents the investment the owner has in the business.

An important consideration when constructing and analyzing balance sheets is the method used to value assets. They can be valued using cost methods, which generally reflect the original investment value or cost, or by using current market valuations. The latter generally results in higher asset values, and therefore higher owner equity, but more accurately reflects the current collateral value of the assets. There are advantages to each method, and the FFSC recommends providing both cost and market values on a farm or ranch balance sheet. This provides full information to the user of the balance sheet.

Two factors, liquidity and solvency, are used to analyze the financial position of a business as shown on a balance sheet. Liquidity considers the ability to generate the cash needed to meet cash requirements over the next year without disrupting the production activities of the business. Solvency measures the debt structure of the business and whether all liabilities could be paid by selling all assets; that is, are assets greater than liabilities. Several ratios are used to measure liquidity and solvency and to analyze the relative strength of the business in these areas.

A statement of owner equity completes the analysis of a balance sheet by showing the sources and amounts of changes in owner equity during the accounting period. Without the detail shown on this statement, it is difficult to identify and explain what caused owner equity to change, and the dollar amount of each change factor.

Questions for Review and Further Thought

1. True or false? If the debt/equity ratio increases, the debt/asset ratio will also increase. Why?
2. True or false? A business with a higher working capital will also have a higher current ratio. Why?
3. Use your knowledge of balance sheets and ratio analysis to complete the following abbreviated balance sheet. The current ratio = 2.0, and the debt/equity ratio = 1.0.

Assets		Liabilities	
Current assets	$80,000	Current liabilities	_____
Noncurrent assets	_____	Noncurrent liabilities	_____
		Total liabilities	_____
		Owner's equity	$100,000
Total assets	_____	Total liabilities and owner equity	_____

4. Can a business be solvent but not liquid? Liquid but not solvent? How?
5. Does a balance sheet show the annual net farm income for a farm business? Why or why not?
6. True or false? Assets + liabilities = equity.
7. Assume you are an agricultural loan officer for a bank, and a customer requests a loan based on the following balance sheet. Conduct a ratio analysis and give your reasons for granting or denying an additional loan. What is the weakest part of this customer's financial condition?

Assets		Liabilities	
Current assets	$ 40,000	Current liabilities	$ 60,000
Noncurrent assets	$240,000	Noncurrent liabilities	$ 50,000
		Total liabilities	$110,000
		Owner's equity	$170,000
Total assets	$280,000	Total liabilities plus equity	$280,000

8. Why are there no noncurrent deferred income taxes on a cost-basis balance sheet?
9. Assume that a mistake was made and the value of market livestock on a balance sheet is $10,000 higher than it should be. How does this error affect the measures of liquidity and solvency? Would the same results occur if land had been overvalued by $10,000?
10. Would the following entries to a farm balance sheet be classified as assets or liabilities? As current or noncurrent?
    a. Machine shed
    b. Feed bill at local feed store
    c. A 20-year farm mortgage contract
    d. A 36-month certificate of deposit
    e. Newborn calves

THE INCOME STATEMENT AND ITS ANALYSIS

CHAPTER OBJECTIVES

1. To discuss the purpose and use of an income statement

2. To illustrate the structure and format of an income statement

3. To define the sources and types of revenue and expenses that should be included on an income statement

4. To show how profit, or net farm income, is computed from an income statement, and what it means and what it measures

5. To analyze farm profitability by computing returns to assets and equity and by studying other measures of profitability

The income statement is a summary of revenue and expenses for a given accounting period. It is sometimes called an operating statement or a profit and loss statement. However, income statement is the preferred term and is used by the Farm Financial Standards Council (FFSC). Its purpose is to measure the difference between revenue and expenses. A positive difference indicates a profit, or a positive net farm income, and a negative value indicates a loss, or a negative net farm income, for the accounting period. Therefore, an income statement answers the question: Did the farm or ranch business have a profit or loss during the last accounting period, and how large was it?

The balance sheet and income statement are two different, yet related, financial statements. A balance sheet shows financial position at a *point in time,* while an income statement is a summary of the revenue and expenses as recorded over a *period of time.* This distinction is shown in Figure 6-1. Using a calendar-year accounting period, a balance sheet is prepared at the end of each year, and it can also serve as the balance sheet for the beginning of the following year. The result is a record of financial position at the beginning and end of each accounting period. However, comparing these balance sheets does not permit a direct calculation of net farm income for the year. This is the purpose of the income statement.

Even though the balance sheet and income statement contain different information and have a different purpose and use, they are both financial statements for the same business. It seems only logical that revenue, expenses, and the resulting profit or loss affect the financial position of the business, and indeed they do. However, an explanation of this relation will be delayed until the end of this chapter, after a complete discussion of the structure and components of an income statement.

IDENTIFYING REVENUE AND EXPENSES

To construct an income statement showing the difference between revenue and expenses, a necessary first step is to identify all revenue and expenses that should be included in the computation of net farm income. The discussion of cash versus accrual accounting in Chapter 3 introduced some of the difficulties that might be encountered in this seemingly simple task. This section will expand on this earlier discussion before introducing the format and construction of an income statement.

Revenue

An income statement should include all business revenue earned during the accounting period but no other revenue. The problem is one of determining when revenue should be recognized; that is, in what accounting period it was earned. This problem is further compounded by the fact that revenue can be either cash or noncash.

When revenue is received in the form of cash for a commodity produced and sold within the same accounting period, recognition is easy and straightforward. However, revenue should also be recognized whenever an agricultural commodity is ready for sale. Inventories of grain and market livestock fit this classification, and changes in these inventories (ending value minus beginning value) are included on an accrual income statement. Accounts receivable represent earned revenue, which should be recognized, but it is revenue for which a cash payment has not yet been received. Any change in their value from the beginning to the end of the year must be included as revenue. Note that these two items represent sources of revenue that may be recognized in one accounting period even though cash will not be received until a later accounting period.

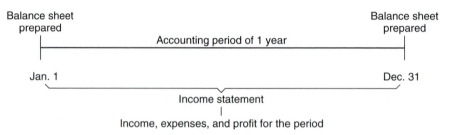

Figure 6-1   **Relation between balance sheet and income statement.**

Box 6-1   Are All Cash Receipts Counted as Revenue?

The answer is NO! Not all receipts of cash result in revenue. Cash received from a gift or inheritance is obviously not business revenue. A new loan from the bank results in the receipt of cash but is also not revenue. Nonfarm income such as salary and investment income would not be counted as farm business revenue. Sales of land or depreciable assets result in revenue only to the extent of any gain or loss realized. Farm business revenue results only from the production of agricultural commodities, services performed, and the gain or loss on the sale of assets used in that production.

When an inventory or account receivable is recognized as revenue, it is noncash revenue at that time, but something for which a cash payment typically will be received at a later date. However, payment may sometimes be received in the form of goods or services instead of cash. This noncash payment should be treated in the same manner as a cash payment. The value of feed or livestock received in payment for custom work should be included in revenue, because a commodity was simply received in lieu of cash.

Gain or Loss on Sale of Capital Assets

The gain or loss on the sale of a capital asset is an entry that often shows up in the revenue section of an income statement. It is the difference between the sale price and the cost of a capital asset such as land. For depreciable assets such as machinery, orchards, and purchased breeding livestock, it is the difference between the selling price and the asset's book value. Gain or loss is recognized only when an asset is actually sold. Before then, the market value or selling price is subject to considerable uncertainty.

Any gain or loss on the sale of a nondepreciable asset such as land is the direct result of an increase or decrease in the market price since its purchase. For depreciable assets, changes in market value affect gains and losses, as well as how accurately the depreciation was estimated. If an asset is sold for exactly its book value, there is no gain or loss. The depreciation over its life has perfectly matched its decline in market value. A sale price higher than book value implies that too much total depreciation has been taken (perhaps due to unanticipated strength in market values) and that past annual depreciation expenses have been too high. Recognizing the gain as revenue adjusts for the fact that depreciation expense was not correct in past years. Selling a depreciable asset for less than its book value means that it lost market value faster than what was accounted for by the annual depreciation. There should have been a higher depreciation expense in the past, and an adjustment is made in the form of a "loss on sale" entry on the income statement.

Given that (1) useful lives and salvage values are only estimates made at the time of purchase, and that (2) the choice of depreciation method will affect the amount of annual depreciation, there will generally be a gain or loss to be recognized when a depreciable asset is sold.

Expenses

Once all revenue for an accounting period has been identified, the next step is to identify all expenses incurred in producing that revenue. They may be either cash or noncash expenses. For example, cash expenses would include purchases of and payment for feed, fertilizer, seed, market livestock, and fuel. Noncash expenses would include depreciation, accounts payable, accrued

---

Box 6-2 **Is Every Expenditure of Cash an Expense?**

The answer is NO! Not all cash expenditures are business expenses. Cash expenditures for food, clothing, gifts, and other personal items are obviously not business expenses. Business expenses are only those items required to produce agricultural commodities and services, and hence, revenue. Principal paid on loans is not an expense, because it only represents the return of borrowed property. Interest is an expense, however, because it is the "rent" paid for the use of the borrowed property. Cash paid to purchase depreciable assets is not an expense in the year of purchase. However, it is converted into an expense over time through an annual depreciation expense, as discussed in Chapter 4.

---

interest, and other accrued expenses. There is also an adjustment for prepaid expenses.

Depreciation is a noncash expense that reflects decreases in the value of assets used to produce the revenue. Accounts payable, accrued interest, and other accrued expenses such as property taxes are expenses that were incurred during the past accounting period but not yet paid. To properly match expenses with the revenue they helped produce, these expenses must be included on this year's income statement. They must also be subtracted from next year's income statement, when the cash will actually be expended for their payment. The difference in timing between the year in which the expense was incurred and the one in which it will be paid for creates the need for these entries.

Accounts payable and accrued expenses are paid for in an accounting period later than the one in which the products or services were used. This is opposite from prepaid expenses which are goods and services paid for in one year but not used to produce revenue until next year. Examples of prepaid expenses would be seed, fertilizer, chemicals, and feed purchased and paid for in December to take advantage of price discounts, income tax deductions, or to assure availability. However, since they will not be used until the next calendar year, the expense should be deferred until then to properly match expenses with the revenue produced by this expense. Because

the timing of paying the expense and using the product is just opposite of an account payable, the accounting procedure is opposite. Prepaid expenses should be subtracted from this year's expenses as they did not produce any revenue this year. They should then be included with next year's expenses, ensuring that the business records properly match expenses with their associated revenue in the same time period.

The FFSC recommends that income taxes be included on an income statement, making the final result *after-tax* net farm income. However, they recognize that taxes due on farm income can be difficult to estimate, particularly when there is also off-farm income to consider. For simplicity and other reasons, the examples used in this text will omit income taxes from the income statement and concentrate on estimating a *before-tax* net farm income. All discussion and use of the term net farm income refers to before-tax net farm income, unless stated otherwise.

INCOME STATEMENT FORMAT

Most of the entries on an income statement have already been discussed, but several items need additional discussion. The outline of an income statement shown in Table 6-1 follows one of the formats recommended by the FFSC. In very condensed form, the basic structure is:

Total revenue

Less total expenses

Equals net farm income from operations

Plus or minus gain/loss on sale of capital assets

Equals net farm income.

Any gain/loss on the sale of culled breeding livestock is shown in the revenue section, but other gains/losses are included in a separate section at the end of the income statement. The reason for including the former in revenue is that the sale of culled breeding livestock is a normal, expected, and usual part of the ongoing production activities of the business. While the FFSC recommends including all gains/losses in net farm income, it recommends that those for machinery and land—along with improvements such as buildings and permanent crops—be shown separately. Any gains/losses from the sale of these assets are less frequent, result from investment rather than production activities, and may be very large. Therefore, net farm income from operations is the profit from normal, ongoing production activities. Net farm income includes this amount, plus any gains/losses on the sale of land and its improvements, buildings, and machinery. Showing these items separately toward the end of the income statement allows the user to determine easily whether they had an unusually large effect on net farm income.

Cash interest paid, accrued interest, and total interest are shown separately from other expenses. While often categorized as an operating expense, the position of the FFSC is that interest is a result of financing activities rather than production activities. Therefore, it should not be included with the direct operating expenses associated with the production of crops and livestock. This separation also makes it easy for the user to find and note the amount of interest expense when conducting an analysis of the income statement.

TABLE 6-1	**Income Statement Format**

Revenue:

Cash crop sales
Cash livestock sales
Inventory changes:
   Crops
   Market livestock
Livestock product sales
Government program payments
Change in value of raised breeding stock
Gain/loss from sale of culled breeding stock
Change in accounts receivable
Other farm income
   Total revenue

Expenses:

Purchased feed and grain
Purchased market livestock
Other cash operating expenses: *deductible*
   Crop expenses
   Livestock expenses
   Fuel, oil
   Labor
   Repairs, maintenance
   Property taxes
   Insurance
   Other: _____
         _____
         _____
Adjustments
   Accounts payable
   Prepaid expenses
Depreciation   *non cash expense*
   Total operating expenses
Cash interest paid
Change in interest payable
   Total interest expense
   Total expenses
Net farm income from operations
Gain/loss on sale of capital assets:
   Machinery
   Land
   Other
Net farm income

The other income statement format recommended by the FFSC computes *value of farm production* as an intermediate step. In condensed form, this format has the following structure

Total revenue

Less cost of purchased feed and grain

Less cost of purchased market livestock

Equals value of farm production

Less all other expenses

Equals net farm income from operations

Plus or minus gain/loss on sale of capital assets

Equals net farm income.

Value of farm production measures the dollar value of all goods and services produced by the farm business. Any purchased feed, grain, and market livestock were produced by some other farm or ranch and should not be credited to this business. Subtracting the cost of these items from total revenue results in a "net" production value. Note that any increase in value obtained by feeding purchased feed and grain to purchased livestock will end up credited to the business being analyzed through sales revenue or an inventory increase.

Value of farm production is a useful measure of farm size when analyzing a farm business and is used to compute other measures as well. Although the FFSC does not require use of the format showing value of farm production, they do recognize its usefulness. Therefore, for the format shown in Table 6-1, they recommend that the cost of purchased feed and grain and purchased market livestock be shown first, separate from other operating expenses. This makes it easy for the user to find the values needed to compute the value of farm production.

ACCRUAL ADJUSTMENTS TO A CASH-BASIS INCOME STATEMENT

While the FFSC encourages the use of accrual accounting, it recognizes that cash accounting is

used currently and will probably continue to be used by a majority of farmers and ranchers for some time. Simplicity and advantages for income tax purposes explain its popularity. However, net farm income from a cash accounting system can be misleading and result in poor decisions when used for management purposes.

The FFSC recommends that anyone using cash accounting convert the resulting net farm income to an *accrual-adjusted* net farm income at the end of each year. This value can then be used for analysis and decision making. Since a cash-basis income statement includes only cash receipts, cash expenses, and depreciation, there are a number of adjustments to be made. Figure 6-2 shows the necessary adjustments and the recommended procedure.

There are two adjustments to cash receipts: change in inventory values, and accounts receivable. Adjustments should be made for inventories of grain, market livestock, and raised breeding livestock. The beginning inventory value is subtracted from cash receipts, because this inventory has been sold (or otherwise consumed) during the year, and the sales have been included as cash receipts. The ending inventory value is added to cash receipts, because this is the current year's production that has not yet been sold.

The process for accounts receivable is the same. Beginning accounts receivable are subtracted from cash receipts, because this amount was collected during the year and recorded as cash receipts. Ending accounts receivable are added, because they represent production during the past year that has not yet shown up in cash receipts. An alternative procedure for both inventories and accounts receivable is to subtract beginning values from ending values and enter only the difference or change. This change can be either a positive or negative adjustment to cash receipts.

There are several adjustments to cash disbursements or expenses, including accounts payable and accrued expenses. They are often the largest expense adjustments, and accrued interest typically will be the largest accrued

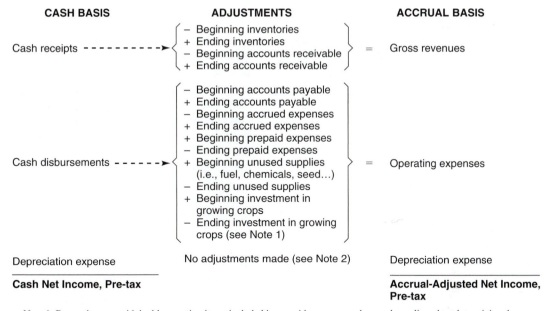

CASH BASIS	ADJUSTMENTS	ACCRUAL BASIS
Cash receipts - - - - - - - - - ➤	− Beginning inventories + Ending inventories − Beginning accounts receivable + Ending accounts receivable	= Gross revenues

Cash disbursements - - - - - - ➤	− Beginning accounts payable
	+ Ending accounts payable
	− Beginning accrued expenses
	+ Ending accrued expenses
	+ Beginning prepaid expenses
	− Ending prepaid expenses
	+ Beginning unused supplies
	  (i.e., fuel, chemicals, seed…)
	− Ending unused supplies
	+ Beginning investment in
	  growing crops
	− Ending investment in growing
	  crops (see Note 1)
		= Operating expenses

Depreciation expense	No adjustments made (see Note 2)	Depreciation expense
Cash Net Income, Pre-tax		**Accrual-Adjusted Net Income, Pre-tax**

Note 1: Remember to avoid double counting items included in prepaid expenses and unused supplies when determining the investment in growing crops.

Note 2: Because depreciation is a non-cash expense, it would technically not be reflected on a cash-basis income statement. Instead the statement would show the cash payments for property, plant, and equipment rather than allocating the cost of the asset over its useful life. However, because the Internal Revenue Code requires capital assets to be depreciated, even for cash-basis taxpayers, the common practice is to record depreciation expense for both cash basis and accrual basis income accounting.

Figure 6-2   Adjustments to get accrual-adjusted net farm income from a cash-basis income statement.

Source: Adapted from *Financial Guidelines for Agricultural Producers,* Recommendations of the Farm Financial Standards Council (Revised), 1997.

expense. Both accounts payable and accrued expenses are expenses that helped produce income during the accounting year but for which no cash has yet been spent. The cash will be spent in the next year. Making the adjustments as shown in Figure 6-2 will get these expenses into the proper year, the year they helped produce revenue.

Prepaid expenses, supplies, and investment in growing crops are other adjustments to cash expenses, but note the different adjustment procedure. The *beginning* values are added to cash disbursements, while the *ending* values are subtracted. This procedure is different from the other two expense adjustments, because the timing of the cash expenditures is different. For prepaid expenses, inventory of supplies, and investment in growing crops, cash was spent during the past accounting period, but it will not produce revenue until the next period. The dif-

ferent signs on the beginning and ending values correctly account for the difference in timing and result in the correct accrual-adjusted operating expenses.

This adjustment process is rather straightforward but does require some effort and attention to detail. It is necessary to know the dollar value of each item at the beginning of the year and again at the end of the year. Because these values are not part of a cash accounting system, they must be measured and recorded in some other way. This adjustment process results in the correct accrual-adjusted net farm income only if all measurements recorded have the same values as they would have had if an accrual accounting system had been used. The most accurate method, and one that will keep all financial records consistent and correlated, is to use values from the beginning and end of the year balance sheets.

TABLE 6-2	Income Statement for I. M. Farmer for Year Ending December 31, 20XX

Revenue:

Cash crop sales	$133,100	
Cash livestock sales	68,400	
Inventory changes:		
Crops	(8,000)	
Market livestock	1,700	
Livestock product sales	0	
Government program payments	3,400	
Change in value of raised		
breeding stock	0	
Gain/loss from sale of culled		
breeding stock	600	
Change in accounts receivable	1,200	
Other farm income	0	
Total revenue		$200,400

Expenses:

Purchased feed and grain	12,000	
Purchased market livestock	28,000	
Other cash operating expenses:		
Crop expenses	44,500	
Livestock expenses	6,500	
Fuel, oil	3,200	
Labor	8,400	
Repairs, maintenance	3,600	
Property taxes	2,800	
Insurance	2,000	
Other: Utilities	2,400	
Adjustments		
Accounts payable	1,000	
Prepaid expenses	1,500	
Accrued expenses	0	
Depreciation	8,200	
Total operating expenses		124,100
Cash interest paid	30,000	
Change in interest payable	(500)	
Total interest expense		29,500
Total expenses		153,600
Net farm income from operations		46,800
Gain/loss on sale of capital assets:		
Machinery	1,100	
Land	0	
Other	0	1,100
Net farm income		47,900

ANALYSIS OF NET FARM INCOME

Table 6-2 contains a complete accrual-basis income statement for I. M. Farmer. It shows a net farm income from operations of $46,800 and a net farm income of $47,900. The business shows a profit for the year, but is it a "profitable" business? Profitability is concerned with the size of the profit relative to the size of the business. Size is measured by the value of the resources used to produce the profit. A business can show a profit but have a poor profitability rating if this profit is small relative to the size of the business. For example, two farms with the same net farm income are not equally profitable if one used twice as much land, labor, and capital as the other to produce that profit.

Profitability is a measure of how efficient the business is in using its resources to produce profit or net farm income. The FFSC recommends four measures of profitability: (1) net farm income, (2) rate of return on assets, (3) rate of return on equity, and (4) operating profit margin ratio. In addition to the rates of return on assets and equity, it is also possible to compute a return to unpaid operator labor and management. This is another common measure of profitability.

Net Farm Income

As shown on the income statement, net farm income is the amount by which revenue exceeds expenses, plus any gain or loss on the sale of capital assets. It can also be thought of as the amount available to provide a return to the operator for the unpaid labor, management, and equity capital used to produce that net farm income. As discussed earlier, it is an absolute dollar amount, making it difficult to use net farm income by itself as a measure of profitability. It should be considered more as a starting point for analyzing profitability than as a good measure of profitability itself.

Rate of Return on Assets

Return on assets (ROA) is the term used by the FFSC, but the same concept has been called return to capital, or return on investment (ROI). It measures profitability with a ratio obtained by dividing the dollar return to assets by the average farm asset value for the year. The latter value is found by averaging the beginning and ending total asset values from the farm's balance sheets. Expressing ROA as a percentage allows an easy comparison with the same values from other farms, over time for the same farm, and with returns from other investments. The equation is

$$\text{Rate of return} \atop \text{on assets (\%)} = \frac{\text{return to assets (\$)}}{\text{average assets (\$)}} \times 100$$

The return on assets is the dollar return to both debt and equity capital, so net farm income from operations must be adjusted. Interest on debt capital was deducted as an expense when calculating net farm income from operations. Any such interest must be added back to net farm income from operations when computing the return to assets. This step eliminates any expense that comes from the type and amount of financing used by the business. The result is equal to what net farm income from operations *would have been* if there had been no debt capital and therefore no interest expense. For I. M. Farmer, the computations are

Net farm income from operations	$46,800
Plus interest expense	29,500
Equals adjusted net farm income from operations	$76,300

This is what Farmer's net farm income from operations would have been if there had been no borrowed capital, or 100 percent equity in the business.

It is also necessary to make an adjustment for the unpaid labor and management provided by the farm operator and the farm family. Without this adjustment, that part of net farm income from operations that is credited to assets would also include the contribution of labor and management toward earning that income. Therefore, a return to unpaid labor and management must be subtracted from adjusted net farm income

from operations to find the actual return to assets in dollars. Because there is no way of knowing exactly what part of that income was earned by labor and management, opportunity costs are used as estimates. Assuming an opportunity cost of $20,000 for the unpaid labor provided by I. M. Farmer and other family members, and $5,000 for management, the calculations are

Adjusted net farm income from operations	$76,300
Less opportunity cost of unpaid labor	−20,000
Less opportunity cost of management	−5,000
Equals return to assets	$51,300

The final step is to convert this dollar return to assets into a percentage of total assets, which is the dollar value of all capital invested in the business. An immediate problem is which total asset value to use, cost or market basis? Although either can be used, the market-basis value is generally used, because it represents the current investment in the business and allows a better comparison of ROAs between farms. It also makes the resulting ROA comparable to the return that these assets could earn in other investments should the assets be converted into cash at current market values and invested elsewhere. However, cost basis provides a better indicator of the return on the actual funds invested and is the better indicator for trend analysis.

I. M. Farmer's balance sheet in Chapter 5 shows a market-basis total asset value of $741,500 on December 31, 20XX. If the total asset value on January 1, 20XX was $710,000, the average asset value for 20XX was $725,750, which is found by averaging the two values. Therefore, the rate of return on Farmer's assets for 20XX was

$$\text{ROA} = \frac{\$51,300}{\$725,750} = 0.0707 \text{ or } 7.07\%$$

This ROA of 7.07 percent represents the return on capital invested in the business after adjusting net farm income from operations for the opportunity cost of labor and management. The "profitability" of the farm can now be judged by comparing this ROA to those from similar farms, the returns from other possible investments, the opportunity cost of the farm's capital, and past ROAs for the same farm.

Several things should be noted whenever ROA is computed and analyzed. First, the FFSC recommends using net farm income from operations rather than net farm income. Any gains or losses on sale of capital assets included in the latter value are sporadic in nature, can be very large, and do not represent income generated by the use of assets in the normal production activities of the business. Therefore, they should not be included in any calculation of how well assets are used in generating profit. Second, the opportunity costs of labor and management are estimates, and changing them will affect ROA. Third, any comparisons of ROAs should be done only after making sure they were all computed the same way and that the same method was used to value assets. Market valuation is recommended for comparison purposes; cost valuation is recommended for checking trends on the same farm. Fourth, the value of any personal or nonfarm assets that might be included on the balance sheet should be subtracted from total assets. Likewise, any nonfarm earnings, such as interest on personal savings accounts, should not be included in the farm income. The computed ROA should be for farm earnings generated by farm assets. Finally, ROA is an average return and not a marginal return. It should not be used when making decisions about investing in additional assets where the marginal return is the important value.

Rate of Return on Equity

The return on assets is the return on all assets or capital invested in the business. On most farms and ranches, there is a mixture of both debt and equity capital. Another important measure of profitability is the return on equity (ROE), which is the return on the owner's share of the capital invested. Should the business be liquidated and

the liabilities paid off, only the equity capital would be available for alternative investments.

The calculation of return on equity begins directly with net farm income from operations. No adjustment is needed for any interest expense. Interest is the payment for the use of borrowed capital, and it must be deducted as an expense before the return on equity is computed. This has already been done when computing net farm income from operations. However, the opportunity cost of operator and family labor and management again must be subtracted, so the ROE does not include their contribution toward earning income.

Continuing with the I. M. Farmer example, the dollar return on equity would be

Net farm income from operations	$46,800
Less opportunity cost of unpaid labor	−20,000
Less opportunity cost of management	−5,000
Equals return on equity	$21,800

The rate of return on equity is computed from the equation

$$\text{Rate of return on equity (\%)} = \frac{\text{return on equity (\$)}}{\text{average equity (\$)}} \times 100$$

Average equity for the year is used in the divisor and is simply the average of beginning and ending market- or cost-basis equity for the year. Return on equity is also expressed as a percent to allow easy comparison among farms and with other investments. Assuming an average market basis equity of $358,565, I. M. Farmer's return on equity would be

$$\text{ROE} = \frac{\$21,800}{\$358,565} = 0.0608 \text{ or } 6.08\%$$

The ROE can be either greater or less than the return on assets, depending on the ROA's relation to the average interest rate on borrowed capital. If the ROA is greater than the interest

---

Box 6-3   **Sources of Comparative Data**

I t is often difficult to judge the strength or weakness of the results of a net farm income analysis without comparing the results to those from similar farms. One source of comparative data is farm business management associations, which exist in many states that have a large number of farms. These associations assist farmers and ranchers with their record keeping during the year and then do a complete farm financial analysis for them at year end. Average values and those for the highest and lowest one-quarter or one-third of all association members are generally available to the public. Average values of farms sorted according to size, region, or major enterprise are often available as well. Information on any associations in your state and published information may be obtained from local county extension directors or from the web site of the agricultural economics department of your state land grant university.

---

rate paid on borrowed capital, this extra margin or return above interest cost accrues to equity capital. Its return becomes greater than the average return on total assets. Conversely, if the return on assets is less than the interest rate on borrowed capital, the return on equity will be less than the return on assets. Some of the equity capital's earnings had to be used to make up the difference when the interest was paid, thereby lowering the ROE. With i being the interest rate on debt, these relations can be summarized as follows:

$$\text{If ROA} > i, \text{ then ROE} > \text{ROA}$$
$$\text{If ROA} < i, \text{ then ROE} < \text{ROA}$$

Given either of the above relations, the absolute difference between the two values depends on another factor: the amount of debt compared to equity. If ROA $> i$, and debt is large relative to equity, ROE can be much greater than ROA. The relatively large amount of debt generates many dollars of return above interest cost, which in turn, accrues to the return earned by the relatively small amount of equity capital itself. This can result in a very large ROE. However, a combination of large relative debt and ROA $< i$ has the opposite effect. All or more of the return on equity may be needed to make up the difference between the return

earned by the debt capital and the interest paid for it. The ROE can become negative, meaning equity itself must be used to pay part of the interest. Several years of this relation can lead to insolvency.

Operating Profit Margin Ratio

This ratio computes operating profit as a percent of total revenue. A higher value means the business is making more profit per dollar of revenue. The first step in computing the operating profit margin ratio is to find the absolute dollar value for operating profit. The process is

Net farm income from operations

Plus interest expense

Less opportunity cost of unpaid labor

Less opportunity cost of management

Equals operating profit

Interest is added back to net farm income from operations to eliminate the effect of debt on operating profit. This allows the operating profit margin ratio to focus strictly on the profit made from producing agricultural commodities without regard to the amount of debt, which can vary substantially from farm to farm. Eliminating this variable permits a valid comparison of this ratio across different farms. To recognize that unpaid labor and management contributed to earning the profit, their opportunity costs are subtracted. This makes the results comparable to those from businesses where all labor and management is hired, as these expenses have already been deducted in the computation of net farm income from operations.

The equation for the operating profit margin ratio is

$$\text{Operating profit margin ratio} = \frac{\text{operating profit}}{\text{total revenue}} \times 100$$

I. M. Farmer's operating profit margin ratio would be

$$\frac{\$46,800 + 29,500 - 20,000 - 5,000}{\$200,400} = \begin{array}{l} 0.256 \text{ or} \\ 25.6\% \end{array}$$

This means that, on the average, for every dollar of revenue, 25.6¢ remained as profit after paying the operating expense necessary to generate that dollar. Farms with a low operating profit margin ratio should concentrate on improving this ratio before expanding production. It does little good to increase total revenue if there is little or no profit per dollar of revenue. Net farm income is related to both the operating profit margin ratio, a measure of profitability, and total revenue, which measures volume of business.

Return to Labor and Management

Net farm income was described as the amount available to provide a return to unpaid labor, management, and equity capital. A rate of return on equity was computed in an earlier section, and in a like manner, it is possible to compute a return to labor and management. This measure of profitability is not one recommended by the FFSC but is widely used. Return to labor and management is a dollar amount that represents the part of net farm income from operations that remains to pay for operator labor and management after all capital (total asset value) is paid a return equal to its opportunity cost.

The procedure is similar to that used to compute returns on assets and equity, except the result is expressed in dollars and not as a ratio or percentage. Return to labor and management is computed as follows, using the same adjusted net farm income from operations as in the ROA section:

Adjusted net farm income from operations

Less opportunity cost of all capital

Equals return to labor and management

If the opportunity cost of I. M. Farmer's capital is assumed to be 8 percent, the opportunity cost on all capital (average asset value) is $725,750 × 8%, or $58,060. Therefore, after all capital is assigned a return equal to its opportunity cost, I. M. Farmer's labor and management earned $76,300 − $58,060 = $18,240.

Assuming that the opportunity costs of I. M. Farmer's labor and management are $20,000 and $5,000, respectively, the $18,240 indicates that capital, labor, management, or some combination of the three did not receive a return equal to their opportunity cost. Unfortunately, there is no way to determine which did or did not have a return equal to or higher than its opportunity cost.

Return to Labor

Return to labor and management can be used to compute a return to labor alone. With the opportunity cost of capital already deducted, the only step remaining to find return to labor is to subtract the opportunity cost of management.

Return to labor and management	$18,240
Less opportunity cost of management	−5,000
Equals return to labor	$13,240

The result is further confirmation that, in this example, net farm income was not sufficient to provide labor, management, and capital with a return at least equal to their opportunity costs.

Return to Management

Management is often considered the residual claimant to net farm income, for several reasons. It is difficult to estimate, and in some ways, it measures how well the manager organized the other resources to generate a profit. Return to management for average or typical farms is often reported in various publications and is computed by subtracting the opportunity cost of labor from the return to labor and management.

Return to labor and management	$18,240
Less opportunity cost of labor	−20,000
Equals return to management	$−1,760

The return to management is highly variable from year to year. A negative return is rather common on many farms and ranches, assuming as this process does, that capital and labor earned their opportunity costs. However, a negative return means only that net farm income was not sufficient to provide a return to capital, labor, and management equal to or higher than their opportunity costs. The net farm income may have been a substantial amount, particularly on a large farm or ranch. It just should have been better to provide labor, management, and capital a return equal to their individual opportunity costs.

CHANGE IN OWNER EQUITY

After computing and analyzing net farm income, several questions may come to mind. What was this profit used for? Where is this money now? Did this profit affect the balance sheet? If so, how? The answers to these questions are related and illustrate the relation between the income statement and the beginning and ending balance sheets.

Any net farm income must end up in one of four uses: (1) owner withdrawals for family living expenses and other uses, (2) payment of income and social security taxes (another reason for withdrawals), (3) increases in cash or other farm assets, or (4) a reduction in liabilities through principal payments on loans or payment of other liabilities. Figure 6-3 shows what happens to net farm income and how it affects the balance sheet. Withdrawals from the business to pay for family living expenses, income and social security taxes, and other purposes reduce the amount of net farm income available for use in the farm business. What remains is called *retained farm earnings.*

As the name implies, retained farm earnings is that part of the farm earnings, after personal withdrawals and taxes, that is retained for use in the farm business. Asset and liability values will have changed, but if retained farm earnings is positive, equity must have increased. Retained farm earnings represent increased assets or decreased liabilities, or some other combination of changes, which will increase the difference between assets and liabilities and therefore increase equity. It is important to note that this change in equity may not be positive. If living expenses, taxes, and other withdrawals are greater than net farm income, retained farm earnings will be negative. In this case, assets had to be withdrawn from the farm business, or additional borrowing was needed to meet living expenses, income taxes, and other withdrawals.

The direct relation between retained farm earnings and change in equity shown in Figure 6-3 applies only to a cost-basis balance sheet. When depreciable assets are valued at book value, their decrease in value matches the depreciation expense on the income statement. Land value is unchanged, and if there are no new farm assets from gifts, inheritance, or nonfarm income, retained farm earnings will equal the change in owner equity for the year. Under these conditions, equity on a cost-basis balance sheet will increase only if retained farm earnings is positive; that is, net farm income is greater than the sum of living expenses, taxes, and other withdrawals. This further emphasizes the necessity of making a substantial profit if the farm

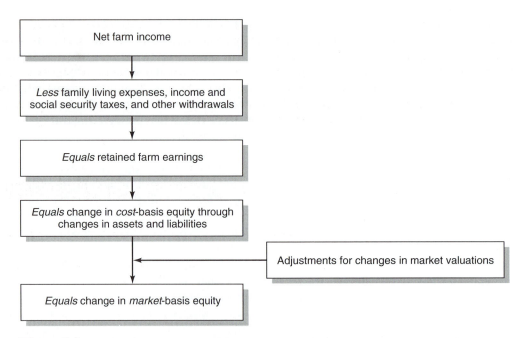

Figure 6-3   **Relation between net farm income and change in equity.**

operator/manager wants to live well and increase business equity at the same time.

On a market-basis balance sheet, changes in equity will come not only from retained farm earnings but also from any changes in the value of assets due to changes in their market value. As shown in Figure 6-3, the change in cost-basis eq-uity must be adjusted for any changes in market valuation of assets and related deferred income taxes. Farms or ranches with a large proportion of their assets in land can experience rapid gains or losses in equity due to fluctuations in land values. Without this adjustment, it is impossible to correlate and account for all changes in equity.

SUMMARY

An income statement organizes and summarizes revenue and expenses for an accounting period and computes net farm income for that period. First, all revenue earned during the period should be identified and recorded. Next, the same thing is done for all expenses incurred in producing that revenue. Net farm income is the amount by which revenue exceeds expenses. If records are kept using cash-basis accounting, accrual adjustments should be made at the end of each accounting period to derive the accrual net farm income. A cash-basis net farm income can be very misleading, and management decisions should be made only on the information obtained from an accrual net farm income.

Net farm income or profit is an actual dollar amount, whereas profitability refers to the size of that profit relative to the resources used to produce it. An analysis of profitability should be done each year to provide a means of comparing results against previous years and against other farm businesses. Net farm income, rate of return on assets, rate of return on equity, and operating profit margin ratio are recommended measures of profitability.

Part of net farm income is withdrawn from the business to pay for family living expenses, income and social security taxes, and other items. The remaining net farm income is called retained farm earnings, because it has been retained for use in the farm business. A positive retained farm earnings ends up as an increase in assets, a decrease in liabilities, or some other combination of changes in assets and liabilities that causes equity to increase. Conversely, a negative retained farm earnings means some farm equity has been used for living expenses, taxes, and other personal withdrawals. Equity based on market values for assets also changes by the amount these market values change during the year.

QUESTIONS FOR REVIEW AND FURTHER THOUGHT

1. Why is inventory change included on an accrual income statement? What effect does an increase in inventory value have on net farm income? A decrease?

2. What are the differences between net farm income computed on a cash basis versus an accrual basis? Which is the better measure of net farm income? Why?

3. What factors determine the change in cost-basis equity? In market-basis equity?

4. Why are changes in land values not included with the inventory changes shown on an income statement?

5. Why are there no entries for the purchase price of new machinery on an income statement? How does the purchase of a new machine affect the income statement?

6. Use the following information to compute values for each of the items listed below.

Net farm income	$ 36,000	Opportunity cost of labor	$16,000
Average equity	$220,000	Opportunity cost of management	$ 8,000
Average asset value	$360,000	Family living expenses	$20,000
Interest expense	$ 11,000	Income and social security taxes	$ 4,000
Total revenue	$109,500	Opportunity cost of capital	10%

a. Rate of return on assets _____ %
b. Rate of return on equity _____ %
c. Operating profit margin ratio _____ %
d. Change in equity $ _____
e. Return to management $ _____

7. Use the following information to compute the values below:

Cash expenses	$110,000	Beginning inventory value	$42,000
Cash revenue	$167,000	Ending inventory value	$28,000
Depreciation	$ 8,500	Cost of new tractor	$48,000
Beginning accounts receivable	$ 2,200	Ending accounts receivable	$ 0
Beginning accounts payable	$ 7,700	Ending accounts payable	$ 1,500

a. Cash-basis net farm income $ _____
b. Total revenue on accrual basis $ _____
c. Accrual-basis net farm income $ _____

DEVELOPING BASIC MANAGEMENT SKILLS

Chapter 2 identified planning as a function of management. Planning is primarily making choices and decisions, selecting the most profitable alternative from among all possible alternatives. The alternative(s) selected form the plan for the next year or longer. Planning might also be thought of as organizing, because a plan is a predetermined way to organize resources to produce some combination and quantity of agricultural commodities. Land, labor, and capital do not automatically produce corn, wheat, vegetables, beef cattle, or any other product. These resources must be organized into the proper combination, the proper amounts, and at the proper time for the desired production to occur. Plans can be developed for an individual enterprise, such as wheat of beef cattle, or for the entire business. Both types of plans are needed.

The alternatives that best meet the manager's goals can be identified only after a thorough analysis of all possible alternatives. "Analyze alternatives" was included as one of the steps in the decision-making process. Performing this important step when developing management plans requires the manager to use some analytical tools and methods. Part Three is designed to introduce the basic concepts, tools, and methods used to analyze alternatives.

Chapters 7, 8, and 9 cover economic principles and cost concepts and their application to decision making in agriculture. Economic principles are important because they provide a systematic and organized procedure to identify the best alternative when profit maximization is the goal. The decision rules derived from these principles require an understanding of some marginal concepts and different types of costs. However, once they are learned, they can be applied to many types of management problems.

Four types of budgeting are explored in Chapters 10, 11, 12, and 13. Budgeting is a formal procedure for comparing alternatives on paper before committing resources to a particular plan or course of action. Knowledge of some marginal and cost concepts is needed to prepare budgets. All budgets are forward-planning tools used to analyze and develop plans for the future. They can be applied to a single enterprise, a small part of the farm business, the whole farm business, or cash flows. Economic principles, combined with enterprise, partial, whole-farm, and cash flow budgeting, provide the farm manager with a powerful set of tools for analyzing and then selecting alternatives to be included in management plans.

ECONOMIC PRINCIPLES— CHOOSING PRODUCTION LEVELS

CHAPTER OUTLINE

CHAPTER OBJECTIVES

1. To explain the concept of marginalism

2. To show the relation between a variable input and an output by use of a production function

3. To describe the calculation of average physical product and marginal physical product

4. To illustrate the law of diminishing returns and its importance

5. To show how to find the profit-maximizing amount of a variable input using the concepts of marginal value product and marginal input cost

6. To show how to find the profit-maximizing amount of output to produce using the concepts of marginal revenue and marginal cost

7. To explain the use of the equal marginal principle for allocating limited resources

A knowledge of economics provides the manager with a set of principles, procedures, and rules for decision making. This knowledge is useful when making plans for organizing and operating a farm or ranch business. It also provides some help and direction when moving through the steps in the decision-making process. Once a problem has been identified and defined, the next step is to acquire data and information. The information needed to apply economic principles provides a focus and direction to this search and prevents time being wasted searching for unnecessary data. Once the data are acquired, economic principles provide some guidelines for processing data into useful information and for analyzing the potential alternatives.

Economic principles consist of a set of rules, which together ensure that the choice or decision made will result in maximum profit. The application of these rules tends to follow the same three steps. The steps are: (1) acquire physical and biological data and process it into useful information, (2) acquire price data and process it into useful information, and (3) apply the appropriate economic decision-making rule to maximize profit. The reader should look for each of these steps as each new economic principle is discussed.

MARGINALISM

Much of the economics of production is related to the concept of marginalism or marginality. Economists and managers are often interested in what changes will result from a change in one or more factors under their control. For example, they may be interested in how cotton yield changes from using an additional 50 pounds of fertilizer, or how 2 additional pounds of grain in the daily feed ration affects milk production from a dairy cow, or how profit changes from raising an additional 20 acres of corn and reducing soybean production by 20 acres.

The term *marginal* will be used extensively throughout this chapter. It refers to incremental changes, increases or decreases, that occur at the edge or the margin. Until the reader thoroughly understands the use of the term, it may be useful to mentally substitute "extra" or "additional" whenever the word marginal is used (remembering that the "extra" can be a negative amount or zero). It is also important to remember that any marginal change being measured or calculated is a result of a marginal change in some other factor.

To calculate a marginal change of any kind, it is necessary to find the difference between an original value and the new value that resulted from the change in the controlling factor. In other words, the change in some value caused by the marginal change in another factor is needed. Throughout this text, a small triangle (actually the Greek letter delta) will be used as shorthand for "the change in." For example, Δ corn yield would be read as "the change in corn yield" and would be the difference in corn yield before and after some change in an input affecting yield, such as seed, fertilizer, or irrigation water. Although other inputs may also be necessary, there is an assumed fixed or constant amount being used. This does not mean they are unimportant, but this assumption serves to simplify the analysis.

THE PRODUCTION FUNCTION

A basic concept in economics is the *production function*. It is a systematic way of showing the relation between different amounts of a resource or input that can be used to produce a product and the corresponding output or yield of that product. In other agricultural disciplines, the same relation may be called a response curve, yield curve, or input/output relation. A production function shows the amount of output that would be produced by using each of a number of different amounts of a variable input. It can be presented in the form of a table, graph, or mathematical equation.

The first two columns of Table 7-1 are a tabular presentation of a production function. Different levels of the input that can be used to produce the product are shown in the first column, assuming that all other inputs are held fixed. The amount of production expected from

	Total physical product (TPP)	Average physical product (APP)	Marginal physical product (MPP)
Input level			
0	0	0	
			12.0
1	12	12.0	
			18.0
2	30	15.0	
			14.0
3	44	14.7	
			10.0
4	54	13.5	
			8.0
5	62	12.4	
			6.0
6	68	11.3	
			4.0
7	72	10.3	
			2.0
8	74	9.3	
			−2.0
9	72	8.0	
			−4.0
10	68	6.8	

TABLE 7-1 | Production Function in Tabular Form

using each input level is shown in the second column, labeled *total physical product.* In economics, output or yield is often called total physical product, abbreviated as TPP.

Average and Marginal Physical Product

The production function provides the basic data that can be used to derive additional information about the relation between the input and TPP. It is possible to calculate the average amount of output, or TPP, produced per unit of input at each input level. This value is called *average physical product* (APP) and is shown in the third column of Table 7-1. APP is calculated by the formula

$$APP = \frac{\text{total physical product}}{\text{input level}}$$

In this example, at 4 units of input, the APP is $54 \div 4 = 13.5$. The production function in Table 7-1 has an APP that increases over a short range and then decreases as more than 2 units of input are used. This is only one example of how APP may change with input level. Other types of production functions have an APP that declines continuously after the first unit of input, which is a common occurrence.

The first marginal concept to be introduced is *marginal physical product* (MPP), shown in the fourth column of Table 7-1. Remembering that marginal means additional or extra, MPP is the additional or extra TPP produced by using an additional unit of input. It requires measuring changes in both output and input.

Marginal physical product is calculated as

$$MPP = \frac{\Delta \text{ total physical product}}{\Delta \text{ input level}}$$

The numerator is the change in TPP caused by a change in the variable input, and the denominator is the actual amount of change in the input. For example, Table 7-1 indicates that using 4 units of input instead of 3 causes TPP to increase by +10 units. Because this change was caused by a 1-unit increase or change in the input level, MPP would be found by dividing +10 by +1 to arrive at an MPP of +10. Using 9 units of input instead of 8, the change in TPP is negative, or −2 units. Dividing this result by the 1-unit increase (+1) in input causing the change in TPP, the MPP is −2. Marginal physical product can be positive or negative. It can also be zero if changing the input level causes no change in TPP. A negative MPP indicates that too much variable input is being used relative to the fixed input(s), and this combination causes TPP to decline.

The example in Table 7-1 has the input increasing by increments of 1, which simplifies the calculation of MPP. Other examples and problems may show the input increasing by increments of 2 or more. In this case, the denominator in the formula for MPP must be the actual total *change* in the input. For example, if the input is changing by increments of four, the denominator in the equation would be 4. The result is not an exact determination of MPP for the last unit of input but an average MPP for the 4 input units in the change. Many times, this calculation will provide sufficient information for decision making, unless the change in input between two possible input levels is fairly large. In this case, either more information should be obtained about expected output levels for intermediate levels of the variable input, or more sophisticated mathematical techniques should be used.[1]

A Graphical Analysis

A production function and its corresponding APP and MPP can also be shown in graphical form. The top portion of Figure 7-1 is a production function with the same general characteristics as the data in Table 7-1. Notice that TPP, or output, increases at an increasing rate as the input level is increased from zero. As the input level is increased further, TPP continues to increase, but at a decreasing rate, and eventually begins to decline absolutely as too much variable input is used relative to the amount of fixed input(s) available.

Table 7-1 and Figure 7-1 illustrate several important relations among TPP, APP, and MPP. Notice that as long as TPP is increasing at an increasing rate, MPP is increasing along with APP. At the point where TPP changes from increasing at an increasing rate to increasing at a decreasing rate, MPP reaches a maximum and then declines continuously, having a value of zero where TPP reaches its maximum. Where TPP is at its maximum, a very small change in the input level neither increases nor decreases output, and therefore, MPP is zero.

Average physical product increases over a slightly longer range than MPP before begin-

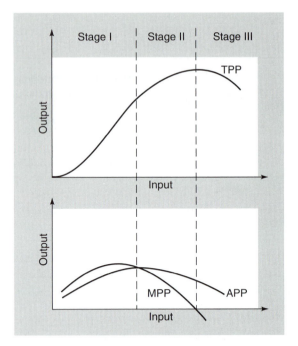

Figure 7-1   **Graphical illustration of a production function.**

ning to decline, which demonstrates an interesting relation between APP and MPP. Notice that whenever MPP is above APP, APP will be increasing, and vice versa. This relation can be explained by remembering that to raise (lower) any average value, the additional or marginal value used in calculating the new average must be above (below) the old average. The only way baseball players can raise their batting averages is to have a daily batting average (the marginal or additional value) that is above their current season average.

The relation among TPP, APP, and MPP often is used to divide this particular type of production function into three stages, as shown in Figure 7-1. Stage I begins at the zero input level and continues to the point where APP is maximum and equal to MPP. Stage II begins where APP is maximum and ends where MPP is zero (or TPP is maximum). Stage III is the range of input levels where MPP is negative and TPP is declining absolutely. The importance of these

[1]The reader who has had a course in calculus will recognize that for a production function that is known and expressed as a continuous mathematical equation, the exact MPP can be found by taking the first derivative of TPP with respect to input.

stages in determining the proper amount of input to use will be discussed later.

Law of Diminishing Marginal Returns

Table 7-1 and Figure 7-1 can be used to illustrate a law that is important from both a theoretical and practical standpoint. The term *diminishing marginal returns* is used to describe what happens to MPP as additional input is used. The law of diminishing marginal returns states that, as additional units of a variable input are used in combination with one or more fixed inputs, marginal physical product will eventually begin to decline. When the production function has three stages, as in Figure 7-1, it is often assumed the decline in MPP begins where Stage II begins. However, a close examination of Figure 7-1 indicates MPP reaches a maximum in Stage I and is declining before the beginning of Stage II. Where MPP begins to decline depends entirely on the characteristics of the production function being analyzed.

Three properties of this law and its definition need to be emphasized. First, for diminishing returns to exist, one or more fixed inputs must be used in the production process in addition to the variable input. One acre of land or one head of livestock is often the fixed input used to define the production function and, therefore, to illustrate diminishing marginal returns. Second, the definition does not preclude diminishing marginal returns from beginning with the first unit of the variable input. This is often the case in agricultural applications of this law. Third, this law is based on the biological

processes found in agricultural production. It results from the inability of plants and animals to provide the same response indefinitely to successive increases in nutrients or some other input.

Numerous examples of diminishing marginal returns exist in agricultural production. As additional units of seed, fertilizer, or irrigation water are applied to a fixed acreage of some crop, the additional output or MPP will eventually begin to decline. The MPP gets smaller and smaller as the crop nears its biological capacity to utilize the input. A similar result is obtained with additional daily feed fed to dairy cows or steers on feed. The importance and practical significance of these examples will become apparent in the next two sections.

HOW MUCH INPUT TO USE

An important use of the information derived from a production function is in determining how much of the variable input to use. Given a goal of maximizing profit, the manager must select from all possible input levels the one that will result in the greatest profit.

Some help in this selection process can be found by referring back to the three stages in Figure 7-1. Any input level in Stage III can be eliminated from consideration, because additional input causes TPP to decrease and MPP to be negative. For any input level in Stage III, the same output can be obtained with less input in one of the other stages. Choosing an input level in Stage III is irrational, and a manager should not use these input levels, even if the input is available at no cost.

Box 7-1   Other Diminishing Returns

Note that the law is specifically described as the law of diminishing *marginal* returns, and is defined in terms of MPP. However, after some input level, both TPP and APP will also begin to decline, and the term "diminishing returns" is sometimes used to describe these results. However, the diminishing MPP is the more interesting and useful concept.

Stage I covers the area where adding additional units of input causes the average physical product to increase. In this stage, adding another unit of input increases the productivity of all previous inputs, as measured by average productivity or APP. If any input is to be used, it seems reasonable that a manager would want to use at least that input level that results in the greatest average physical product per unit of input. This point is the boundary of Stage I and Stage II and represents the greatest efficiency in the use of the variable input. However, as will be shown soon, profit often can be increased further by using more input even though APP is declining along with MPP.

The above discussion of Stages I and III assumes positive and constant input and output prices. Other less common price situations can cause the profit-maximizing input level to be in Stage I or III. For example, a negative input price (someone pays you to use more input) can result in a Stage III solution. However, given the more usual and normal price relation, Stage I and Stage III can be eliminated from consideration in determining the profit-maximizing input level. This leaves only Stage II. Using only the physical information available from TPP, APP and MPP, it is not possible to determine which input level in Stage II will actually maximize profit. More information is needed, specifically price information plus some way to incorporate it into the decision-making process.

Marginal Value Product

Table 7-2 contains the same production function data as Table 7-1 but with APP eliminated. As will be shown, APP is of little or no value in determining the profit-maximizing input level once Stage I is eliminated from consideration. Two additional columns of information needed to determine the optimum input level have been added to the table, using an input price of $12 per unit and an output price of $2 per unit. Marginal value product (MVP) is the additional or marginal income received from using an additional unit of input. It is calculated by using the equation

$$\text{MVP} = \frac{\Delta \text{ total value product}}{\Delta \text{ input level}}$$

Total value product (TVP) for each input level is found by simply multiplying the quantity of output (TPP) times its selling price.[2] For example, the MVP for moving from 2 to 3 units of input is found by subtracting the total value product at 2 units of input ($60) from the total value product at 3 units ($88) and dividing by the change in input (1) to get an MVP of $28. The other MVP values are obtained in a similar manner. A short-cut method to find MVP is to multiply the MPP by the output selling price; that is, $\text{MVP} = \text{MPP} \times \text{output price}$. However, this method works only if the selling price remains constant regardless of output level.

Marginal Input Cost

Marginal input cost (MIC) is defined as the change in total input cost, or the addition to total input cost caused by using an additional unit of input. It is calculated from the equation

$$\text{MIC} = \frac{\Delta \text{ total input cost}}{\Delta \text{ input level}}$$

Total input cost is equal to the quantity of input used times its price. For example, the MIC for moving from 2 to 3 units of input is found by subtracting the total input cost at 2 units ($24) from the cost at 3 units ($36) and dividing by the change in input (1) to find the MIC of $12.

Table 7-2 indicates that MIC is constant for all input levels. This should not be surprising if the definition of MIC is reviewed. The additional cost of acquiring and using an additional unit of input is equal to the price of the input, which is equal to MIC. This MIC = input price relation is a handy calculation short-cut. It will always hold, provided the input price does not change as more or less input is purchased.

---

[2]Total value product is the same as total income or gross income. However, TVP is the term used in economics when discussing input levels and input use.

TABLE 7-2	Marginal Value Product, Marginal Input Cost, and the Optimum Input Level *(Input Price = $12 per Unit; Output Price = $2 per Unit)*				
Input level	Total physical product (TPP)	Marginal physical product (MPP)	Total value product (TVP) ($)	Marginal value product (MVP) ($)	Marginal input cost (MIC) ($)
0	0		0		
		12		24	12
1	12		24		
		18		36	12
2	30		60		
		14		28	12
3	44		88		
		10		20	12
4	54		108		
		8		16	12
5	62		124		
		6		12	12
6	68		136		
		4		8	12
7	72		144		
		2		4	12
8	74		148		
		−2		−4	12
9	72		144		
		−4		−8	12
10	68		136		

The Decision Rule

Both MVP and MIC are dollar values which can now be compared and used to determine the optimum input level. Notice that the first few lines of Table 7-2 show MVP to be greater than MIC. Here the additional income received from using one more unit of input exceeds the additional cost of that input. Therefore, additional profit is being made with each additional unit of input. These relations exist until the input level reaches 6 units. At this point, the additional income from and additional cost of another unit of input are just equal. Using more than 6 units of input causes the additional income (MVP) to be less than the additional cost (MIC), which causes profit to decline as more input is used.

This discussion leads to the decision rule for determining the profit-maximizing input level. This input level is the point where

marginal value product = marginal input cost

or

$$MVP = MIC$$

If MVP is greater than MIC, additional profit can be made by using more input. When MVP is less than MIC, more profit can be made by using

---

Box 7-2   Negative Profit?

It is important to note that using input at a level where MIC is greater than MVP does *not* imply a *negative* profit. Profit is just less than it was at the previous input level. There may still be a substantial profit accumulated from using the previous units of inputs.

---

less input. Note that unless the input is free, the profit-maximizing point is not at the input level that maximizes TVP or output. Profit is maximized at a lower level of input and output.

Obviously, the profit-maximizing input level can also be found by calculating the total value product (total income) and total input cost at each input level and then subtracting cost from income to find the level where profit is greatest. This procedure has two disadvantages. First, it conceals the marginal effects of changes in the input level. Notice how diminishing marginal returns affect MVP as well as MPP, causing MVP to decline until it finally equals MIC. Second, it takes more time to find the new optimum input level if the price of either the input or the output changes. Using the MVP and MIC procedure requires only recalculating MVPs if output price changes before the MVP = MIC rule is applied. Similarly, a change in input price requires only substituting the new MIC values for the old. The reader should experiment with different input and output prices to determine their effect on MVP, MIC, and the optimum input level.

An Alternative View

The relation between prices and the production function can be viewed another way. Remember that, given constant prices, MVP = MPP × P_o, where P_o is the price of the output, and MIC = P_i, where P_i is the price of the input. Substituting these equalities for MVP and MIC in the profit-maximizing rule results in

$$MPP \times P_o = P_i$$

Next, divide each side of this equation by P_o.

The final result is

$$MPP = \frac{P_i}{P_o}$$

which says that MPP will equal the ratio of the input and output prices at the profit-maximizing input level.

What happens if this ratio changes? If it decreases (P_i decreases or P_o increases), MPP must be smaller to maximize profit. More input should be used as MPP decreases with higher input levels. Conversely, if the price ratio increases (P_i increases or P_o decreases), less input should be used to reach a higher MPP, which is needed to maintain the equality. These results show the importance of prices, specifically the relationship between input and output prices, in determining the profit-maximizing input level.

HOW MUCH OUTPUT TO PRODUCE

The discussion in the previous section concentrated on finding the input level that maximized profit. There is also a related question. How much output should be produced to maximize profit? To answer this question directly requires the introduction of two new marginal concepts utilizing input and output prices.

Marginal Revenue

Table 7-3 is the same as the previous table, except the last two columns are now marginal revenue and marginal cost. Marginal revenue (MR) is defined as the change in income or the additional income received from selling one more unit of output. It is calculated from the equation

TABLE 7-3	**Marginal Revenue, Marginal Cost, and the Optimum Output Level** *(Input Price = $12 per Unit; Output Price = $2 per Unit)*				
Input level	**Total physical product (TPP)**	**Marginal physical product (MPP)**	**Total revenue (TR) ($)**	**Marginal revenue (MR) ($)**	**Marginal cost (MC) ($)**
0	0		0		
		12		2.00	1.00
1	12		24.00		
		18		2.00	0.67
2	30		60.00		
		14		2.00	0.86
3	44		88.00		
		10		2.00	1.20
4	54		108.00		
		8		2.00	1.50
5	62		124.00		
		6		2.00	2.00
6	68		136.00		
		4		2.00	3.00
7	72		144.00		
		2		2.00	6.00
8	74		148.00		
		−2		2.00	
9	72		144.00		
		−4		2.00	
10	68		136.00		

$$MR = \frac{\Delta \text{ total revenue}}{\Delta \text{ total physical product}}$$

where total revenue is the same as total value product (output quantity × price). Total revenue (TR) is used in place of total value product when discussing output levels. It is apparent from Table 7-3 that MR is constant and equal to the output selling price. This should not be surprising given the definition of MR. Provided the output price does not change as more or less output is sold, which is typical for an individual farmer or rancher, MR will always equal the price of the output. The additional income received from selling one more unit of output will equal the price received for that output. However, if the selling price varies with changes in

the quantity of output sold, MR must be calculated using the equation above.

Marginal Cost

Marginal cost (MC) is defined as the change in cost, or the additional cost incurred from producing another unit of output. It is calculated from the equation

$$MC = \frac{\Delta \text{ total input cost}}{\Delta \text{ total physical product}}$$

where total input cost is the same as previously defined. In Table 7-3, marginal cost decreases slightly and then begins to increase as additional input is used. Notice the inverse relation between MPP and MC. When MPP is declining

---

Box 7-3   Which of the Two Rules Should Be Used?

Theoretically, the MVP = MIC rule should be used to find the profit-maximizing input level, and the MR = MC rule to find the profit-maximizing output level. However, once either of these levels has been found, the other can be found by referring to the corresponding value on the production function. What is important is to use the correct marginal concepts. Always use MVP and MIC together, because both are computed per unit of *input*. Similarly, always use MR and MC together, because both are computed per unit of *output*. A third possibility is to use the MPP = price ratio relation.

(diminishing marginal returns), MC is increasing, because it takes relatively more input to produce an additional unit of output. Therefore, the additional cost of another unit of output is increasing.

The Decision Rule

In a manner similar to analyzing MVP and MIC, MR and MC are compared to find the profit-maximizing output level. When MR is greater than MC, the additional unit of output increases profit, because the additional income exceeds the additional cost. Conversely, if MR is less than MC, producing the additional unit of output will decrease profit. The profit-maximizing output level is therefore at the output level where

marginal revenue = marginal cost

or

MR = MC

In Table 7-3, this condition occurs at 68 units of output.

It should be no surprise that the optimum output level of 68 units is produced by the optimum input level of 6 units found in the previous section. There is only one profit-maximizing combination of input and output for a given production function and a given set of prices. When either the optimum input or output level

is found, the other value can be determined from the production function.

APPLYING THE MARGINAL PRINCIPLES

A common management decision is how much fertilizer, seed, or irrigation water to use per acre for a certain crop. Table 7-4 contains information for a hypothetical irrigation problem where the manager must select the amount of water to apply to an acre of corn. Given prices for water and corn, the principles from the last two sections can be used to determine the amount of water and the corresponding amount of output that will maximize profit. The first two columns of Table 7-4 contain the production function data, which often can be obtained from experimental trials. Marginal physical product is computed from these data and represents the additional bushels of corn produced by an additional acre-inch of water within each 2-inch increment.

An examination of the MVP and MIC data shows they are never exactly equal. This will happen often when using a tabular production function with an input level that increases by several units at each step. If the MVP = MIC rule cannot be met exactly, then the next best alternative is to get as close to this point as possible but never go past it. Never select an input

TABLE 7-4	Determining the Profit-Maximizing Irrigation Level for Corn Production *(Water at $3.00 per acre-inch, and corn at $2.50 per bushel)*					
Irrigation water (acre-inch)	Corn yield per acre (bu)	Marginal physical product (MPP) (bu)	Marginal value product (MVP) ($)	Marginal input cost (MIC) ($)	Marginal revenue (MR) ($)	Marginal cost (MC) ($)
10	104.0					
		6.40	16.00	3.00	2.50	0.47
12	116.8					
		5.90	14.75	3.00	2.50	0.51
14	128.6					
		4.80	12.00	3.00	2.50	0.63
16	138.2					
		3.30	8.25	3.00	2.50	0.91
18	144.8					
		2.10	5.25	3.00	2.50	1.43
20	149.0					
		1.40	3.50	3.00	2.50	2.14
22	151.8					
		0.90	2.25	3.00	2.50	3.33
24	153.6					
		0.30	0.75	3.00	2.50	10.00
26	154.2					

level where the MVP is less than MIC, because that step results in less profit than the next lower input level.

Table 7-4 assumes that water can be applied only in increments of 2 acre-inches. For up to and including 22 acre-inches, MVP is greater than MIC, and each additional acre-inch causes profit to increase. This makes 22 acre-inches the profit-maximizing amount to apply. Increasing the amount of water applied to 24 inches would result in an MVP less than MIC, and profit would be reduced. Profit would be reduced by $0.75 for each of the two additional inches, for a total reduction of $1.50. The $0.75 comes from the difference between the additional cost of $3.00 per inch and the additional income of only $2.25.

The same logic can be applied to marginal revenue and marginal cost. Profit will be maximized at 151.8 bushels of output, where MR is

still greater than MC. Increasing production to 153.6 bushels causes MR to be less than MC, which will reduce profit. As before, the actual reduction in profit can be found by taking the difference between MR and MC ($0.83), which is the reduction in profit per bushel of additional output, and multiplying it by the additional output of 1.8 bushels. The result of $1.50 is the same as that found in the previous paragraph. This helps verify that 22 acre-inches and 151.8 bushels are the profit-maximizing combination of input and output. Moving to the combination of 24 acre-inches and 153.6 bushels will reduce profit by $1.50.

What would happen to the profit-maximizing levels if one or more of the prices changed? A change in the price of water would cause MIC and MC to change, and a change in the price of corn would cause MVP and MR to change. Such

---

Box 7-4   New Technology and Production Decisions

In the past, farmers had little choice but to optimize input and output levels using marginal values computed from a field's "average" response to changes in input levels. Yet most fields have areas with different soil types, fertility levels, slope, drainage, pH, weed problems, and so forth. New technology is now available, including global positioning systems (GPS) to find exact locations in a field, yield monitors that use GPS to record yield by actual location in a field, and precision application equipment that can adjust seed, fertilizer, and pesticide application rates "on the go" based on field location. Farmers now have new information and the equipment needed to apply the marginal principles and optimize input and output levels on each different area in a field. This can be done by soil type, fertility level, soil pH, or any other factor which affects productivity.

---

changes are very likely to change the profit-maximizing solution. Assuming that the corn price is constant at $2.50, the MVP and MIC columns indicate that the price of water would have to fall to $2.25 per inch to make 24 inches the new profit-maximizing level. It would take a price of $3.50 or greater (but not more than $5.25) to make 20 inches the correct amount to use.

The same sort of analysis can be done with MR and MC. With the price of water constant at $3.00, the price of corn would have to increase to $3.33 per bushel or higher to make 153.6 bushels the new profit-maximizing amount of output. Similarly, if the price of corn dropped to $2.14 per bushel or less (but not less than $1.43), 149.0 bushels would then become the optimum output level.

Price changes are common in agriculture, and they affect the optimum input and output levels, as shown in this example. An increase in the input price or a decrease in the output price tends to lower the profit-maximizing input and output levels. Price increases for the output or price decreases for the input tend to increase the profit-maximizing input and output levels. However, it is important to note that there must be a change in the *relative* prices. Doubling both prices or decreasing both by 50 percent will change the numbers in Table 7-4 but will not change the choice of input and output levels. One price must move proportionately more than the other, remain constant, or move in the opposite direction to cause a change in the relative prices.

EQUAL MARGINAL PRINCIPLE

The discussion so far in this chapter has assumed that sufficient input is available or can be purchased to set MVP = MIC for each and every acre or head of livestock. Another, possibly more likely, situation is a limited amount of input that prevents reaching the MVP = MIC point for all possible uses. Now the manager must decide how the limited input should be allocated or divided among several or many possible uses or alternatives. Decisions must be made on the best allocation of fertilizer among many acres, fields, and different crops; irrigation water between fields and crops; and feed between different types and weights of livestock. If capital is limited, it first must be allocated somehow among the purchases of the fertilizer, water, feed, and any other inputs.

The equal marginal principle provides the guidelines and rules to ensure that the allocation is done in such a way that profit is maximized from the use of any limited input. This principle can be stated as follows: *A limited input should be allocated among alternative uses in such a way that the marginal value products of the last unit used on each alternative are equal.*

TABLE 7-5	Application of the Equal Marginal Principle to the Allocation of Irrigation Water[*]

	Marginal value products ($)		
Irrigation water (acre-inch)	Wheat (100 acres)	Grain sorghum (100 acres)	Cotton (100 acres)
0			
	1,200	1,600	1,800
4			
	800	1,200	1,500
8			
	600	800	1,200
12			
	300	500	800
16			
	50	200	400
20			

[*]Each application of 4 acre-inches on a crop is a total use of 400 acre-inches (4 acre-inches times 100 acres).

Table 7-5 is an application of this principle, in which irrigation water must be allocated among three crops in three fields of equal size. The MVPs are obtained from the production functions relating water use to the yield of each crop, and from the respective crop prices. Assume that a maximum of 2,400 acre-inches of water is available and can be applied only in increments of 4 acre-inches. The limited supply of water would be allocated among the three crops in the following manner, using the MVPs to make the decisions. The first 400 acre-inches (4 acre-inches on 100 acres) would be allocated to cotton, because it has the highest MVP. The second 400 acre-inches would be allocated to grain sorghum, because it has the second highest MVP. In a similar manner, the third 400 acre-inches would be used on cotton, and the fourth, fifth, and sixth 400-acre-inch increments on wheat, grain sorghum, and cotton, respectively.

Each successive 400-acre-inch increment is allocated to the field that has the highest MVP remaining after the previous allocations.

The final allocation is 4 acre-inches on wheat, 8 on grain sorghum, and 12 on cotton. Each final 4-acre-inch increment on each crop has an MVP of $1,200, which satisfies the equal marginal principle. If more water was available, the final allocation obviously would be different. For example, if 3,600 acre-inches were available, it would be allocated 8 acre-inches to wheat, 12 to grain sorghum, and 16 to cotton. This equates the MVPs of the last 4-acre-inch increment on each crop at $800, which again satisfies the equal marginal principle. When inputs must be applied in fixed increments, it may not be possible to exactly equate the MVPs of the last units applied to all alternatives. However, the MVP of the last unit allocated should always be equal to or greater than the MVP available from any other alternative use.

When allocating an input that is thought to be limited, care must be taken not to use it past the point where MVP = MIC for any alternative. This would result in something less than maximum profit. The input is not really limited if a sufficient amount is available to reach or exceed the point where MVP = MIC for every alternative.

The profit-maximizing property of this principle can be demonstrated for the 2,400-acre-inch example above. If the 4 acre-inches on wheat were allocated to grain sorghum or cotton, $1,200 of income would be lost and $800 gained, for a net loss of $400. The same loss would be incurred if the last 4-acre-inch increment was removed from either grain sorghum or cotton and reallocated to another crop. When the MVPs are equal, profit cannot be increased by a different allocation of the limited input.

The equal marginal principle can also be presented in graphical form, as in Figure 7-2, where there are only two alternative uses for the limited input. The problem is to allocate the input between the two uses, keeping the MVPs equal, so that input quantity 0a plus quantity 0b just equals the total amount of input available. If

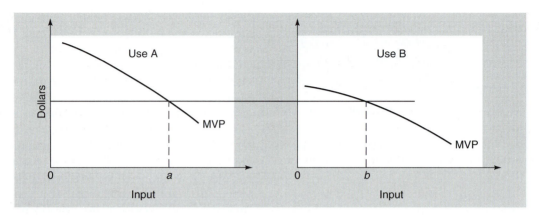

Figure 7-2 **Illustration of the equal marginal principle.**

$0a$ plus $0b$ is less than the total input available, more should be allocated to each use, again keeping the MVPs equal, until the input is fully used. There would have to be a decrease in the input used on both alternatives if $0a$ plus $0b$ exceeded the total input available.

The equal marginal principle applies not only to purchased inputs but also to those that are already owned or available, such as land, the manager's time, and machinery time. It also prevents making the mistake of maximizing profit from one enterprise by using the input until MVP = MIC and not having enough to use on other enterprises. Maximizing profit from the *total* business requires the proper allocation of limited inputs, which will not necessarily result in maximizing profit from any *single* enterprise.

SUMMARY

Economic principles using the concept of marginality provide useful guidelines for managerial decision making. They have direct application to the basic decisions of how much to produce, how to produce, and what to produce. The production function that describes the relation between input levels and corresponding output level provides some basic technical information. When this information is combined with price information, the profit-maximizing input and output levels can be found.

Marginal value product and marginal input cost are equated to find the proper input level, and marginal revenue and marginal cost are equated to find the profit-maximizing output level. These are all marginal concepts that measure the changes in revenue and cost that result from small changes in either input or output. When a limited amount of input is available and there are several alternative uses for it, the equal marginal principle provides the rule for allocating the input and maximizing profit under these conditions.

A manager will seldom have sufficient information to fully use the economic principles discussed in this chapter. This does not detract from the importance of these principles, but their application and use are often hindered by insufficient physical and biological data. Prices must also be estimated before the product is available for sale, which adds more uncertainty to the decision-making process. However, a complete understanding of the economic principles permits making changes in the right direction in response to price and other changes. A manager should continually search for better information to use in refining the decisions made through the use of these basic economic principles.

QUESTIONS FOR REVIEW AND FURTHER THOUGHT

1. In Table 7-2, assume that the prices of both the input and output have doubled. Calculate the new MVPs and MICs, and determine the profit-maximizing input level for the new prices. Now assume that both prices have been cut in half and repeat the process. Explain your results.

2. Use several other price combinations for the production function data in Tables 7-2 and 7-3, and find the profit-maximizing input and output for each price combination. Pay close attention to what happens to these levels when the input price increases or decreases, or when the output price increases or decreases.

3. In Table 7-2, what would the profit-maximizing input level be if the input price was $0? Notice the TPP and MPP at this point. Can you think of any other situation where you would maximize profit by producing at this level?

4. Does the law of diminishing marginal returns have anything to do with MVP declining and MC increasing? Why?

5. Find an extension service or experiment station publication for your state that shows the results of a seeding, fertilizer, or irrigation rate experiment. Do the results exhibit diminishing returns? Use the concepts of this chapter to find the profit-maximizing amount of the input to use.

6. Does the equal marginal principle apply to personal decisions when you have limited income and time? How do you allocate a limited amount of study time when faced with three exams on the same day?

7. Freda Farmer can invest capital in $100 increments and has three alternative uses for the capital, as shown in the following table. The values in the table are the marginal value products for each successive $100 of capital invested.

Capital invested	Fertilizer ($)	Seed ($)	Chemicals ($)
First $100	400	250	350
Second $100	300	200	300
Third $100	250	150	250
Fourth $100	150	105	200
Fifth $100	100	90	150

  a. If Freda has an unlimited amount of her own capital available and no other alternative uses for it, how much should she allocate to each alternative?
  b. If Freda can borrow all the capital she needs for 1 year at 10 percent interest, how much should she borrow and how should it be used?
  c. Assume that Freda has only $700 available. How should this limited amount of capital be allocated among the three uses? Does your answer satisfy the equal marginal principle?
  d. Assume that Freda has only $1,200 available. How should this amount be allocated? What is the total income from using the $1,200 this way? Would a different allocation increase the total income?

ECONOMIC PRINCIPLES— CHOOSING INPUT AND OUTPUT COMBINATIONS

CHAPTER OBJECTIVES

1. To explain the use of substitution in economics and decision making

2. To demonstrate how to compute a substitution ratio and a price ratio for two inputs

3. To use the input substitution and price ratios to find the least-cost combination of two inputs

4. To describe the characteristics of competitive, supplementary, and complementary enterprises

5. To show the use of the output substitution and price ratios to find the profit-maximizing combination of two enterprises

Substitution takes place in the daily lives of most people. It occurs whenever one product is purchased or used in place of another, or whenever personal income is spent for one type or class of product rather than another. Steak replaces hamburger on the dinner table, a large car replaces the economy model in the garage, and a new brand of soap or toothpaste is purchased instead of the old brand. Some substitutions or replacements are made due to an increase (or decrease) in one's personal income. The motivation for others comes from changes in relative prices or perceived differences in quality.

Substitution also takes place in the production of goods and services. Often there is more than one way to produce a product or provide a service. Machinery, computers, and robots can substitute for labor, and one ingredient can often be replaced by another. Prices, specifically relative prices, play a major role in determining whether and how much substitution should take place. In many production activities, including those on farms and ranches, the substitution is not an all-or-nothing decision. It is often a matter of making some relatively small changes in the ratio or mix of two or more inputs being used. Substitution is therefore another type of marginal analysis that considers the changes in costs and/or income when making the substitution decision.

INPUT COMBINATIONS

One of the basic decisions a farm or ranch manager must make is how to produce a given amount of some product. Most products require two or more inputs in the production process, but the manager can often choose the input combination or ratio to be used. The problem is one of determining whether more of one input can be substituted for less of another and thereby reduce total input cost. This leads to the least-cost combination of inputs to produce a given amount of output.

Substitution of one input for another occurs frequently in agricultural production. One grain can be substituted for another, or forage for grain in a livestock ration. Herbicides substitute for mechanical and hand cultivation, and machinery can replace labor. The manager must select the combination of inputs that will produce a given amount of output or perform a certain task for the least cost. In other words, the problem is to find the least-cost combination of inputs, because this combination will maximize the profit from producing a given amount of output. The alert manager will always be looking for a different input combination that will do the same job for less cost.

Input Substitution Ratio

The first step in analyzing a substitution problem is to determine whether it is physically possible to make a substitution, and at what rate. Figure 8-1 illustrates the more common types of substitution between two inputs. In Figure 8-1a, the line PP′ is an isoquant (from isoquantity, meaning the same quantity) and shows a number of possible combinations of corn and barley in a feed ration. This line is called an isoquant because any of these combinations will produce the same quantity of output or weight gain in this example.

The input substitution ratio, or the rate at which one input will substitute for another, is determined from the equation

$$\text{Input substitution ratio} = \frac{\text{amount of input replaced}}{\text{amount of input added}}$$

where both the numerator and denominator are the differences or changes in the amount of inputs being used between two different points on the isoquant PP′.

In Figure 8-1a, moving from point A to point B means 4 pounds of corn are being replaced by 5 additional pounds of barley in order to maintain the same weight gain or output. The input substitution ratio is $4 \div 5 = 0.8$, which means 1 pound of barley will replace 0.8 pounds of corn. Since PP′ is a straight line, the input

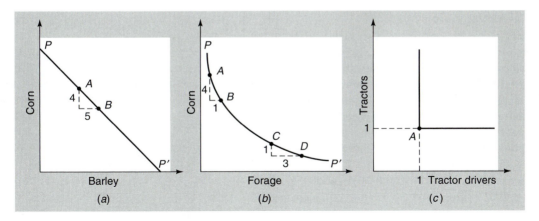

Figure 8-1   Three possible types of substitution.

substitution ratio will always be 0.8 between any two points on this isoquant. This is an example of a constant rate of substitution between two inputs. Whenever the input substitution ratio is equal to the same numerical value over the full range of possible input combinations, the inputs exhibit a constant rate of substitution. This occurs most often when the two inputs contribute the same or nearly the same factor to the production process. Corn and barley, for example, both contribute energy to a feed ration.

Another, and perhaps more common, example of physical substitution is shown in Figure 8-1b. Here the isoquant PP′ shows the different combinations of corn and forage that might produce the same weight gain on a steer, or the same milk production from a dairy cow. This is an illustration of a decreasing rate of substitution. The input substitution ratio is $4 \div 1 = 4$ when moving from point A to point B on isoquant PP′, but it is $1 \div 3 = 1/3$ when moving from point C to point D. In this example, the input substitution ratio depends on the location on the isoquant and will decline with any movement down and to the right on the curve.

Many agricultural substitution problems have a decreasing input substitution ratio, which occurs when the two inputs are dissimilar or contribute different factors to the production process. As more of one input is substituted for another, it becomes increasingly difficult to make any further substitution and still maintain the same level of output. More and more of the added input is needed to substitute for a unit of the input being replaced, which causes the input substitution ratio to decrease. This is an indication that the inputs work together best when used at some combination containing a relatively large proportion of each. At low levels of one, there is a near surplus of the other, and it is not an efficient or productive combination.

One other possible substitution situation is where no substitution is possible. Figure 8-1c illustrates one example with tractors and tractor drivers. It takes one of each to make a working combination as shown by point A. The right-angled isoquant indicates that more that one tractor with just one driver does not increase production but would increase costs. The same is true for more than one driver. Therefore, the only efficient, and the least-cost combination, is one tractor and one driver. Other examples might be the input combinations of computers and computer programmer, fence posts and wire for a given fence type, and many chemical reactions which require a fixed proportion of chemicals to obtain the desired chemical reaction and output.

The Decision Rule

Identifying the type of physical substitution that exists and calculating the input substitution ratio are necessary steps, but they alone do not permit a determination of the least-cost input combination. Input prices are also needed, and the ratio of the input prices must be compared to the input substitution ratio. The input price ratio is computed from the equation

$$\text{Input price ratio} = \frac{\text{price of input being added}}{\text{price of input being replaced}}$$

It is sometimes called the *inverse* price ratio, because it is the ratio of the *added* input price divided by the *replaced* input price. This is inverse to the input substitution ratio, which was the amount of *replaced* input divided by the amount of *added* input.

With this ratio, the least-cost combination can be found. The decision rule for finding this combination is where the

$$\text{Input substitution ratio} = \text{input price ratio}$$

Table 8-1 is an application of this procedure. Each of the feed rations is a combination of grain and forage that will put the same pounds of gain on a feeder steer. The problem is to select the ration that produces this gain for the least cost. The fourth and fifth columns of Table 8-1 contain the input substitution ratio and input price ratio for each feed ration. The input substitution ratio is declining, and the input price ratio is constant for the given prices. If the substitution ratio is greater than the price ratio, the total cost of the feed ration can be reduced by moving to the next lower ration in the table. The converse is true if the substitution ratio is less than the price ratio. This can be verified by calculating the total cost of two adjacent rations for each situation.

The decision rule states that the least-cost combination is where the input substitution ratio equals the input price ratio. However, in Table 8-1 and many other problems where the choices involve a number of discrete rather than an infi-

| TABLE 8-1 | Selecting a Least-Cost Feed Ration *(Price of grain = 4.4¢ per lb.; price of hay = 3.0¢ per lb.)*[*] | | | |

Feed ration	Grain (lb.)	Hay (lb.)	Input substitution ratio	Input price ratio
A	825	1,350		
			2.93	1.47
B	900	1,130		
			2.60	1.47
C	975	935		
			2.20	1.47
D	1,050	770		
			1.93	1.47
E	1,125	625		
			1.33	1.47
F	1,200	525		
			1.07	1.47
G	1,275	445		

[*]Each ration is assumed to put the same weight gain on a feeder steer with a given beginning weight.

nite number of combinations, there is no combination where the two ratios are equal. In these cases, select the last combination where the input substitution ratio is greater than the input price ratio. Ration E is therefore the least-cost combination for this problem. At the next choice, Ration F, the substitution ratio is less than the price ratio and therefore represents a higher-cost ration—actually $0.30 higher, $68.55 versus $68.25 for Ration E.

It may not be apparent why this rule works. Using the input substitution and price ratios is just a convenient method of comparing the additional cost of using more of one input versus the reduction in cost from using less of the other. Whenever the additional cost is less than the cost reduction, the total cost is lower and the substitution should be made. This is the case whenever the input substitution ratio is greater than the input price ratio.

Box 8-1 — Prices, Price Ratio, and Profit

The least-cost combination always depends on the *relative* prices, as shown in the input price *ratio,* not the absolute prices. Doubling or halving both prices results in the same price ratio and the same least-cost combination. However, doubling input prices obviously will reduce profit, and halving prices will increase profit, even though the least-cost input combination is unchanged.

In any input substitution problem, the least-cost combination depends on both the substitution ratio and the price ratio. The substitution ratios will remain the same over time provided the underlying physical or biological relations do not change. However, the price ratio will change whenever the relative input prices change, which will result in a different input combination becoming the new least-cost combination. As the price of one input increases relative to the other, the new least-cost input combination will tend to have less of the higher-priced input and more of the now relatively less expensive input.

When two inputs exhibit a decreasing input substitution ratio, such as in Table 8-1, the least-cost combination will generally include at least some of both inputs. If Table 8-1 was expanded to include additional combinations, it would likely show that even a very high or very low input price ratio would still result in a least-cost ration that includes some of each input. However, different results are obtained when the input substitution ratio is constant.

Panel (a) in Figure 8-1 is an example of a constant input substitution ratio, in this case 0.8 between any two points on the isoquant. Assume an input price ratio of 0.6. Beginning at the top of the isoquant and moving downward, any point would have a substitution ratio greater than the price ratio. The rule says to make this substitution and move to the next point, where the result is exactly the same. Once the horizontal axis is reached, the end result is a ration of all barley, no corn. If the input price ratio should be greater than 0.8, say 1.0, all points on the iso-

quant have a substitution ratio less than the price ratio. Every move down the isoquant represents a higher-cost ration than the one before. Therefore, the least-cost solution is all corn, no barley.

These results, and those for a decreasing input substitution ratio, can be summarized as follows:

1. Given a constant rate of substitution between two inputs, the least-cost combination will be all of one input or all of the other. The only exception occurs when the input substitution and price ratios happen to be exactly equal. Then any combination is least-cost, because they all have the same total cost.
2. With a decreasing rate of substitution between two inputs, the least-cost combination will generally include some of each input. First, the types of input that have a decreasing rate of substitution generally must be used in some combination to produce any output. All of one or the other may not even be an option. Second, even if it is physically and biologically possible to produce the output with only one of the inputs, the input substitution ratio at that point will be either so high or so low that an input price ratio probably would never reach these extremes.

ENTERPRISE COMBINATIONS

The third basic decision to be made by a farm or ranch manager is what to produce, or what combination of enterprises will maximize profit. A choice must be made from among all possible enterprises, which may include vegetables, wheat, soybeans, cotton, beef cattle, hogs, poultry, and others. Climate, soil, range vegetation, and limits on other inputs may restrict the list of possible enterprises to only a few on some farms and ranches. On others, the manager may have a large number of possible enterprises from which to select the profit-maximizing combination. In all cases, it is assumed that one or more inputs are limited, which places an upper limit on how much can be produced from a single enterprise or any combination.

Competitive Enterprises

The first step in determining profit-maximizing enterprise combinations is to determine the physical relationship among the enterprises. Given a limited amount of land, capital, or some other input, the production from one enterprise can generally be increased only by decreasing the production from another enterprise. Some of the limited input must be switched from use in one enterprise to use in another, causing the changes in production levels. There is a trade-off, or substitution, to be considered when changing the enterprise combination. These are called competitive enterprises, because they are competing for the use of the same limited input at the same time.

Figure 8-2 illustrates two types of competitive enterprises. In the first graph, corn and soybeans are competing for the use of the same 100 acres of land. Planting all corn would result in the production of 12,000 bushels of corn, and planting all soybeans would produce a total of 4,000 bushels. Other combinations of corn and soybeans totaling 100 acres would produce the combinations of corn and soybeans shown on the line connecting these two points. This line is called a *production possibility curve* (PPC); it shows all possible combinations of corn and soybeans that can be produced from the given 100 acres.

Beginning with producing all corn, replacing an acre of corn with an acre of soybeans results in a loss of 120 bushels of corn and a gain of 40 bushels of soybeans. The trade-off, or output substitution ratio, is 3, because three bushels of corn must be given up to gain one bushel of soybeans. With a straight-line production possibility curve, this substitution ratio is the same between any two points on the curve. This is an example of competitive enterprises with a constant substitution ratio.

Over a period of time, a combination of crop enterprises may benefit each other because of better weed, disease, and insect control, soil fertility, erosion control, and improved timeliness in planting and harvesting large acreages. This situation is shown in the second graph in Figure 8-2. The curved production possibility

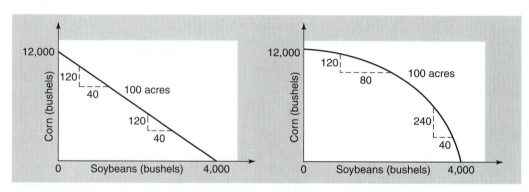

Figure 8-2   **Production possibility curves for competitive enterprises.**

curve indicates that total corn production increases at a slower rate, as a higher proportion of the land is used for corn production. A similar situation holds true for soybeans. This causes the output substitution ratio to be different for different combinations of the two enterprises. The substitution ratio is $120 \div 80 = 1.5$ near the top of the curve and increases to $240 \div 40 = 6.0$ near the bottom of the curve. These enterprises are still competitive, but they have an increasing output substitution ratio.

The most profitable combination of two competitive enterprises can be determined by comparing the output substitution ratio and the output price ratio. Each substitution ratio is calculated from the equation

$$\text{Output substitution ratio} = \frac{\text{quantity of output lost}}{\text{quantity of output gained}}$$

where the quantities gained and lost are the *changes* in production between two points on the PPC. The output price ratio is found from the equation

$$\text{Output price ratio} = \frac{\text{price of output gained}}{\text{price of output lost}}$$

The decision rule for finding the profit-maximizing combination of two competitive enterprises is the point where the[1]

output substitution ratio = output price ratio

Table 8-2 is an example of two competitive enterprises, corn and wheat, that have an increasing output substitution ratio. The method for determining the profit-maximizing enterprise combination is basically the same as for determining the least-cost input combination but with one important exception. For enterprise combi-

TABLE 8-2	Selecting a Profit-Maximizing Enterprise Combination, with Land the Fixed Input *(Price of corn = $2.80 per bu; price of wheat = $4.00 per bu)*

Combination number	Corn (bu)	Wheat (bu)	Output substitution ratio	Output price ratio
1	0	6,000		
			0.20	0.70
2	2,000	5,600		
			0.30	0.70
3	4,000	5,000		
			0.45	0.70
4	6,000	4,100		
			0.55	0.70
5	8,000	3,000		
			0.65	0.70
6	10,000	1,700		
			0.85	0.70
7	12,000	0		

nations, when the output price ratio is greater than the output substitution ratio, substitution should continue by moving downward and to the right on the PPC and down to the next combination in Table 8-2. Conversely, a substitution ratio greater than the price ratio means that too much substitution has taken place, and the adjustment should be upward and to the left on the PPC and up to the next combination in the table.

As often happens when studying a discrete rather than an infinite number of possible combinations, Table 8-2 does not contain a combination where the output substitution ratio is equal to the output price ratio. In these cases, continue to move down the table, accepting each new combination as long as the price ratio is greater than the substitution ratio. Profit will be increasing. The last combination where this relation exists will be the profit-maximizing combination.

---

[1]This rule assumes that total production costs are the same for any combination of the two enterprises. If this is not true, the ratio of the *profit* or *gross margin* per unit of each enterprise should be used instead of the output price ratio.

Never choose a combination where the output substitution ratio is greater than the output price ratio. Even though the ratio values may be closer to being equal, the profit will be less than that in the previous combination. Given only the seven possible combinations in Table 8-2, the profit-maximizing combination is number six (10,000 bu. of corn and 1,700 bu. of wheat). It appears that the two ratios are equal somewhere between combination six and seven in this example, but we do not have sufficient information to determine the exact point.

The decision rule of output substitution ratio equals output price ratio is a shortcut method for comparing the income gained from producing more of one enterprise versus the income lost from producing less of the other. Whenever the price ratio is greater than the substitution ratio, the additional income is greater than the income lost, and the substitution will increase total income. Profit will also increase, because total cost is assumed to be constant for any enterprise combination.

When the enterprises have a constant output substitution ratio, the profit-maximizing solution will be to produce all of one or all of the other enterprise, not a combination. This is because the price ratio will be either greater than or less than the substitution ratio for all combinations. An increasing substitution ratio generally will result in the production of a combination of the enterprises, with the combination dependent on the current price ratio. Any change in the price(s) of the output that changes the price ratio will affect the profit-maximizing enterprise combination when there is an increasing output substitution ratio. As with input substitution, it is the price *ratio* that is important, not just the price level.

Supplementary Enterprises

While competitive enterprises are the most common, other types of enterprise relationships do exist. One of these is supplementary, and an example is shown in the left diagram of Figure 8-3. Two enterprises are supplementary if the production from one can be increased without affecting the production level of the other.

The example in Figure 8-3 shows that beef production from stocker steers run on winter wheat pasture can be increased over a range without affecting the amount of wheat produced. The PPC shows a relationship that eventually becomes competitive as beef production cannot be increased indefinitely without affecting wheat production. In states where landowners can lease hunting rights to hunters, this leasing enterprise can be supplementary with livestock and crop production.

A manager should take advantage of supplementary relationships by increasing production from the supplementary enterprise. This should continue at least up to the point where the enterprises become competitive. If the supplementary enterprise shows any profit, however small, total profit will be increased by producing some of it, because production from the primary enterprise does not change. Two general conclusions can be drawn about the profit-maximizing combination of supplementary enterprises. First, this combination will not be within the supplementary range. It will be at least at the point where the relationship changes from supplementary to competitive. Second, it will in all likelihood be in the competitive portion of the production possibility curve. The exact location will depend, as it always does with competitive enterprises, on the output substitution and price ratios.

Complementary Enterprises

Another possible type of enterprise relationship is complementary. This type of relationship exists whenever increasing the production from one enterprise causes the production from the other to increase at the same time. The right-hand graph in Figure 8-3 illustrates a possible complementary relationship between wheat production and land left fallow.

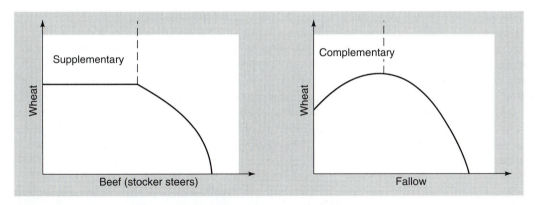

Figure 8-3   **Supplementary and complementary enterprise relationships.**

In many dryland wheat production areas with limited rainfall, some land is left fallow, or idle, each year as a way to store part of one year's rainfall to be used by a wheat crop the following year. Leaving some acres fallow reduces the acres in wheat, but the per-acre yield may increase because of the additional moisture available. This yield increase may be enough that total wheat production is actually greater than from planting all acres to wheat every year.

A complementary enterprise should be increased to at least the point where production from the primary enterprise (wheat in this example) is at a maximum. This is true even if the complementary enterprise has no value because production from the primary enterprise is increasing at the same time. Enterprises will generally be complementary only over some limited range, and they will then become competitive. As with supplementary enterprises, two general conclusions can be drawn about the profit-maximizing combination of complementary enterprises. First, this combination will not be within the complementary range. It will be at least at the point where the relationship changes from complementary to competitive. Second, assuming that both enterprises produce some revenue, it will be somewhere in the competitive range. The correct combination can be found using the substitution ratio and price ratio decision rule for competitive enterprises.

SUMMARY

This chapter continues the discussion of economic principles begun in Chapter 7. Here the emphasis was on the use of substitution principles to provide a manager with a procedure to answer the questions of how and what to produce. The question of how to produce concerns finding the least-cost combination of inputs to produce a given amount of output. Calculation and use of the input substitution and price ratios was shown, as well as the decision rule of finding the point where these two values are equal. This decision rule determines the least-cost combination of two inputs. Marginality is still an important concept here, because input substitution ratios are computed from small or marginal changes in the two inputs.

The question of what to produce is one of finding the profit-maximizing combination of enterprises when the quantity of one or more inputs is limited. That combination will depend first on the type of enterprise relationship that exists—competitive, supplementary, or complementary. The profit-maximizing combination for competitive enterprises is found by computing output substitution ratios and the output price ratio. Finding the point where they are equal determines the correct combination. Supplementary and complementary enterprises have unique properties over a limited range, but they will eventually become competitive when they begin competing for the use of some limited input. The profit-maximizing combination for these two types of enterprise relationships will not be within these ranges, but generally will be within their competitive range.

QUESTIONS FOR REVIEW AND FURTHER THOUGHT

1. First double and then halve both prices used in Table 8-1, and find the new least-cost input combination in each case. Why is there no change? But what would happen to profit in each case?

2. Is there always one "best" livestock ration? Or does it depend on prices? Would you guess that feed manufacturers know about input substitution and least-cost input combinations? Explain how they might use this knowledge.

3. Explain how the slope of the isoquant affects the input substitution ratio.

4. Carefully explain the difference between an isoquant and a production possibility curve.

5. Increase the price of one of the enterprises in Table 8-2, calculate the new output price ratio, and find the new profit-maximizing combination. Do you produce more or less of the enterprise with the now higher price? Why? What would be the effect of doubling both prices? Halving both prices?

6. Why is the profit-maximizing combination of even supplementary and complementary enterprises generally in the competitive range?

7. Where climate and rainfall permit, most farms produce two or more crops. Explain the reason for this production practice in terms of the shape of the PPC and the output price ratio.

8. For two very similar inputs, such as soybean oil meal and cottonseed oil meal in a livestock feed ration, would you expect the substitution ratio to be nearly constant or to decline sharply as one input is substituted for another? Why?

COST CONCEPTS IN ECONOMICS

CHAPTER OBJECTIVES

1. To explain the importance of opportunity cost and its use in managerial decision making

2. To clarify the difference between short run and long run

3. To discuss the difference between fixed and variable costs

4. To identify the fixed costs and show how to compute them

5. To show how to compute the different average costs

6. To demonstrate the use of fixed and variable costs in making short-run and long-run production decisions

7. To explore economies of size and how they help explain changes in farm size and profitability

A number of costs are important and useful in economics and for making management decisions. Costs can be classified in different ways, depending on whether they are fixed or variable, cash or noncash, included in record books as an accounting expense, or not. Opportunity cost is a type of cost that is not included in the accounting expenses for a business but is an important economic cost. It will be widely used in later chapters and will be the first cost discussed in this chapter.

OPPORTUNITY COST

Opportunity cost is an economic concept, not a cost that can be found in an accountant's ledger. It does not affect accounting profit. However, it is an important and basic concept that needs to be considered when making managerial decisions. Opportunity cost is based on the fact that every input or resource has an alternative use, even if the alternative is nonuse. Once an input is committed to a particular use, it is no longer available for any other alternative, and the income from the alternative must be foregone.

Opportunity cost can be defined in one of two ways:

1. the value of the product not produced because an input was used for another purpose, or
2. the income that would have been received if the input had been used in its most profitable alternative use.

The latter definition is perhaps the more common, but both should be kept in mind as a manager makes decisions on input use. The real cost of an input may not be its purchase price. Its real cost, or its opportunity cost, in any specific use is the income it would have earned in its next best alternative use. If this is greater than the income expected from the planned use of the input, the manager should reconsider the decision. The alternative appears to be a more profitable use of the input.

The concept can be illustrated by referring to Table 9-1, which is the same example used in discussing the equal marginal principle in Chapter 7. If the first 400 acre-inches of water is put on the 100 acres of wheat, the opportunity cost is $1,800. This income would be lost by not using the water for its best alternative use, which is cotton. Because the expected return from wheat is $1,200, the higher opportunity cost indicates that this first unit of water is not earning its greatest possible income. If the first 400 acre-inches is used on cotton, the opportunity cost is $1,600 (from grain sorghum), indicating that there is no more profitable use for the water. The expected income exceeds the opportunity cost. This same line of reasoning can be used to determine the most profitable use of each successive 400-acre-inch increment of water. A little thought will reveal this to be just another way of applying the equal marginal principle. It also illustrates the relation between these two concepts.

Opportunity costs are used widely in economic analysis. For example, the opportunity costs of a farm operator's labor, management, and capital are used in several types of budgeting and when analyzing farm profitability. The opportunity cost of a farm operator's labor (and perhaps that of other unpaid family labor) would be what that labor would earn in its next best

TABLE 9-1	Opportunity Cost of Applying Irrigation Water Among Three Uses[*]

	Marginal value products ($)		
Irrigation water (acre-inch)	Wheat (100 acres)	Grain sorghum (100 acres)	Cotton (100 acres)
4	1,200	1,600	1,800
8	800	1,200	1,500
12	600	800	1,200
16	300	500	800
20	50	200	400

[*]Each application of 4 acre-inches on a crop is a total use of 400 acre-inches (4 acre-inches times 100 acres).

alternative use. That alternative use could be nonfarm employment, but depending on skills, training, and experience, it might also be employment on someone else's farm.

Opportunity cost of management is difficult to estimate. For example, what is the value of management per crop acre? Some percentage of the gross income per acre is often used, but what is the proper percent? In other cases, an annual opportunity cost of management is needed. Here and elsewhere, the analyst must be careful to exclude labor from the estimate. For example, if the opportunity cost of labor is estimated at $20,000 per year, and the individual could get a job in "management" paying $30,000 per year, the opportunity cost of management is estimated at the difference, or $10,000 per year. The opportunity cost of labor plus that of management cannot be greater than the total salary in the best alternative job. Because it is difficult to estimate the opportunity cost of management, the opportunity costs of labor and management are often left combined as one value.

Capital presents a different set of problems when estimating opportunity costs. There are many uses of capital and generally a wide range of possible returns. However, the higher the average return expected on an investment, the higher the expected risk as measured by the variability of the return. To avoid the problem of identifying a comparable level of risk, the opportunity cost of farm capital is often set equal to the interest rate on savings accounts or the current cost of borrowed capital. This represents a minimum opportunity cost and is a somewhat conservative approach. If comparable levels of risk can be determined, it would be appropriate to use higher rates when analyzing farm investments with above-average levels of risk.

A special problem is how to determine the opportunity cost on the annual service provided by long-lived inputs such as land, buildings, breeding stock, and machinery. It may be possible to use rental rates for land and custom rates for machinery services. However, this does not work well for all such items, and the opportunity cost of these inputs often should be determined

by the most profitable alternative use outside the farm or ranch business. This is the real or true cost of using inputs to produce agricultural products.

It is difficult to evaluate directly the opportunity cost of these long-lived inputs in terms of nonagricultural alternative uses. Land, fences, buildings, and machinery may have a very low opportunity cost based on nonagricultural use, but they represent a large capital investment. This capital investment is generally used to estimate their opportunity cost. The total dollar value of the input is estimated and then multiplied by the opportunity cost of capital. This is an indirect procedure for estimating the annual opportunity cost. However, it has the advantages of converting everything into a common unit, its capital equivalent, and then using an opportunity cost that is easier to estimate, that of capital.

COSTS

A number of cost concepts are used in economics. Marginal input cost was studied in Chapter 7. Seven additional cost concepts and their abbreviations are

1. Total fixed cost (TFC)
2. Average fixed cost (AFC)
3. Total variable cost (TVC)
4. Average variable cost (AVC)
5. Total cost (TC)
6. Average total cost (ATC)
7. Marginal cost (MC)

All these costs are *output* related. Marginal cost, which was also studied in Chapter 7, is the additional cost of producing an additional unit of output. The others are either the total or average costs for producing a given amount of output.

Short Run and Long Run

Before these costs are discussed in some detail, it is necessary to distinguish between what economists call the short run and the long run. These are time concepts, but they are not defined as fixed periods of calendar time. The short run is that period of time during which the available quantity of one or more production inputs is fixed and cannot be changed. For example, at the beginning of the planting season, it may be too late to increase or decrease the amount of crop-land owned or rented. The current crop production cycle would be a short-run period, as the amount of available land is fixed.

Over a longer period of time, land may be purchased, sold, or leased, or leases may expire and the amount of land available may increase or decrease. The long run is defined as that period of time during which the quantity of all necessary productive inputs can be changed. In the long run, a business can expand by acquiring additional inputs or go out of existence by selling all its inputs. The actual calendar length of the long run as well as the short run will vary with the situation and circumstances. Depending on which input(s) is fixed, the short run may be anywhere from several days to several years. One year or one crop or livestock production cycle are common short-run periods in agriculture.

Fixed Costs

The costs associated with owning a fixed input are called fixed costs. These are the costs that are incurred even if the input is not used, and there may be additional costs if it is actually used in producing some product. Fixed costs do not change as the level of production changes in the short run but can change in the long run as the quantity of the fixed input changes. Since there need not be any fixed inputs owned in the long run, fixed costs exist only in the short run and are equal to zero in the long run.

Another characteristic of fixed costs is that they are not under the control of the manager in the short run. They exist, and at the same level, regardless of how much or how little the resource is used. The only way they can be avoided is to sell the item, which can be done in the long run.

Total fixed cost (TFC) is simply the summation of the several types of fixed costs. Depreciation, insurance, taxes (property taxes, not income taxes), and interest are the usual components of TFC. Repairs and maintenance may also be included as a fixed cost. (See Box 9-1.)

Computing the average annual TFC for a fixed input requires some method of finding the average annual depreciation and interest. The straight-line depreciation method discussed in Chapter 4 provides the average annual depreciation from the equation

$$\text{Annual depreciation} = \frac{\text{cost} - \text{salvage value}}{\text{useful life}}$$

where cost is equal to the purchase price, useful life is the number of years the item is expected to be owned, and salvage value is its expected value at the end of that useful life.

Because capital invested in a fixed input has an opportunity cost, interest on that investment is included as part of the fixed cost. This interest is always included and always computed the same way, regardless of whether or not any money was borrowed to finance the purchase. However, it is not correct to charge interest on the purchase price or original cost of a depreciable asset every year, because its value is decreasing over time. Therefore, the interest component of total fixed cost is commonly computed from the formula

$$\text{Interest} = \frac{\text{cost} + \text{salvage value}}{2} \times \text{interest rate}$$

where the interest rate is the opportunity cost of capital.[1] This equation gives the interest charged for the average value of the item over its life (as shown by the first part of the equation) and reflects the fact that it is decreasing in value over time. Depreciation is being charged to account for this decline in value.

As an example, assume the purchase of a machine for $20,000, with a salvage value of $5,000 and a useful life of 5 years. Annual

---

[1]This equation is widely used but is only a close approximation of the true opportunity cost. The capital recovery methods discussed in Chapter 17 can be used to find the true charge, which combines depreciation and interest into one value. This value recovers the investment in the asset plus compound interest over its useful life.

| **Box 9-1** | **Repairs as a Fixed Cost?** |

Repairs can be found in some lists of fixed costs. However, repairs typically increase as use of the asset increases, which does not fit the definition of a fixed cost. The argument for including repairs in fixed costs is that some minimum level of maintenance is needed to keep an asset in working order, even if it is not being used. Therefore, this maintenance expense is more like a fixed cost. The practical problem with this is calculating what part of an asset's total repair and maintenance expense should be a fixed cost.

In practice, all machinery repairs are generally considered variable costs, because most repairs are necessitated by use. Building repairs are more likely to be included as a fixed cost. These repairs are often more from routine maintenance caused by age and less from use. This is particularly true for storage facilities and many general-purpose buildings. For convenience, all repair costs will be classified as variable costs when developing budgets in later chapters of this text.

---

property taxes are estimated to be $25, annual insurance is $50, and the opportunity cost of capital is 10 percent. Using these values and the two preceding equations results in the following annual total fixed cost:

$$\text{Depreciation} = \frac{\$20,000 - 5,000}{5 \text{ years}} = \$3,000$$

$$\text{Interest} = \frac{\$20,000 + 5,000}{2} \times 10\% = \$1,250$$

Taxes	25
Insurance	50
Annual total fixed cost	$4,325

Notice that annual total fixed costs are more than 20 percent of the purchase price. Total annual fixed costs are often 20 to 25 percent of the purchase price for a depreciable asset.

Fixed cost can be expressed as an average per unit of output. Average fixed cost (AFC) is found using the equation

$$\text{AFC} = \frac{\text{TFC}}{\text{Output}}$$

where output is measured in physical units such as bushels, bales, or hundredweights. Acres or hours are often used as the measure of output for machinery even though they are not actually units of production. By definition, TFC is a fixed or constant value, so regardless of the quantity of output, AFC will decline continuously as output increases (see Figure 9-2). One way to lower the cost of producing a given commodity is to get more output from the fixed resource. This will always lower the AFC per unit of output.

Fixed costs can be either *cash* or *noncash* expenses. They can be easily overlooked or underestimated, because a large part of total fixed cost can be noncash expenses, as shown in the following chart.

Expense item	**Cash expense**	**Noncash expense**
Depreciation		X
Interest on the investment	X	X
Repairs	X	
Taxes, property	X	
Insurance	X	X

Depreciation is always a noncash expense, because there is no annual cash outlay for this fixed cost. Repairs and property taxes are always cash expenses, and interest and insurance may be either. If money is borrowed to purchase the asset, there will be some cash interest expense.

Box 9-2 Cash and Noncash Expenses

$\mathbf{A}$ny distinction between cash and noncash expenses does not imply that noncash expenses are any less important than cash expenses. In the short run, noncash expenses do mean that less cash is needed to meet current expenses. However, income must be sufficient to cover all expenses in the long run if the business is to survive, prosper, replace capital assets, and make an economic profit.

When the item is purchased entirely with the buyer's own capital, the interest charge would be the opportunity cost on this capital, and there is no cash payment to a lender. Insurance would be a cash expense if it is carried with an insurance company, or noncash if the risk of loss is assumed by the owner. In the latter case, there would be no annual cash outlay, but the insurance charge should still be included in fixed costs to cover the possibility of damage to or loss of the item from fire, windstorm, theft, and so forth.

Variable Costs

Variable costs are those over which the manager has control at a given point in time. They can be increased or decreased at the manager's discretion and will increase as production is increased. Items such as feed, fertilizer, seed, chemicals, fuel, and livestock health expenses are examples of variable costs. A manager has control over these expenses in the short run, and they will tend to change as total production changes.

Total variable cost (TVC) can be found by summing the individual variable costs, each of which is equal to the quantity of the input purchased times its price. Average variable cost (AVC) is calculated from the equation

$$AVC = \frac{TVC}{Output}$$

where output again is measured in physical units. Average variable cost may be either in-

creasing or decreasing, depending on the underlying production function and the output level. For the production function illustrated in Figure 7-1, AVC will initially decrease as output is increased and then will increase beginning at the point where average physical product begins to decline.

Variable costs exist in both the short run and the long run. All costs are considered to be variable costs in the long run, because there are no fixed inputs. The distinction between fixed and variable costs also depends on the exact point in time where the next decision is to be made. Fertilizer is generally a variable cost. Yet once it has been purchased and applied, the manager no longer has any control over the size of this expenditure. It must be considered a fixed cost for the remainder of the crop season, which can affect future decisions during that crop season. Labor cost and cash rent for land are similar examples. After a labor or lease contract is signed, the manager cannot change the amount of money obligated, and the salary or rent must be considered a fixed cost for the duration of the contract.

Total and Marginal Costs

Total cost (TC) is the sum of total fixed cost and total variable cost (TC = TFC + TVC). In the short run, it will increase only as TVC increases, because TFC is a constant value. Average total cost (ATC) can be found by one of two methods. For a given output level, it is equal to AFC + AVC. It can also be calculated from the equation

$$ATC = \frac{TC}{Output}$$

which will give the same result. Average total cost will typically be decreasing at low output levels, because AFC is decreasing rapidly and AVC may be decreasing also. At higher output levels, AFC will be decreasing less rapidly, and AVC will eventually increase and be increasing at a rate faster than the rate of decrease in AFC. This combination causes ATC to increase.

Marginal cost (MC) is defined as the change in total cost divided by the change in output:

$$MC = \frac{\Delta\, TC}{\Delta\, Output} \text{ or } MC = \frac{\Delta\, TVC}{\Delta\, Output}$$

It is also equal to the change in total variable cost divided by the change in output. Since TC = TFC + TVC, and TFC is constant, the only way TC can change is from a change in TVC. Therefore, MC can be calculated either way with the same result.

Cost Curves

Relations among the seven output-related cost concepts can be illustrated graphically by a series of curves. The shape of these cost curves depends on the characteristics of the underlying production function. Figure 9-1 contains cost curves that represent the general production function shown in Figure 7-1. Other types of production functions would have cost curves with different shapes.

The relations among the three total costs are shown in Figure 9-1. Total fixed cost is constant and unaffected by output level. Total variable cost is always increasing, first at a decreasing rate and then at an increasing rate. Because it is the sum of total fixed cost and total variable cost, the total cost curve has the same shape as the total variable cost curve. However, it is always higher by a vertical distance equal to total fixed cost.

The general shape and relation of the average and marginal cost curves are shown in Fig-

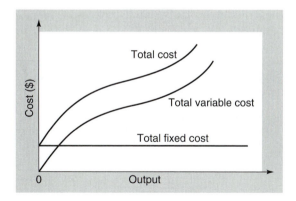

Figure 9-1   **Typical total cost curves.**

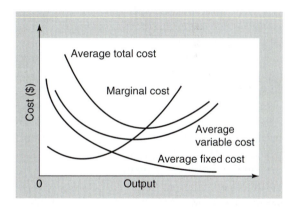

Figure 9-2   **Average and marginal cost curves.**

ure 9-2. Average fixed cost is always declining but at a decreasing rate. The other two average curves are U-shaped, declining at first, reaching a minimum, and then increasing at higher levels of output. Notice that they are not an equal distance apart. The vertical distance between them is equal to average fixed cost, which changes with output level. This accounts for their slightly different shape and for the fact that their minimum points are at two different output levels.

The marginal cost curve will generally be increasing. However, for this particular production function, it decreases over a short range before starting to increase. Notice that the marginal cost curve crosses both average curves at their minimum points. As long as the marginal value is below the average value, the latter will be decreasing, and vice versa. For this reason, the marginal cost curve will always cross the average cost curves at their minimum point, if such a point exists.

Other Possible Cost Curves

As stated earlier, the shape of the cost curves is directly related to the underlying production function. The cost curves in Figures 9-1 and 9-2 are all derived from the shape of the generalized production function in Figure 7-1. Other types of production functions exist in agriculture, in particular, those which increase at a *decreasing* rate from the first unit of input. They do not have a Stage I and therefore begin with diminishing marginal returns.

Figure 9-3 shows the total, average, and marginal cost curves for this type of production function. Since the production function effectively begins in Stage II with diminishing marginal returns, the TVC curve increases at an increasing rate from the beginning. This in turn causes the AVC curve to increase at an increasing rate throughout. However, since the ATC curve is the sum of AVC and AFC, it begins high due to the high AFC. Initially, average fixed costs will be decreasing at a rapid rate and faster than AVC is increasing. This combination re-

sults in an ATC which decreases at first but which eventually increases as the AVC curve begins increasing at a more rapid rate than the AFC curve is declining.

Cost curves with a different shape can result when output is measured in something other than the usual agricultural commodities. The output from machinery services is one example. It is difficult if not impossible to measure machinery output in bushels, pounds, or tons, particularly if the machine is used in the production of several outputs. Therefore, the output from tractors in particular, and often other machinery, is measured in hours of use or acres covered during a year. There is no declining marginal product in this case as another hour is another hour of the same length and the same amount of work can be performed in that hour. A constant marginal physical product results from each additional hour of use, that is, another hour of work performed.

Figure 9-4 illustrates the cost curves for this example. With a constant marginal physical product, TVC increases at a constant rate which in turn causes AVC to be constant *per hour* of use. However, AFC is decreasing as hours of use increases. Since ATC is the sum of AVC and AFC, it will also be continually decreasing as annual hours of use increases. (See Chapter 22 for more on machinery costs.)

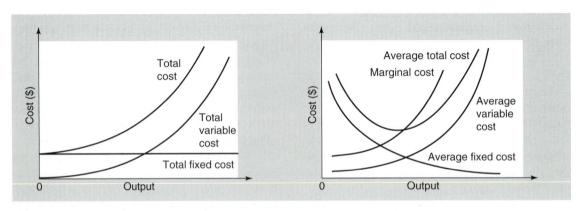

Figure 9-3   **Cost curves for a diminishing marginal returns production function.**

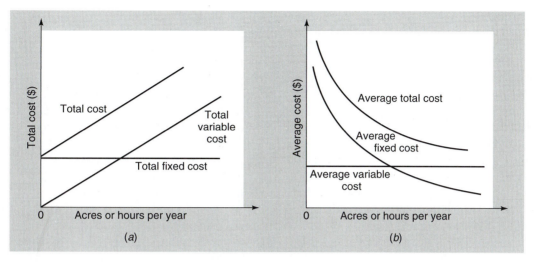

Figure 9-4   **Cost curves for a production function with constant marginal returns.**

APPLICATION OF COST CONCEPTS

Table 9-2 is an example of some cost figures for the common problem of determining the profit-maximizing stocking rate for a pasture of fixed size. It illustrates many similar problems where an understanding of the different cost concepts and relations will help a manager in planning and decision making. Total fixed costs are assumed to be $5,000 per year in the example. This would cover the annual opportunity cost on the land and any improvements, depreciation on fences and water facilities, and insurance. Variable costs are assumed to be $495 per steer (steers are the variable input in this example). This includes the cost of the steer, transportation, health expenses, feed, interest on the investment in the steer, and any other expenses that increase along with the number of steers purchased.

Because the size of the pasture and amount of forage available are both limited, running more and more steers will eventually cause the average weight gain per steer to decline. This is reflected in diminishing returns and a declining MPP when more than 30 steers are placed in the pasture. Total hundredweights of beef sold off the pasture still increases, but at a decreasing rate as more steers compete for the limited forage.

The cost figures in Table 9-2 have the usual or expected pattern as production or output is increased. Total fixed cost remains constant by definition, while both TVC and TC are increasing. Average fixed cost declines rapidly at first and then continues to decline but at a slower rate. Average variable cost, average total cost, and marginal cost all decline initially, reach a minimum at different output levels, and then begin to increase.

The profit-maximizing output level was defined in Chapter 7 to be where MR = MC, but this exact point does not exist in Table 9-2. However, MC is less than MR when moving from 50 to 60 steers, but more than MR when moving from 60 to 70 steers. This makes 60 steers and 420 hundredweights of beef the profit-maximizing levels for this example. When MC is greater than MR, the additional cost per hundredweight of beef produced from the additional 10 steers is greater than the additional income per hundredweight. Therefore, placing 70 steers on the pasture would result in less profit than 60 steers.

TABLE 9-2	Illustration of Cost Concepts Applied to a Stocking Rate Problem*

Number of steers	Output (hundred weight of beef)	MPP	TFC ($)	TVC ($)	TC ($)	AFC ($)	AVC ($)	ATC ($)	MC ($)	MR ($)
0	0		5,000	0	5,000	—	—	—		
		7.2							68.75	87.50
10	72		5,000	4,950	9,950	69.44	68.75	138.19		
		7.6							65.13	87.50
20	148		5,000	9,900	14,900	33.78	66.89	100.68		
		7.7							64.29	87.50
30	225		5,000	14,850	19,850	22.22	66.00	88.22		
		7.0							70.71	87.50
40	295		5,000	19,800	24,800	16.95	67.12	84.07		
		6.5							76.15	87.50
50	360		5,000	24,750	29,750	13.89	68.75	82.64		
		6.0							82.50	87.50
60	420		5,000	29,700	34,700	11.90	70.71	82.62		
		5.5							90.00	87.50
70	475		5,000	34,650	39,650	10.53	72.95	83.47		
		5.0							99.00	87.50
80	525		5,000	39,600	44,600	9.52	75.43	84.95		
		4.5							110.00	87.50
90	570		5,000	44,550	49,550	8.77	78.16	86.93		
		4.0							123.75	87.50
100	610		5,000	49,500	54,500	8.20	81.15	89.34		

*Total fixed cost is $5,000, and variable costs are $495 per steer. Selling price of the steers is assumed to be $87.50 per hundredweight.

The profit-maximizing point will depend on the selling price, or MR, and MC. Selling prices often change, and MC can change with changes in variable cost (primarily from a change in the cost of the steer in this example). The values in the MC column indicate that the most profitable number of steers would be 70 if the selling price is over $90.00 but less than $99.00. That number would drop to 50 steers if the selling price drops to between $76.15 and $82.50. As shown in Chapter 7, the profit-maximizing input and output levels will always depend on prices of both the input and output.

For the selling price of $87.50 per hundredweight and a stocking rate of 60 steers, the TR is $36,750 (420 hundredweight × $87.50) and TC is $34,700, leaving a profit of $2,050. For a price of $76.15, the MR = MC rule indicates a profit-maximizing point of 50 steers and 360 hundredweights. However, the ATC per hundredweight at this point is $82.64, which means a loss of $6.49 per hundredweight ($76.15 − $82.64 = −$6.49) or a total loss of $2,336 ($6.49 × 360 hundredweights). What should a manager do in this situation? Should any steers be purchased if the expected selling price is less than ATC and a loss will result?

The answer to this question is "yes" for some situations and "no" for others. Data in Table 9-2 indicate that there would be a loss equal to the TFC of $5,000 if no steers were purchased. This loss would exist in the short run as long as the land is owned. It could be avoided in the long run by selling the land, which would eliminate the fixed costs. However, the fixed costs cannot be avoided in the short run, and the relevant question is: Can a profit be made or the loss reduced to less than $5,000 in the short run by purchasing some steers? Obviously, steers should not be purchased if it would result in a loss greater than $5,000, because the loss can be minimized at $5,000 by not purchasing any.

Variable costs are under the control of the manager and can be reduced to zero by not purchasing any steers. Therefore, no variable costs should be incurred unless the expected selling price is at least equal to or greater than the minimum AVC. This will provide enough total revenue to cover total variable costs. If the selling price is greater than the minimum AVC but less than the minimum ATC, the income will cover all variable costs, with some left over to pay part of the fixed costs. There would be a loss, but it would be less than $5,000 in this example. To answer the question above, yes, steers should be purchased when the expected selling price is less than the minimum ATC, but only if it is above the minimum AVC. This action will result in a loss, but it will be the minimum loss possible given these cost and price relations.

If the expected selling price is less than the minimum AVC, total revenue will be less than TVC: there will be a loss, and it will be greater than $5,000. Under these conditions, no steers should be purchased, which will minimize the loss at $5,000. In Table 9-2, the lowest AVC is $66.00, and the lowest ATC is $82.62. The loss would be minimized by not purchasing steers when the expected selling price is less than $66.00, and by purchasing steers when the expected selling price is between $66.00 and $82.62. In the last situation, the loss-minimizing output level is where MR = MC.

Production Rules for the Short Run

The preceding discussion leads to three rules for making production decisions in the short run. They are:

1. Expected selling price is greater than minimum ATC (or TR greater than TC). A profit can be made and is maximized by producing where MR = MC.
2. Expected selling price is less than minimum ATC but greater than minimum AVC (or TR is greater than TVC but less than TC). A loss cannot be avoided but will be minimized by producing at the output level where MR = MC. The loss will be somewhere between zero and the total fixed cost.
3. Expected selling price is less than minimum AVC (or TR less than TVC). A loss cannot be avoided but is minimized by not producing. The loss will be equal to TFC.

The application of these rules is illustrated graphically in Figure 9-5. With a selling price equal to MR_1, the intersection of MR and MC is well above ATC, and a profit is being made. When the selling price is equal to MR_2, the income will not be sufficient to cover total costs but will cover all variable costs, with some left over to pay part of fixed costs. In this situation, the loss is minimized by producing where MR = MC, because the loss will be less than TFC. Should the selling price be as low as MR_3, income would not even cover variable costs, and the loss would be minimized by stopping production. This would minimize the loss at an amount equal to TFC.

Production Rules for the Long Run

The three rules above apply only to the short run where fixed costs exist. What about the long run where there are no fixed costs? Continual losses incurred by producing in the long run will eventually force the firm out of business.

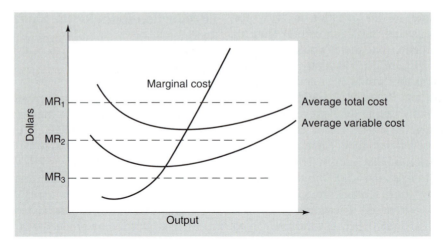

Figure 9-5   Illustration of short-run production decisions.

There are only two rules for making production decisions in the long run.

1. Selling price is greater than ATC (or TR greater than TC). Continue to produce, because a profit is being made. This profit is maximized by producing at the point where MR = MC.
2. Selling price is less than ATC (or TR less than TC). There will be a continual loss. Stop production and sell the fixed asset(s), which eliminates the fixed costs. The money received should be invested in a more profitable alternative.

This does not mean that assets should be sold the first time a loss is incurred. Short-run losses will occur when there is a temporary drop in the selling price. Long-run rule number two should be invoked only when the drop in price is expected to be long lasting or permanent.

ECONOMIES OF SIZE

Economists and managers are interested in farm size and the relation between costs and size, for a number of reasons. The following are examples of questions being asked that relate to farm size and costs. What is the most profitable farm size? Can larger farms produce food and fiber more cheaply? Are large farms more efficient? Will family farms disappear and be replaced by large corporate farms? Will the number of farms and farmers continue to decline? The answers to all these questions depend at least in part on what happens to costs and the cost per unit of output as farm size increases.

First, how is farm size measured? Number of livestock, number of acres, number of full-time workers, equity, total assets, profit, and other factors have all been used to measure size, and all have some advantages and disadvantages. For example, number of acres is a common and convenient measure of farm size but should be used only to compare farm sizes in a limited geographical area where farm type, soil type, and climate are very similar. It is obvious that 100 acres of irrigated vegetables in California is not the same size operation as 100 acres of arid range land in neighboring Arizona or Nevada.

Gross farm income, or total revenue, is a common measure of farm size. It has the advantage of converting everything into the common denominator of the dollar. This and other measures that are in dollar terms are better than any physical measure for determining and comparing farm size across different farming regions.

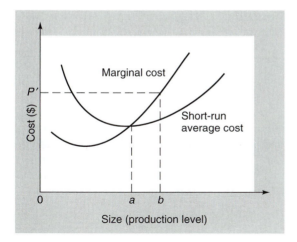

Figure 9-6 **Farm size in the short run.**

Size in the Short Run

In the short run, the quantity of one or more inputs is fixed, with land often being the fixed input. Given this fixed input, there will be a short-run average total cost curve, as shown in Figure 9-6. Short-run average cost curves typically will be U-shaped, with the average cost increasing at higher levels of production, because the limited fixed input makes additional production more and more difficult and therefore increases the average cost per unit of output.

For simplicity, size is measured as the output of a single product in Figure 9-6. The product can be produced at the lowest average cost per unit by producing the quantity 0a. However, this may not be the profit-maximizing quantity, as profit is maximized at the output level where marginal revenue is equal to marginal cost. Since marginal revenue is equal to output price, a price of P' would maximize profit by producing the quantity 0b. A higher or lower price would cause output to increase or decrease to correspond with the point where the new price is equal to marginal cost.

Because of a fixed input such as land, output can be increased in the short run only by intensifying production. This means more variable inputs such as fertilizer, chemicals, irrigation water, labor, and machinery time must be used. However, the limited fixed input tends to increase average and marginal costs as production is increased past some point and an absolute production limit is eventually reached. Additional production is possible only by acquiring more of the fixed input, which is a long-run problem.

Size in the Long Run

The economics of farm size is more interesting when analyzed in a long-run context. This gives the manager time to adjust all inputs to the level that will result in the desired farm size. One measure of the relation between output and costs as farm size increases is expressed in the following ratio:

$$\frac{\text{Percent change in costs}}{\text{Percent change in output value}}$$

Both changes are calculated in monetary terms to allow combining the cost of the many inputs and the value of several outputs into one figure. This ratio can have three possible results called decreasing costs, constant costs, or increasing costs.

Ratio value	Type of costs
< 1	Decreasing
= 1	Constant
> 1	Increasing

These three possible results are also called, respectively, increasing returns to size, constant returns to size, and decreasing returns to size. Decreasing costs means increasing returns to size, and vice versa. These relations are shown in Figure 9-7 using a long-run average cost curve that is the average cost per unit of output. When decreasing costs exist, the average cost per unit of output is decreasing, so that the average profit per unit of output is increasing. Therefore, increasing returns to size are said to exist. The same line of reasoning explains the relation

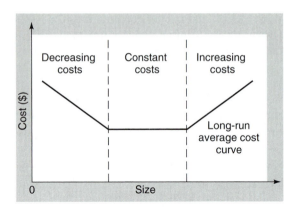

Figure 9-7   **Possible size-cost relations.**

between constant costs and constant returns and between increasing costs and decreasing returns.

Economies of size exist whenever the long-run average cost curve is decreasing over some range of output (decreasing costs and increasing returns to size). *Diseconomies of size* exist when long-run average costs are increasing (increasing costs and decreasing returns to size). Many studies have been done to verify the existence of economies and diseconomies of size and their causes. The existence or nonexistence of each and the range of farm sizes over which each occur help explain and predict farm size. Farms of a size where long-run average costs are declining have an incentive to get larger. Any farms which may be so large that their long-run average costs are increasing will have less incentive to grow and will soon reach their long-run profit-maximizing size if they haven't already done so.

Causes of Economies of Size

Studies of U.S. farms and ranches have documented the existence of economies of size over at least some initial size range. What is not always clear are the exact causes of these lower average costs as size increases. The following have been identified as possible causes with several likely contributing to lower long-run average costs on any given farm at any given time.

Full Utilization of Existing Resources

One of the basic causes of economies of size is the full use of existing labor, machinery, capital, and management. These resources have fixed costs whether they are used or not. Their full use does not increase total fixed costs but does lower average fixed cost per unit of output. Consider the farmer who rents some additional land and is able to farm it with existing labor and machinery. Average fixed cost per unit of output is now lower not only for the additional production but also for all of the original production.

Technology

New technology is often expensive but it can reduce the average cost per unit of output by the combination of replacing some current inputs and increasing production per acre or per head. However, because of the high initial investment, new technology must often be used over a large number of acres or head to obtain these lower costs. This produces economies of size for larger farms and ranches but leaves smaller ones with older and more costly technology.

Use of Specialized Resources

Labor and equipment on smaller farms must often be used for many different tasks and for several different enterprises, perhaps both crops and livestock. No single enterprise is large enough to justify specialized equipment which could do the job more efficiently and at less cost. Labor must perform so many different tasks that individuals do not have time to get sufficient experience to become skilled at any task. So little time may be spent on any single task during the year that it is difficult to justify the job training necessary to become more proficient. Larger farms can make full-time use of specialized equipment and individual workers can work full time in one enterprise and perhaps at only one task. A large dairy where workers can work full time in the milking parlor is one example. Specialization often increases efficiency and reduces costs per unit of output.

Input Prices

Price discounts when purchasing large amounts and for buying in bulk are common in most industries including agriculture. Large farms and ranches can obtain substantial price discounts, for example, by buying feed in bulk by the truckload rather than a few bags at a time. Chemicals, fertilizer, seed, fuel, and supplies such as baling twine and spare parts may also have price discounts for volume buying. Even if there is no price discount for bulk buying, the labor saved by the convenience, ease, and speed of handling the material can easily make up for additional storage and handling equipment needed.

Output Prices

Large volume producers may also have a price advantage over smaller ones when selling their production. Grain producers may earn a premium price if they can deliver a large quantity at one time or guarantee the delivery of a fixed amount each month to a feedlot or feed mill. Ranchers who can deliver a full truckload of uniform weight cattle directly to a feedlot will usually receive a higher net price than one who sells a few at a time through a local auction barn.

In recent years many new, special-use grain varieties have been developed. Farmers receive a price premium for raising these specialized grains but they require special handling and separate storage away from all other varieties. Combines and trucks often need to be thoroughly cleaned before and after harvesting and hauling these varieties. This is another case where only larger volume producers can justify the expense of separate and adequate storage and other related expenses in order to receive the premium for growing the crop.

Management

Many of the functions and tasks of management may contribute to economies of size. Purchasing inputs, marketing products, accounting, gathering information, planning, and, at least up to some point, labor supervision, are examples.

Doubling the size of the business may increase the time spent on each of these activities but not to the extent of doubling it.

Causes of Diseconomies of Size

It is clear that economies of size exist over some initial range of farm sizes for nearly all types of farms and ranches. The existence of diseconomies of size is less clear as is the size at which diseconomies may begin. Diseconomies may result from any or all of the following.

Management

Limited management capacity has always been considered a prime cause of diseconomies of size. As a farm or ranch business becomes larger, it becomes more difficult for a manger to be knowledgeable about all aspects of all enterprises and to properly organize and supervise all activities. Timeliness of operations and attention to detail begin to decline. The business becomes less efficient and average costs increase through some combination of increasing expenses and declining production.

Labor Supervision

Related to management is the increased difficulty of managing a larger labor force as size increases. In many types of agriculture this is often further complicated by individuals working alone or in small groups spread over an entire field or many different fields on several different farms. This may require additional supervisors or considerable unproductive travel time by one supervisor.

Geographical Dispersion

Agricultural operations such as greenhouse production, cattle feedlots, confinement swine facilities, and poultry production can have a large operation concentrated in a relatively small area. However, increasing the size of a field crop operation requires additional land which may not always be available nearby. This increases the time and expense needed to move labor and equipment from field to field and the products from field to centralized storage or processing facilities.

Large Livestock Operations

Odor control, manure disposal, and increased risk of disease in large concentrations of livestock are potential sources of diseconomies in large livestock operations. State and Federal regulations often require operations above a certain size to implement strict odor control and manure disposal procedures which can increase costs compared to a smaller operation not subject to the regulations. Control of additional surrounding land may be required to keep odors from reaching neighbors and for manure disposal.

SHAPE OF THE LRAC CURVE

Whether long run average costs (LRAC) are decreasing, constant, or increasing depends on the size of the business and what types of economies and diseconomies exist at that size. Some of both may exist at some sizes and the combined effect on average cost will depend on the relative number and strength of each.

Two possible LRAC curves are shown in Figure 9-8. Figure 9-8a is the typical, U-shaped curve found in most textbooks. It assumes economies of size will dominate as a small business becomes larger. At some size all economies have been realized and LRAC is at a minimum. It is possible that this minimum cost may be constant or nearly so over a range of sizes. (See Figure 9-7) However, diseconomies will eventually dominate as size grows beyond some point and LRAC will begin to increase. Management as the limiting input is often cited as the reason for the increasing costs.

The LRAC curve in Figure 9-8b is often referred to as an L-shaped curve. It better describes the results found in a number of cost studies on crop farms of different sizes. Average costs usually fall rapidly reaching a minimum at a size often associated with a full-time family farmer who hires at least some additional labor, either full- or part-time. As size increases beyond this point average costs remains constant or nearly so over a wide range of sizes. These studies show little or no increase in average costs over the range of sizes studied. However, this does not mean diseconomies do not exist, as most studies do not include enough very large farms to make a definite conclusion.

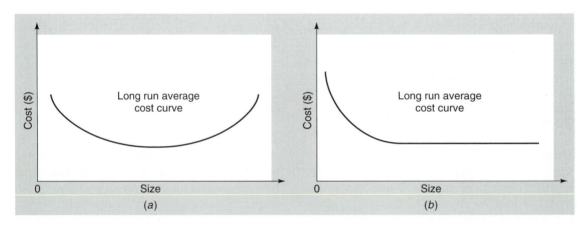

Figure 9-8   **Two Possible LRAC curves.**

SUMMARY

This chapter discussed the different economic costs and their use in managerial decision making. Opportunity costs are often used in budgeting and farm financial analysis. This noncash cost stems from inputs that have more than one use. Using an input one way means it cannot be used in any other alternative use, and the income from that alternative must be foregone. The income given up is the input's opportunity cost.

An analysis of costs is important for understanding and improving the profitability of a business. The distinction between fixed and variable costs is important and useful when making short-run production decisions. In the short run, production should take place only if the expected income will exceed the variable costs. Otherwise, losses will be minimized by not producing. Production should take place in the long run only if income is high enough to pay all costs. If all costs are not covered in the long run, the business will eventually fail or will be receiving less than opportunity cost on one or more inputs.

An understanding of costs is also necessary for analyzing economies of size. The relation between cost per unit of output and size of the business determines whether there are increasing, decreasing, or constant returns to size. If unit costs decrease as size increases, there are increasing returns to size, and the business would have an incentive to grow, and vice versa. The type of returns that exist for an individual farm will determine in large part the success or failure of expanding farm size. Future trends in farm size, number of farms, and form of business ownership and control will be influenced by economies and diseconomies in farm and ranch businesses.

QUESTIONS FOR REVIEW AND FURTHER THOUGHT

1. How would you estimate the opportunity cost for each of the following items? What do you think the actual opportunity cost would be?
   a. Capital invested in land
   b. Your labor used in a farm business
   c. Your management used in a farm business
   d. One hour of tractor time
   e. The hour you wasted instead of studying for your next exam
   f. Your time earning a college degree

2. For each of the following, indicate whether it is a fixed or variable cost, and a cash or noncash expense. (Assume short run.)

	Fixed or Variable		Cash or Noncash	
a. Gas and oil	____	____	____	____
b. Depreciation	____	____	____	____
c. Property taxes	____	____	____	____
d. Salt and minerals	____	____	____	____
e. Labor hired on an hourly basis	____	____	____	____
f. Labor contracted for 1 year in advance	____	____	____	____
g. Insurance premiums	____	____	____	____
h. Electricity	____	____	____	____

3. Assume that Freda Farmer has just purchased a new combine. She has calculated total fixed cost to be $22,500 per year and estimates a total variable cost of $9.50 per acre.
   a. What will her average fixed cost per acre be if she combines 1,200 acres per year? 900 acres per year?
   b. What is the additional cost of combining an additional acre?
   c. Assume that Freda plans to use the combine only for custom work on 1,000 acres per year. How much should she charge per acre to be sure all costs are covered? If she would custom harvest 1,500 acres per year?

4. Assume the purchase of a combine for $152,500 that is estimated to have a salvage value of $42,000 and a useful life of 8 years. The opportunity cost of capital is 10 percent. When computing annual total fixed cost, how much should be included for depreciation? For interest?

5. Using the data in Table 7-3 and a TFC of $50, compute the three total costs and the three average costs.
   a. What is the maximum profit that can be made with the given prices?
   b. To continue production in the long run, the output price must remain equal to or above $_____ .
   c. In the short run, production should stop whenever the output price falls below $_____ .

6. Why is interest included as a fixed cost even though no money was borrowed to purchase the item?

7. What will profit be if production takes place where MR = MC just at the point where ATC is minimum? At the point where MR = MC, and AVC is at its minimum?

8. Explain why and under what conditions it is rational for a farmer to produce a product at a loss.

9. Imagine a typical farm or ranch in your local area. Assume it doubles in size to where it is producing twice as much of each product as before.
   a. If total cost also doubles, is the result increasing, decreasing, or constant returns to size? What if total cost increases by only 90 percent?
   b. Which individual costs would you expect to exactly double? Which might increase by more than 100 percent? By less than 100 percent?
   c. Would you expect this farm or ranch to have increasing, decreasing, or constant returns to size? Economies or diseconomies of size? Why?

10

ENTERPRISE BUDGETING

CHAPTER OBJECTIVES

1. To define an enterprise budget and discuss its purpose and use

2. To illustrate the different sections of an enterprise budget

3. To learn how to construct a crop enterprise budget

4. To outline additional problems and steps to consider when constructing a livestock enterprise budget

5. To show how data from an enterprise budget can be analyzed and used for computing cost of production and break-even prices and yields

The previous three chapters discussed some basic economic principles and cost concepts. One practical application of these concepts is their use in various types of budgets. Enterprise, partial, whole-farm, and cash flow budgets are important and useful tools for the farm and ranch manager, and each provides a framework for the use of these concepts. These budgeting techniques are the subject of the next four chapters.

Budgeting is often described as "testing it out on paper" before committing resources to a plan or to a change in an existing plan. It is a forward-planning tool, or a way to estimate the profitability or feasibility of a plan, a proposed change in a plan, or an enterprise, before making the decision and implementing it. Budgets reflect the manager's best estimate as to what will happen in the future if a certain plan is followed. Paper, pencils, calculators, and computers are the tools of budgeting and, as such, may be the most important tools a manager uses.

The combination of budgeting and economic principles provides some powerful, practical, and useful techniques for the manager to use when analyzing alternatives. This step is an important one in the decision-making process, because only a proper and correct analysis will lead to making the right decision. The discussion of budgeting begins with enterprise budgets, because they often provide much of the data needed for developing the other types of budgets.

ENTERPRISE BUDGETS

An enterprise budget provides an estimate of the potential revenue, expenses, and profit for a single enterprise. Each crop or type of livestock that can be grown is an enterprise. Therefore, there can be enterprise budgets for cotton, corn, wheat, beef cows, dairy cows, farrow-finish hogs, watermelons, soybeans, peanuts, and so forth. They can be created for different levels of production or types of technology, so there can be more than one budget for a given enterprise. The base unit for enterprise budgets is typically one acre for crops. For livestock, some managers develop enterprise budgets on a per head basis, while others pick a "typical" size operation (e.g. a 30-head cow-calf operation) as the basis for a budget. Using common units permits an easy and fair comparison across different enterprises. Enterprise budgets for many enterprises are usually available from the county or state Cooperative Extension office. These budgets, developed to represent a "typical" situation, may or may not be accurate for a particular operation and often need to be adjusted.

The primary purpose of enterprise budgets is to estimate the projected costs, returns, and profit per unit for the enterprises. Once this is done, the budgets have many uses. They help identify the more profitable enterprises to be included in the whole-farm plan. Because a whole-farm plan often consists of several enterprises, enterprise budgets are often called the "building blocks" in a whole-farm plan and budget. (See Chapter 12 for more on whole-farm budgeting.) Although constructing an enterprise budget requires a large amount of data, once completed, it is a source of data for other types of budgeting. A manager will refer often to enterprise budgets for information and data when making many types of decisions. An illustration of other uses for enterprise budgets will be postponed until after discussing their organization and construction.

Table 10-1 is an example that will be used to discuss the content, organization, and structure of a crop enterprise budget. Although no single organization or structure is used by everyone, most budgets contain the sections or parts included in Table 10-1. This example does not include all the detail on physical quantities and prices usually found on an enterprise budget, but it will serve to illustrate the basic organization and content.

The name of the enterprise being budgeted and the budgeting unit are shown first. Most enterprise budgets cover a year or less, but for some enterprises, with a long production process, a multi-year budget is more useful. If the time period is longer than a year, it is useful to state the time period being covered. Income or revenue from the enterprise is typically shown next. Quantity, unit, and price should all be included to provide full information to the user. The cost section comes next and is generally divided into two parts: variable (or operating) costs, and fixed (or ownership) costs. Some budgets further divide variable costs into preharvest variable costs, or those that occur before harvest, and harvest costs, or those that are a direct result of harvesting. Income or revenue above variable costs is an intermediate calculation and shows the revenue remaining to be

TABLE 10-1	Example Enterprise Budget for Watermelon Production (One Acre)

Item		Value per acre
Revenue		
250 cwt @ $5.50 per cwt		$1,375.00
Variable Costs		
Seed	$ 80.00	
Fertilizer	95.50	
Pesticides	97.75	
Machinery fuel, lube, and repairs	35.15	
Custom spraying	8.00	
Harvesting and hauling	500.00	
Labor	320.00	
Interest @ 10% for 6 months	56.82	
Total variable cost		$1,193.22
Income above variable cost		$ 181.78
Fixed Costs		
Machinery depreciation, interest, taxes, and insurance	$ 62.00	
Land charge	100.00	
Total fixed costs		$ 162.00
Total costs		$1,355.22
Estimated profit (return to management)		$ 19.78

applied to fixed costs. Income above variable costs is sometimes called the *gross margin* of an enterprise.

Fixed costs in a crop enterprise budget typically include the fixed costs for the machinery needed to produce the crop, and a charge for land use. These costs are probably the most difficult to estimate, and discussion of the procedures to be used will be postponed until the next section.

The estimated profit per unit is the final value and is found by subtracting total costs from total revenue. Everyone is interested in this value, and it is important that it be interpreted correctly. Most enterprise budgets such as the one in Table 10-1 are economic budgets. This means that, in addition to cash expenses and depreciation, opportunity costs are also included. Typically, there would be opportunity costs for operator labor, capital used for variable costs, capital invested in machinery, and possibly for that invested in land. Therefore, the profit or return shown on an enterprise budget is an estimated *economic* profit. This is different from the concept of accounting profit, where opportunity costs are not recognized. When working with enterprise budgets developed by others, managers should be careful to check whether opportunity costs have been included in the cost figures.

Enterprise and other types of budgets often use a slightly different terminology to describe the different types of costs. Economic variable costs may be called *operating* or *direct costs,* emphasizing that they arise from the actual operation of the enterprise. These costs would not exist except for the production from this enterprise.

Fixed costs may be called *ownership* or *indirect costs.* The term ownership costs refers to the fixed costs that arise from *owning* machinery, buildings, or land. They result from owning assets and would exist even if they were not used for this enterprise. They include such things as the usual economic fixed costs, and other farm expenses such as property and liability insurance, legal and accounting fees, pickup truck expenses, and subscriptions to agricultural publications. These expenses are necessary and appropriate expenses but they are not directly tied to any single enterprise. It is therefore difficult to assign these expenses correctly when making enterprise budgets. They are often prorated on some basis to all enterprises to make sure all farm expenses, direct and indirect, are charged to the farm enterprises. Machinery expenses, for example, can be prorated based on the number of hours a machine is used for a particular enterprise. Other general operating expenses are sometimes assigned based on that enterprise's share of either the farm's total variable costs or its total gross revenue.

Box 10-1   **Opportunity Cost of Management**

The opportunity cost of management is often omitted from an enterprise budget. While it would be appropriate to include it, management is a difficult opportunity cost to estimate. This may explain its frequent omission. If no opportunity cost for management is shown in the budget, the estimated profit should be interpreted as "estimated return to management and profit." This terminology is used on some enterprise budgets.

CONSTRUCTING A CROP ENTERPRISE BUDGET

The first step in constructing a crop enterprise budget is to determine tillage and agronomic practices, input levels, type of inputs, and so forth. What seeding rate? Fertilizer levels? Type and amount of herbicide? Number and type of tillage operations? All of the agronomic, production, and technical decisions must be made before beginning work on the enterprise budget.

Table 10-2 is an enterprise budget for wheat that will be used to discuss the steps in constructing a crop enterprise budget. This budget assumes an owner/operator who is paying all the expenses and receiving all the crop.

Revenue

The revenue section should include all cash and noncash revenue from the crop. Some crops have two sources of revenue, such as cotton lint and cottonseed, oat grain and oat straw. All cash revenue should be included, and the concept of opportunity cost should be used to value any noncash sources of revenue such as crop residue used by livestock.

The accuracy of the projected profit for the enterprise may depend more on the estimates made in this section than in any other. It is important that both yield and price estimates be as accurate as possible. Projected yield should be based on historical yields, yield trends, and the type and amount of inputs to be used. The appropriate selling price will depend somewhat on whether the budget is for only the next year or for long-run planning purposes. In either case, a review of historical price levels, price trends, and outlook for the future should be conducted before estimating a price.

Operating or Variable Expenses

This section includes those costs that will be incurred only if this crop is produced. The amount to be spent in each case is under the control of the decision maker and can be reduced to $0 by not producing this crop.

Seed, Fertilizer, and Pesticides

Costs for these items are relatively easy to determine once the levels of these inputs have been selected. Prices can be found by contacting input suppliers, and the total per-acre cost for each item is found by simply multiplying quantity times price.

Fuel and Lubricants

Fuel and lubricants expense is related to the type and size of machinery used and to the number and type of machinery operations performed for this crop. A quick and simple way to obtain this value is to divide the total farm expense for fuel and lubricants by the number of crop acres. However, this method is not accurate if some machinery is also used for livestock production and if some crops take more machinery time than others.

A more accurate method is to determine fuel consumption per acre for each machine

TABLE 10-2	Enterprise Budget for Wheat (One Acre)

Item	Unit	Quantity	Price	Amount
Revenue				
Wheat grain	bu	48	$ 3.00	$144.00
Total revenue				$144.00
Operating expenses				
Seed	lb	75	$ 0.20	$ 15.00
Fertilizer: Nitrogen	lb	60	0.15	9.00
Phosphate	lb	30	0.20	6.00
Potash	lb	30	0.15	4.50
Pesticides	acre	1	7.50	7.50
Fuel & Lubricants	acre	1	9.50	9.50
Machinery repairs	acre	1	6.40	6.40
Labor	hr	2	8.00	16.00
Interest (operating expenses for 6 months)	$	36.95	10%	3.70
Total operating expense				$ 77.60
Income above variable costs				$ 66.40
Ownership expenses				
Machinery depreciation	acre	1	14.20	14.20
Machinery interest	acre	1	10.60	10.60
Machinery taxes & insurance	acre	1	2.50	2.50
Land charge	acre	1	50.00	50.00
Misc. overhead	acre	1	4.00	4.00
Total ownership expenses				$ 81.30
Total expense				$158.90
Profit (return to management)				$ (14.90)

operation and then simply sum the fuel usage for all the operations scheduled for this crop. The result can be multiplied by the price of fuel to find the per-acre cost. Another method is to compute fuel consumption per hour of tractor use and then determine how many hours will be needed to perform the machine operations. This method is used by many computer programs written to calculate enterprise budget costs.

Machinery Repairs

Estimating per-acre machinery repairs has many of the same problems as estimating fuel use. A method must be devised that allocates repair expense relative to the type of machinery used and the amount of use. Any of the methods discussed for estimating fuel expense can also be used to estimate machinery repair expense. Chapter 22 contains more detailed methods for estimating repair costs for all types of machinery.

Labor

Some enterprise budgets go into enough detail to divide labor requirements into that provided by the farm operator and that provided by hired labor. However, most use one estimate for labor with no indication of the source. Total labor needed for crop production is heavily influenced by the size of the machinery used and the number of machine operations. Besides the labor needed to operate machinery in the field, care should be taken to include time needed to get to and from fields, adjust and repair machinery, and perform other tasks related directly to the crop in the enterprise budget.

The opportunity cost of farm operator labor is generally used to value labor. If some hired labor is used, this value should certainly be at least equal to the cost of hired labor, including any fringe benefits. When estimating the opportunity cost of labor, it is important to include only the cost of labor, not management. A management charge should be shown as a separate item or included in the estimated net return.

Interest

This interest expense is for the capital tied up in operating expenses. For annual crops, it is generally less than a year from the time of the expenditure until harvest when income is or can be received. Therefore, interest is charged for a period of less than a year. This time should be the average length of time between when the operating costs are incurred and harvest. Interest is charged on operating expenses without regard to how much is borrowed or even if any is borrowed. Even if no capital is borrowed, there is an opportunity cost on the farm operator's capital. If the amount of borrowed capital and equity capital that will be used is known, a weighted average of the interest rate on borrowed money and the opportunity cost of equity capital can be used.

Income above Variable Costs

This value, which may also be called *gross margin,* shows how much an acre of this enterprise will contribute toward payment of fixed or ownership expenses. It also shows how much revenue could decrease before this enterprise could no longer cover its variable or operating expenses.

Ownership or Fixed Expenses

This section includes those costs that would exist even if the specific crop were not grown. These are the costs incurred due to ownership of machinery, equipment, and land used in crop production.

Machinery Depreciation

The amount of machinery depreciation to charge to a crop enterprise will depend on the size and type of machinery used and the number and type of machine operations. As with machinery operating expenses, the problem is the need to properly allocate the total machinery depreciation to a specific enterprise. The first step is to compute the average annual depreciation on each machine. This can be quickly and easily done with the straight-line depreciation method. Annual depreciation on each machine can then be converted to a per-acre or per-hour value based on acres or hours used per year. Next, it can be prorated to a specific crop enterprise based on use.

Machinery Interest

Interest on machinery is based on the average investment in the machine over its life and is computed the same way no matter how much, if any, money was borrowed to purchase it. The equation used in Chapter 9 to compute the interest component of total fixed costs should be used to find the average annual interest charge. Next, this interest charge should be prorated to each enterprise using the same method as was used for depreciation.

Machinery Taxes and Insurance

Machinery is subject to personal property taxes in some cases, and most farmers carry some type of insurance on their machinery. The annual expense for these items should be computed and then allocated again using the same method as for the other machinery ownership costs.

Land Charge

There are several ways to calculate a land charge: (1) What it would cost to cash rent similar land, (2) the net cost of a share-rent lease for this crop on similar land and, (3) for owned land, the opportunity cost of the capital invested; that is, the value of an acre multiplied by the opportunity cost of the owner's capital. The three methods can give greatly different values.

This is particularly the case for the third method if it is used during periods of rapidly increasing land values. During inflationary periods, land values reflect both the appreciation potential of land and its value for crop production.

Most enterprise budgets use one of the rental charges even if the land is owned. Assuming a short-run enterprise budget, the land owner/operator could not sell the acre and invest the resulting capital. As long as the land is owned, if it is not farmed by the owner, the alternative is to rent it to another farm operator. The rental amount then becomes the short-run opportunity cost for the land charge.

Miscellaneous Overhead

Many enterprise budgets contain an entry for miscellaneous overhead expense. This entry can be used to cover many expenses such as a share of pickup truck expenses, farm liability insurance, farm shop expenses, and so on. These are expenses that cannot be directly associated with a single enterprise but are necessary and important farm expenses. These expenses are often allocated based on the enterprise's share of either total revenue or total variable costs.

Profit and Return to Management

The estimated profit is found by subtracting total expenses from total revenue. If a charge for management has not been included in the budget, this value should be considered the return to management *and* profit. Management is an economic cost and should be recognized in an economic budget either as a specific expense or as part of the residual net return or loss.

Other Considerations for Crop Enterprise Budgets

The wheat example used to describe the process of constructing a crop enterprise budget is a fairly simple example. Other crops may have many more variable inputs or revenue and expenses specific to that crop. Special problems that may be encountered when constructing crop enterprise budgets include:

1. Double cropping, where two crops are grown on the same land in the same year. In this case, budgets should be developed for each crop, and the annual ownership costs for land divided between the two crops.

2. Storage, transportation, and marketing expenses may be important for some crops. Most enterprise budgets assume sale at harvest and do not include storage costs. They are assumed to be part of a marketing decision and not a production decision. However, there may still be transportation and marketing expenses even with a harvest sale. If storage costs are included in the budget, then the selling price should be that expected at the end of the storage period, not the harvest price.

3. Establishment costs for perennial crops, orchards, and vineyards present another problem. These and other crops may take a year or more to begin production, but their enterprise budgets are typically for a year of full production. It is often useful to develop separate budgets for the establishment phase and the production phase. The latter budget must include an expense for a prorated share of the establishment costs and any other costs incurred before receiving any income. This is done by accumulating costs for all years before the onset of production and then determining the present value of these costs. This value can be used to determine the annual equivalent, which is included as an annual expense on the enterprise budget. (See Chapter 17 for a discussion of present value and annual equivalent.)

4. The methods for computing machinery depreciation and interest discussed earlier are easy to apply and widely used. However, they do not result in the exact amounts needed to cover depreciation and provide interest on the machine's remaining value each year. The capital recovery method, while more complex, does provide the correct amount. Capital

---

| **Box 10-2** | **Computers and Enterprise Budgets** |

Computers and spreadsheets are well adapted and widely used to develop enterprise budgets. Many agricultural universities and extension services use computers to develop annual enterprise budgets for the major enterprises in their state. Their computer programs often calculate machinery and building costs based on typical costs, sizes, use, and other agricultural engineering factors. Many individual farmers and ranchers have developed their own enterprise budgeting spreadsheet templates.

Enterprise budgets are the starting point in some large farm financial planning software programs. Once these budgets are prepared, a farm plan is produced by selecting the number of acres or head for each enterprise. The program then calculates income and expenses from the budgets and transfers the results to a cash flow budget, projected income statement, and other projected financial statements. It then becomes easy to make small changes to the farm plan and observe the results on cash flows and profit before selecting a final farm plan.

---

recovery combines depreciation and interest into one value.

The general equation for the annual capital recovery amount is

$$(TD \times \text{amortization factor})$$
$$+ (\text{salvage value} \times \text{interest rate})$$

where TD is the total depreciation over the life of the machine or the capital that must be recovered. An amortization factor corresponding to the interest rate used and the life of the machine can be found in Table 1 in the Appendix or from a financial calculator. The capital recovery amount obtained from this equation is money required at the end of each year to pay interest on the remaining value of the machine *and* recover the capital lost through depreciation. Because the salvage value will be recovered at the end of the machine's useful life, only interest is charged on this amount. (See Chapter 17 and Chapter 22 for discussions of capital recovery.)

CONSTRUCTING A LIVESTOCK ENTERPRISE BUDGET

Livestock enterprise budgets include many of the same entries and problems as crop enterprise budgets. However, livestock budgets can also have some unique and particular problems, which will be discussed in this section. The cow/calf budget in Table 10-3 will be used as the basis for the discussion.

Unit

The budgeting unit for livestock is usually one head, but different units such as one litter for swine, one cow unit, or 100 birds for poultry can also be used. In some instances, livestock budgets may be developed for several different typical sizes of the enterprise (e.g. 30 head, 50 head, and so on), to reflect economies of size.

Time Period

Although many livestock enterprises are budgeted for one year, some feeding and finishing enterprises require less than a year. Some types of breeding livestock, such as swine, produce offspring more than once a year. Whatever the time period chosen, it is important that all costs and revenue in the budget be calculated for the same time period.

Multiple Products

Many livestock enterprises will have more than one product producing revenue. For example, a dairy would have revenue from cull cows,

TABLE 10-3	Example of a Cow/Calf Budget for One Cow Unit*			

Item	Unit	Quantity	Price	Amount
Revenue				
Cull cow (0.10 head)	cwt	10.00	$50.00	$ 50.00
Heifer calves (0.33 head)	cwt	5.20	80.00	137.28
Steer calves (0.45 head)	cwt	5.50	88.00	217.80
Total revenue				$405.08
Operating expenses				
Hay	ton	1.50	60.00	90.00
Grain & supplement	cwt	8.00	7.00	56.00
Salt, minerals	cwt	0.40	6.00	2.40
Pasture maintenance	acre	3.00	7.50	22.50
Veterinary & health expense	head	1.00	10.00	10.00
Livestock facilities	head	1.00	8.00	8.00
Machinery & equipment	head	1.00	5.00	5.00
Breeding expenses	head	1.00	5.00	5.00
Labor	hours	5.00	6.00	30.00
Miscellaneous	head	1.00	10.00	10.00
Interest (on half of operating expenses)	$	119.45	10%	11.95
Total operating expense				$250.85
Ownership expenses				
Interest on breeding herd	$	750.00	10%	75.00
Livestock facilities				
Depreciation & interest	head	1.00	10.00	10.00
Machinery & equipment				
Depreciation & interest	head	1.00	6.50	6.50
Land charge	acre	3.00	35.00	105.00
Total ownership expenses				$196.50
Total expenses				$447.35
Profit (return to management)				($ 42.27)

*(1 cow unit = 1 cow, 0.04 bull, 0.9 calf, 0.12 replacement heifer)

calves, and milk, while a sheep flock would have revenue from cull breeding stock, lambs, and wool. All sources of revenue must be identified and then prorated correctly to an average individual animal in the enterprise.

The cow/calf budget in Table 10-3 shows average revenue from cull cows, heifer calves, and steer calves *per cow unit* in the producing herd. A cow unit includes the cow and a por-

tion of the calves, bulls, and replacement heifers. There is an implied 10 percent replacement rate each year based on the 0.10 cull cows sold per producing cow. Less than one calf is sold per cow unit due to several assumptions: (1) calving percentage is less than 100 percent, (2) some heifer calves are retained for replacing cull cows, and (3) some death loss is incurred.

Breeding Herd Replacement

An important consideration in enterprise budgets for breeding herds is properly accounting for replacements for the producing animals. The example in Table 10-3 assumes that female replacements are raised rather than purchased. With a 90 percent calving rate, there are 0.45 steer calves available for sale per cow, and there should also be 0.45 heifer calves. However, at least 0.10 heifer calves per cow must be retained as replacements. This example shows that 0.12 (0.45 available less 0.33 actually sold) are actually retained. The additional heifer calves retained would cover a 2 percent death loss among the replacements and the producing herd. Whenever replacement animals are raised, the number of female offspring sold must be adjusted to reflect the percentage retained for replacing animals culled from the breeding herd and death loss.

If replacements are purchased, the enterprise budget would show revenue from selling all female offspring as calves. Including the net cost of the replacements can be done in one of two ways. First, an annual depreciation charge, computed using the straight-line method, can be included as an expense. This will account for the decline in value from purchase price to market value at time of culling for the typical cow in the herd. No revenue from cull animals nor expense for purchasing replacements would be shown. As an alternative, the purchase cost of a replacement divided by the useful life can be shown as an expense, and the revenue from culled animals divided by useful life can be shown as revenue. The difference between these two values should be the same value as the annual depreciation in the first method.

Feed and Pasture

Many livestock enterprises will consume both purchased feed and farm-raised feed. Purchased feed is easily valued at cost. Farm-raised feed should be valued at its opportunity cost or what it would sell for if marketed off the farm. Care should be taken to include expenses for salt and minerals, any annual charges for establishment costs, fertilizer, spraying or mowing of pastures, and for the feed needed to maintain the replacement herd if replacements are being raised. Pasture costs would also include a charge for the land used, based either on the cost of renting the pasture or, if owned, its opportunity cost.

Livestock Facilities

Livestock facilities include buildings, fences, pens, working chutes, feeders, waterers, wells, windmills, feed storage, milking equipment, and other specialized items used for livestock production. Operating expenses for these items include repairs and any fuel or electricity required to operate them.

These items also incur fixed costs and present some of the same computational and allocation problems outlined when discussing crop machinery. Annual depreciation, interest, taxes, and insurance should be computed for each livestock facility. For specialized items used only by one livestock enterprise, the total annual fixed costs can simply be divided by the number of head that use it each year to get the per-head charge. When more than one livestock enterprise uses the item, some method must be used to allocate the fixed costs among the enterprises.

Machinery and Equipment

Tractors, pickups, and other machinery and equipment may be used for both crop and livestock production. Both operating and ownership expenses should be divided between crop and livestock enterprises according to the proportional use of the item.

GENERAL COMMENTS ON ENTERPRISE BUDGETS

Several factors need to be considered when constructing and using enterprise budgets. The economic principles of marginal value product (MVP) equals marginal input cost (MIC) and least-cost input combinations should be considered when selecting input levels for a budget.

> ### Box 10-3   Third-Party Enterprise Budgets
>
> It is often convenient to use enterprise budgets prepared by someone else. However, no budget prepared by a third party is likely to fit any individual farm situation exactly. Most prepared budgets use typical or average values for a given geographic area. However, there are differences in yields, input levels, and other management practices from farm to farm. It is also important to know the assumptions and equations used in any budget, because they may not fit all situations. Different state extension services and other orga-
>
> nizations often use slightly different formats for organizing and presenting budget data. However, the major headings and items discussed in this chapter should be included no matter what format is chosen.
>
> Prepared budgets should be considered only as a guide and viewed as something that will need adjusting to fit any individual farm or ranch. Many prepared budgets include a column labeled "Your Values" or something similar, so users can make individual adjustments to fit their farm.

However, it should not be assumed that all third-party published budgets were constructed using these principles. Even if they were, these levels may not be correct for a specific, individual situation. The typical or average input levels used in many published budgets may not be the profit-maximizing levels for any individual farm.

Because it is possible to use many different input levels and input combinations, even on a single farm, there is not a single budget for each enterprise. There are as many potential budgets as there are possible input levels and combinations. This again emphasizes the importance of selecting the profit-maximizing input levels for the individual situation, or if capital is limited, selecting the input levels that will satisfy the equal marginal principle.

The fixed-cost estimates for enterprise budgets are usually based on an assumed farm size or level of use. Different estimates may be needed if fixed costs are spread over significantly more or fewer output units than the level assumed in the budget.

Enterprise budgets require a large amount of data. Past farm or ranch records are an excellent source if they are available in sufficient detail and for the enterprise being budgeted. Many states publish an annual summary of the average income and expenses for farms participating in a statewide record-keeping service.

These summaries may contain sufficient detail to be a useful source of data for enterprise budgeting. Research studies conducted by agricultural universities, the U.S. Department of Agriculture, and agribusiness firms are reported in bulletins, pamphlets, special reports, and farm magazines. This information often includes average yields and input requirements for individual enterprises.

The appropriate price and yield data may depend on the purpose of the budget. A budget to be used only for making adjustments in next year's crop and livestock plan should contain estimates of next year's price and expected yield. Estimates of long-run prices and yields should be used in budgets constructed to assist in developing a long-run plan for the business. The appropriate yield for a particular enterprise budget will also depend on the types of inputs included in the budget, and on the input levels. Higher input levels should be reflected in higher yields, and vice versa.

INTERPRETING AND ANALYZING ENTERPRISE BUDGETS

Any economic enterprise budget must be interpreted correctly. An economic budget includes opportunity costs on labor, capital, land, and perhaps management as expenses. The resulting profit (or loss) is the revenue remaining after covering all expenses, including opportunity costs. This can be thought of as an *economic* profit, which will not be the same as *accounting* profit. The latter would not include any opportunity costs as operating expenses. Rather, accounting profit is the revenue remaining to pay management, unpaid labor, and equity capital for their use. A projected economic profit of zero is not as bad as it might seem. This result simply means all labor, capital, and land are just earning their opportunity costs—no more, no less. A positive projected profit means one or more of these factors are earning more than their opportunity cost.

The data in an enterprise budget can be used to perform several types of analyses. These include calculating cost of production and computing break-even prices and yields.

Cost of Production

Cost of production is a term used to describe the average cost of producing one unit of the commodity. It is the same as average total cost discussed in Chapter 9, provided the same costs and production level are used to compute each. The cost of production equation for crops is

$$\text{Cost of production} = \frac{\text{total cost}}{\text{yield}}$$

which is the same as for average total cost, with "output" and "yield" being interchangeable terms. For the example in Table 10-2, the cost of production for wheat is $158.90 divided by 48 bushels, or $3.31 per bushel. Notice that cost of production will change if either costs or yield change.

Cost of production is a useful concept, particularly when marketing the product. Any time the product can be sold for more than its cost of production, a profit is being made. If opportunity costs are included in the expenses, the profit is an economic profit. The resulting accounting profit will be even higher.

Break-Even Analysis

The data contained in an enterprise budget can be used to do a break-even analysis for prices and yields. The formula for computing the *break-even yield* is

$$\text{Break-even yield} = \frac{\text{total cost}}{\text{output price}}$$

This is the yield necessary to cover all costs at a given output price. For the example in Table 10-2, it would be $158.90 divided by $3.00, or 53.0 bushels per acre. The output price is only an estimate, so it is often useful to compute the break-even yield for a range of possible prices, as shown below:

Price per bushel ($)	Break-even yield (bu)
2.50	63.6
2.75	57.8
3.00	53.0
3.25	48.9
3.50	45.4

This approach often provides some insight into how sensitive the break-even yield is to changes in the output price. As shown in this table, the break-even yield can be very sensitive to changes in the output price.

The *break-even price* is the output price needed to just cover all costs at a given output level, and it can be found from the equation

$$\text{Break-even price} = \frac{\text{total cost}}{\text{expected yield}}$$

Again using the example in Table 10-2, the break-even price would be $158.90 divided by 48 bushels, or $3.31. Notice that the break-even price is the same as the cost of production. They are only two different ways of looking at the same value.

The break-even price can also be computed for a range of possible yields, as in the following table. Different yields cause different break-even prices (and cost of production), and these prices can vary widely depending on the yield level.

Yield (bu)	Break-even price ($)
36.0	$4.41
42.0	3.78
48.0	3.31
54.0	2.94
60.0	2.65

Because both the yield and output price in an enterprise budget are estimated rather than actual values, the calculation of break-even yields and prices can aid managerial decision making. By studying the various combinations of break-even yields and prices, managers can form their own expectations about the probability of obtaining a price and yield combination that would just cover total costs. Break-even prices and yields can also be calculated from total variable costs rather than total costs. These results can help managers make the decisions discussed in Chapter 9 concerning continuing or stopping production to minimize losses in the short run.

SUMMARY

Enterprise budgets are an organization of projected income and expenses for a single enterprise. They are constructed for a single unit of the enterprise, such as one acre for a crop and one head for a livestock enterprise. Most enterprise budgets are economic budgets and, as such, include all variable or operating expenses, all fixed or ownership expenses, and opportunity costs on factors such as operator labor, capital, and management.

Enterprise budgets can be used to compare the profitability of alternative enterprises and are particularly useful when developing a whole-farm plan. They can also be used to make minor year-to-year adjustments in the farm plan in response to short-run price and yield changes. Once completed, an enterprise budget contains the data needed to compute cost of production, break-even yield, and break-even price.

QUESTIONS FOR REVIEW AND FURTHER THOUGHT

1. Suggest how the "return above variable costs" value found on most enterprise budgets can be used to make some short-run production decisions.

2. Should the economic principles for determining profit-maximizing input levels be applied before or after completing an enterprise budget? Why?

3. An enterprise budget for soybeans shows a yield of 36 bushels, a selling price of $5.85 per bushel, and total cost of $220.00 per acre. What is the cost of production? The break-even yield? The break-even price?

4. Why should the opportunity costs of a farmer's labor, capital, and management be included on an enterprise budget?

5. If the land is owned, should a land charge be included on an enterprise budget? Why?

6. Would you expect two farms of widely different size to have the same fixed costs on their enterprise budgets for the same enterprise? Might economies or diseconomies of size explain any differences?

7. "There are potentially many different enterprise budgets for a single enterprise." Defend or refute that statement.

8. How might an agricultural loan officer use enterprise budgets? A farm real estate appraiser? A farmer when ordering input supplies for the coming year?

9. How would an enterprise budget for a perennial or long-term crop differ from one for an annual crop?

11

PARTIAL BUDGETING

CHAPTER OBJECTIVES

1. To discuss the purpose of a partial budget

2. To emphasize the many possible uses of a partial budget

3. To illustrate the format of a partial budget

4. To show what types of entries are made on a partial budget

5. To note the importance of including only changes in revenue and expenses on a partial budget

6. To demonstrate the use of partial budgeting with several examples

Enterprise budgets are very useful, but they do have limitations. Analyzing a possible change involving several enterprises and any interactions among them requires a different budgeting technique. A *partial budget* is often the appropriate way to analyze these types of problems using information obtained from enterprise budgets.

Many of the day-to-day management decisions made by farmers and ranchers are really adjustments to, or fine-tuning of, an existing farm plan. Even the best farm plan will need some occasional fine-tuning as changes occur and new information becomes available. These adjustment decisions often affect revenue and expenses. A convenient and practical method for analyzing the profit potential of these partial changes in the overall whole-farm plan is the use of a partial budget.

USES OF A PARTIAL BUDGET

Examples of decisions that can be analyzed with a partial budget are whether to increase the size of, or to eliminate, a small herd of beef cows, to own harvesting equipment or custom hire harvesting, and to plant more barley and less wheat. Most of these decisions could be evaluated by comparing two whole-farm budgets, but time and effort would be wasted collecting and organizing information that will not change and therefore does not affect the decision.

A partial budget provides a formal and consistent method for calculating the expected change in profit from a proposed change in the farm business. It compares the profitability of one alternative, typically what is being done now, with a proposed change or new alternative. Throughout the discussion of partial budgeting, the emphasis will be on *change*. The analysis is for a relatively small change in the farm plan, so only changes in revenue and expenses are included on a partial budget. The final result is the expected change in profit.

Because it is designed to analyze relatively small changes in the farm business, partial budgeting is really a form of marginal analysis. Figure 11-1 illustrates this point by showing how typical changes analyzed by partial budgeting relate to a production function, an isoquant, and a production possibility curve. The production function in the first panel of Figure 11-1 shows,

assuming that the current input/output combination is point A, possible increases or decreases in that combination. Examples would be using more or less fertilizer, irrigation water, labor, or capital and analyzing the effects on output, revenues, expenses, and profit.

The second panel of Figure 11-1 shows possible movements up or down an isoquant, or different combinations of two inputs to produce a given amount of output. Possible changes in input combinations can be analyzed easily with a partial budget. Substituting larger machinery for less labor would be an example. Another typical use of a partial budget is to analyze the change in profit from substituting more of one enterprise for another. This adjustment is shown in the third panel of Figure 11-1 by possible movements up or down the production possibility curve from the current combination at point A. A fourth general type of alternative adapted to partial budget analysis is expanding or contracting one or more enterprises. This would be illustrated by moving to a higher or lower isoquant or a higher or lower production possibility curve.

PARTIAL BUDGETING PROCEDURE

Steps in the decision-making process discussed in Chapter 2 included: identify and define the problem, identify alternatives, collect data and

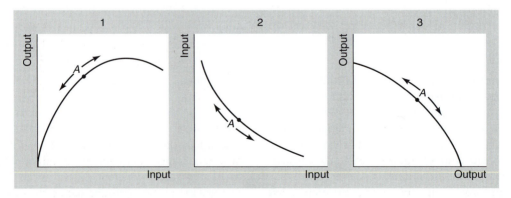

Figure 11-1   Partial budgeting and marginal analysis.

information, and analyze alternatives. Partial budgeting fits this process, with one modification. It is capable of analyzing only two alternatives at a time: the current situation and a single proposed alternative. Of course, several partial budgets can be used to evaluate a number of alternatives. Identifying the alternative to be analyzed before gathering any information reduces the amount of information needed. The only information required is changes in costs and revenues *if* the proposed alternative is implemented. There is no need for information on any other alternative or for information about costs and returns that will not be affected by the proposed change.

The changes in costs and revenues needed for a partial budget can be identified by considering the following four questions. They should be answered on the basis of what would happen *if* the proposed alternative was implemented.

1. What new or additional costs will be incurred?
2. What current costs will be reduced or eliminated?
3. What new or additional revenue will be received?
4. What current revenue will be lost or reduced?

For many problems, it will be easier to first identify all physical changes that would result if the alternative was adopted. These can then be assigned a dollar value to use in the partial budget.

THE PARTIAL BUDGET FORMAT

The answers to the preceding questions are organized within one of the four categories shown on the partial budget form in Table 11-1. There are different partial budgeting forms, but all have these four categories arranged in some manner. It is important to remember that, for each category, only the *changes* are included, not all costs or revenues.

Additional Costs

These are costs that do not exist at the current time with the current plan. A proposed change may cause additional costs because of a new or expanded enterprise that requires the purchase of additional inputs. Other causes would be increasing the current level of input use or substituting more of one input for another for an existing enterprise. Additional costs may be either variable or fixed, because there will be additional fixed costs whenever the proposed alternative requires additional capital investment. These additional fixed costs would include depreciation, interest (opportunity cost), taxes, and insurance for a new depreciable asset.

Reduced Revenue

This is revenue currently being received but which will be lost or reduced should the alternative be adopted. Revenue may be reduced if an enterprise is eliminated or reduced in size, if the change causes a reduction in yields or production levels, or if the selling price will decrease. Estimating reduced revenue requires careful attention to information about yields, livestock birth and growth rates, and output selling prices.

Additional Revenue

This is revenue to be received only if the alternative is adopted. It is not being received under the current plan. Additional revenue can be received if a new enterprise is added, if there is an increase in the size of a current enterprise, or if the change will cause yields, production levels, or selling price to increase. As with reduced revenue, accurate estimates of yields and prices are important.

Reduced Costs

Reduced costs are those now being incurred that would no longer exist under the alternative being considered. Cost reduction can result from eliminating an enterprise, reducing the size of an enterprise, reducing input use, substituting more

TABLE 11-1	**Partial Budget Form**

PARTIAL BUDGET

Alternative:

Additional Costs:	**Additional Revenue:**
Reduced Revenue:	**Reduced Costs:**
A. Total additional costs and reduced revenue $_____	B. Total additional revenue and reduced costs $_____
	$_____
	Net Change in Profit (B minus A) $_____

Box 11-1	**Can Fixed Costs Really "Change"?**

It may seem strange to talk about computing changes in fixed costs when Chapter 9 emphasized that fixed costs do not change. The explanation is in the difference between short run and long run. Fixed costs do not change in the short run. However, analyzing the purchase or sale of a capital asset is a long-run decision with respect to that asset, and fixed costs can change from one point in time to another. Computing the fixed costs that would exist at one point in time and those that would exist at another after a purchase or sale would indicate a difference. It is that difference, or change, that should be included on a partial budget.

TABLE 11-2	Partial Budget For Owning Combine Versus Custom Hiring

PARTIAL BUDGET

Alternative: Purchase combine to replace custom hiring (1,000 acres of wheat)

Additional Costs:		Additional Revenue:	
Fixed costs		None	
Depreciation	$10,000		
Interest	8,000		
Taxes	100		
Insurance	300		
Variable costs			
Repairs	2,500		
Fuel and Lubricants	1,300		
Labor	550		
Reduced Revenue:		**Reduced Costs:**	
None		Custom combining charge	
		1,000 acres @ $20.00 per acre	$20,000
A. Total additional costs and reduced revenue	$22,750	B. Total additional revenue and reduced costs	$20,000
			$22,750
		Net Change in Profit (B minus A)	$(2,750)

of one input for another, or being able to purchase inputs at a lower price. Reduced costs may be either fixed or variable. A reduction in fixed costs will occur if the proposed alternative will reduce or eliminate the current investment in machinery, equipment, breeding livestock, land, or buildings.

The categories on the left-hand side of the partial budget are the two that reduce profit—additional costs and reduced revenue. On the right-hand side of the budget are the two categories that increase profit—additional revenue and reduced costs. Entries on the two sides of the form are summed and then compared to find the net change in profit. Whenever opportunity costs are included on a partial budget, the result is the estimated change in "economic profit." This will not be the same as the change in "accounting profit."

PARTIAL BUDGETING EXAMPLES

Two examples will illustrate the procedure and possible uses of partial budgeting. Table 11-2 is a relatively simple budget analyzing the profitability of purchasing a combine to replace the current practice of hiring a custom combine operator to harvest 1,000 acres of wheat. All values in the budget are average annual values, and the result is net change in annual economic profit. Costs or revenues for wheat production are not included in the budget, because these will be the same under either alternative.

Purchasing the combine will increase fixed costs, as well as incurring the additional variable costs associated with operating the combine. Depreciation and interest are computed from the equations used to compute fixed costs in Chapter 9. The values used were a purchase cost of $120,000, a salvage value of $40,000, an 8-year useful life, and a 10 percent opportunity cost for capital. The additional annual fixed costs total $18,400. Additional variable costs such as repairs, fuel and lubricants, and labor to operate the combine were estimated to be $4,350. No opportunity cost on capital tied up in these additional variable costs was included, because little time elapses between their occurrence and when income from the wheat can or will be received. Assuming the same yield, there is no reduced revenue, so the profit-reducing effects total $22,750.

No additional revenue is shown in the example, although some might be expected if the timeliness of harvesting would be improved or field losses reduced. These factors are difficult to estimate but should be included if a significant change is expected. The only reduced cost is the elimination of the payment for custom combining, which is $20,000. This example shows that purchasing the combine instead of continuing to hire the custom operator would *reduce* annual economic profit by $2,750.

A second and more detailed example of a partial budget is shown in Table 11-3. The proposed change is the addition of 50 beef cows to an existing herd. However, not enough forage is available, and 100 acres currently in grain production must be converted to forage production.

This proposed change will cause additional fixed costs. There will be additional interest

on the increased investment in beef cows, depreciation on additional bulls, and additional property taxes on the new animals. Herd replacements are assumed to be raised, so there is no depreciation included on the cows. Variable or operating costs will also increase as shown, including labor, feed, and health costs for the additional livestock, as well as fertilizer for the new 100 acres of pasture and opportunity cost interest on all additional variable costs. Income from the grain now being produced on the 100 acres will no longer be received, and this reduced revenue is estimated to be $15,000. The sum of additional costs and reduced revenue equals $23,220.

Additional revenue will be received from the sale of cull cows, steer calves, and heifer calves from the additional 50 cows. Several items are important in estimating this revenue, in addition to carefully estimating prices and weights. First, it is unrealistic to assume that every cow will wean a calf every year. This example assumes 46 calves from the 50 cows, or a 92 percent weaned calf crop. Second, herd replacements are raised rather than purchased, so 5 heifer calves must be retained each year to replace the 5 cull cows that are sold. This assumption results in the sale of only 18 heifer calves each year, compared with 23 steer calves. The reduced costs include all expenses from planting the 100 acres of grain that would no longer exist. Opportunity cost interest on these expenses is also included as a reduced cost. No reduction in machinery fixed costs is included, because the machinery complement is assumed to be no different after the proposed change than before. However, should the reduction in crop acres be large enough that some machinery could be sold or a smaller size used, reduced costs should include the change in related fixed costs.

The total additional revenue and reduced costs are $27,053, or $3,833 higher than the total additional costs and reduced revenue. This positive difference indicates that the proposed change would increase profit.

TABLE 11-3	**Partial Budget for Adding 50 Beef Cows**

PARTIAL BUDGET

Alternative: Add 50 beef breeding cows and convert 100 acres from grain to forage production

Additional Costs:		Additional Revenue:	
Fixed costs			
Interest on cows/bulls	$ 2,500	5 cull cows	$ 2,500
Bull depreciation	200	23 steer calves	
Taxes	100	500 lb. @ 85¢	9,775
		18 heifer calves	
		460 lb. @ 78¢	6,458
Variable costs			
Labor	600		
Vet and health	500		
Feed and hay	2,000		
Hauling	300		
Miscellaneous	200		
Pasture fertilizer	1,500		
Interest on variable costs	320		
Reduced Revenue:		**Reduced Costs:**	
Grain production			
5,000 bu. @ $3.00	$15,000	Fertilizer	$ 2,750
		Seed	1,400
		Pesticides	1,200
		Labor	1,500
		Machinery	1,000
		Interest on variable costs	470
A. Total additional costs		B. Total additional revenue	
and reduced revenue	$23,220	and reduced costs	$27,053
			$23,220
		Net Change in Profit (B minus A)	$ 3,833

FACTORS TO CONSIDER WHEN COMPUTING CHANGES IN REVENUE AND COSTS

In addition to the usual problem of acquiring good information and data, there are several other potential problems when doing a partial budget. The first is nonproportional changes in costs and revenue. This problem will occur more often with costs, but it is possible for revenues. Assume the proposed change is a 20 percent increase (decrease) in the size of an enterprise. It would be easy to take the totals for each existing expense and revenue and assume that each will be 20 percent higher (lower). This could be wrong for two reasons. Fixed costs would not change unless the 20 percent change caused an increase or decrease in capital investment. Many relatively small changes will not. Even variable costs may not change proportionally. For example, adding 20 cows to an existing beef or dairy herd of 100 cows will increase labor requirements but probably by something less than 20 percent. Economies and diseconomies of size must be considered when estimating cost and revenue changes.

Opportunity costs are other easily overlooked items. They should be included on a partial budget to permit a fair comparison of the alternatives. This is particularly true if the difference in capital or labor requirements is large. Additional variable costs represent capital that could be invested elsewhere, so opportunity cost on them should be included as another additional cost. The reverse of this argument holds true for reduced variable costs, so an opportunity cost on these should be included as a reduced cost. Opportunity cost on any additional capital investment becomes part of additional fixed costs and likewise should be a part of reduced fixed costs if the capital investment will be reduced.

Opportunity cost on the farm operator's labor may also be needed on a partial budget. However, several things should be considered when estimating this opportunity cost. Is there really an opportunity cost for using additional labor if it is currently unused? Free or leisure time would be given up, and there may be an opportunity cost on this time. Alternatively, the farm operator may desire some minimum return before using any excess labor in a new alternative. The same question exists in reverse if the alternative will reduce labor requirements. Is there a productive use for an additional 50 or 100 hours of labor, or will it just be additional leisure time? What will it earn in the alternative use, or what is the value of an additional hour of leisure time? Answers to these questions help determine the appropriate opportunity cost of operator labor on a partial budget.

Another consideration is the unit of change used in the partial budget. Is the budget based on changes in total farm revenue and expenses or is it for one acre or one head of livestock? In other words, is the unit the whole farm or some smaller unit? Some alternatives can be analyzed either way if they have a common physical unit such as acres. Others where the alternative involves both changes in acres of crops and head of livestock have no common unit of measurement. They must be budgeted on a whole farm basis. Budgeting on a whole farm basis is always the safer method to prevent any confusion about the budgeting unit.

SENSITIVITY ANALYSIS

It is often difficult to estimate the average prices and yields needed in a partial budget. Estimation is particularly difficult if the budget projects well into the future. Yet, the accuracy of the analysis and of the resulting decision directly depends on these values. A sensitivity analysis of the budget can provide some additional information on just how dependent the results are on the prices and yields used.

Sensitivity analysis consists of doing the budget computations several times, each with a different set of prices or yields. The results show how sensitive the estimated change in profit is to changes in these values. One way to perform a basic sensitivity analysis is to use low, average, and high prices each in a different partial budget. The same could be done for low, average, and high yields if appropriate. A comparison of the results will show how sensitive the expected change in profit is to price or yield variation. This will provide the manager with some idea of the risk involved in the proposed change.

Another way to do sensitivity analysis would be to look at prices which, for example, are 10, 20 and 30 percent higher and lower than the expected average price. Using this method, one of the prices may result in an expected change in profit of somewhere near zero meaning it is close to the break-even price. For some types of partial budgeting problems it is possible to compute the break-even price or yield directly, which simplifies the calculations. Once a break-even value is calculated the manager can decide if the future price or yield is more likely to be above or below that value. This information can help make the final decision.

The example in Table 11-2 can be used to illustrate sensitivity analysis. The key price in this problem is the custom combining charge, as the price of the combine would generally be known and most of the additional costs flow directly from that value. We would expect higher custom charges to favor purchasing the combine and vice versa. For example, a custom combining charge of $18.00 per acre would lower reduced costs by $2,000 making the net change in profit a *negative* $4,750. Increasing the charge to $22.00 per acre would increase the reduced costs by $2,000 and the estimated net change in profit would be a *negative* $750.

The break-even custom combining charge is the value which makes the net change in profit just equal to zero. This requires a $2,750 increase in the total custom charge or $2.75 per acre. Therefore, a custom rate of $20.00 + $2.75 = $22.75 per acre is the break-even value. Any custom rate above this value would make it profitable to purchase the combine.

Another question might be: What if the purchased combine was used to do custom work for others? Then the partial budget would need to show additional revenue from this source as well as the additional variable costs that would increase directly with the number of acres custom harvested. The fixed costs would not change unless the additional use decreased the useful life or salvage value of the combine. Because repair costs might increase more than proportionally, a variable cost of $5.50 per acre was assumed. With a custom charge of $20.00 per acre, this would provide an additional net return of $14.50 for each acre of custom work done. Dividing this value into the $2,750 reduction in profit indicates that custom combining 190 acres would be the break-even amount. If more than 190 acres can be custom harvested, purchasing the combine becomes the more profitable alternative.

Sensitivity analysis and break-even calculations can also be done on the budget in Table 11-3 but it is more difficult because of the greater number of prices and yields. However, it can be done by holding all price and yield values

constant but the one of most interest. For example, if all livestock prices were held constant, break-even grain prices or yields could be computed. The budget indicates that the value of the grain production (either from price or yield) must increase by $3,833 before the net change in profit would become zero. Keeping the production constant at 5,000 bu. and dividing $3,833 by 5,000 bu. indicates the price could increase by $0.77 to $3.77 per bu. before the additional beef cows would not increase profit. Similarly, with price constant at $3.00, the total production must increase by $3,833 ÷ $3.00 = 1,278 bu. (to 6,278 bu.) before the decision would change. If these two break-even values are considerably higher than the manager believes to be likely, it provides a level of confidence that choosing to add 50 beef cows to the existing herd will increase profits.

Performing a sensitivity analysis and computing break-even values can require numerous calculations. However, a partial budget is relatively easy to set up on a computer spreadsheet. Once this is done, it is quick and easy to change a value and observe the result.

LIMITATIONS OF PARTIAL BUDGETING

Partial budgets are easy to use, require minimal data, and are readily adaptable to many types of management decisions. However, partial budgeting does have some limitations. It can only compare the present management plan with one alternative at a time. This requires many budgets when there are many alternatives to consider. Partial budgeting can still be used in this situation but it can be time consuming.

The data in partial budgets are expected average annual changes in economic revenue and expenses. While an alternative may increase profit based on average changes there are other factors to consider when the changes are not constant from year to year. An example would be the planting of an orchard or other crop where no revenue is expected for several years and then expected to increase annually for several years before reaching a maximum level. Though it may be a profitable alternative based on *average* annual values, it may be difficult to meet the crop expenses in the early years and to make any loan payments. In other words, the cash shortage in the early years is not reflected in the partial budget. Any type of change requiring a large capital investment and revenues which vary over time should be analyzed by more detailed procedures which should include a cash flow projection. (See Chapters 13 and 17)

A partial budget should include appropriate opportunity costs to account for all economic costs. However, they are not included as accounting costs so the expected change in net profit shown on a partial budget should not be interpreted as expected change in accounting profit. The expected change in net profit must be adjusted for any opportunity costs included in its calculation to find the expected change in accounting profit.

FINAL CONSIDERATIONS

The partial budget in Table 11-3 can be used to illustrate two additional factors to be considered in the decision. Before adopting a proposed change that appears to increase profit, any additional risk and capital requirements should be carefully evaluated. If risk is measured in terms of annual variability in profit, is the profit from the additional 50 cows more variable than the profit from the 100 acres of grain? If the answer is yes, then the decision maker must evaluate the potential effects of additional risk on the financial stability of the business. Is the additional average profit worth the additional risk or variability of profit? Chapter 15 discusses risk in more detail.

Adding 50 cows to the herd will require an additional capital investment. Is the capital for acquiring the cows available or can it be borrowed? If it is borrowed, how will this affect the financial structure of the business, risk, cash flow requirements, and repayment ability? Will this additional investment cause a capital shortage in some other part of the business? These questions need to be evaluated carefully before making the final decision to adopt the change. A profitable change may not be adopted if the increase in profit is relatively small, it increases risk, or an additional capital investment is required.[1]

SUMMARY

A partial budget is an extremely useful type of budget. It can be used to analyze many of the common, everyday problems and opportunities that confront the farm and ranch manager. Partial budgets are intended to analyze the profitability of proposed changes in the operation of the business where the change affects only part of the farm plan or organization. The current situation is compared to the expected situation after implementing a proposed change.

Data requirements are rather small, because only changes in costs and revenues are included on a partial budget. The sum of additional costs and reduced revenue is subtracted from the sum of additional revenue and reduced costs to find the estimated change in profit. A positive result indicates that the proposed change would increase profit. However, any additional risk and capital requirements should be considered before making the final decision.

[1]Potential changes requiring additional capital investment can also be analyzed other ways. See Chapter 17 for capital budgeting and other more comprehensive investment analysis methods.

QUESTIONS FOR REVIEW AND FURTHER THOUGHT

1. Will all partial budgets contain some fixed ownership costs? Why or why not?

2. List the types of changes that would appear in a partial budget for determining the profitability of participating in a government farm program. The program requires that 10 percent of your crop land be left idle in exchange for a lump-sum payment.

3. Why are opportunity costs included in partial budgets?

4. Assume that a proposed change would reduce labor requirements by 200 hours. If this was the farm operator's labor rather than hired hourly labor, would you include a reduced cost for labor? What factors would determine the value to use?

5. True or false and explain. Partial budgeting can be used to develop a whole-farm plan.

6. Besides additional profit, what other factors should a farm operator take into account when evaluating a proposed change?

Whole-Farm Planning

CHAPTER OBJECTIVES

1. To show how whole-farm planning differs from the planning of individual enterprises

2. To learn the steps and procedures to follow in developing a whole-farm plan

3. To understand the various uses for a whole-farm plan and budget

4. To compare the assumptions used for short-run and long-run budgeting

5. To introduce linear programming as a tool for choosing the most profitable combination of enterprises

Once management has developed a strategic plan for a farm or ranch business, the next logical step is to develop a tactical plan to carry it out. Every manager has a plan of some kind—about what to produce, how to produce, and how much to produce—even if it has not been fully developed on paper. However, a systematic procedure for developing a whole-farm plan may well result in one that will increase farm profits or come closer to attaining other goals.

WHAT IS A WHOLE-FARM PLAN?

As the name implies, a *whole-farm plan* is an outline or summary of the type and volume of production to be carried out on the entire farm, and the resources needed to do it. It may contain sufficient detail to include fertilizer, seed, and chemical application rates and actual feed rations for livestock, or it may simply list the enterprises to be carried out and their desired levels of production. When the expected costs and returns for each part of the plan are organized into a detailed projection, the result is a *whole-farm budget.*

Chapters 10 and 11 discussed enterprise and partial budgeting. Enterprise budgets can be used as building blocks for the development of a whole-farm plan and its associated budget. Partial budgets are particularly useful for making minor adjustments or fine-tuning a previously developed whole-farm plan. However, a major change that affects several different enterprises may require a new whole-farm plan and budget to adequately compare the profitability and feasibility of the proposed change against the current plan.

The whole-farm plan can be designed specifically for the current or upcoming year, or it may reflect a typical year over a longer time period. In some instances, a transitional plan may be needed for the period it takes to fully implement a major change in the operation. The effects of alternative plans on the financial position and risk exposure of the business also need to be considered.

THE PLANNING PROCEDURE

The development of a whole-farm plan can be divided into six steps, as shown in Figure 12-1: (1) review goals and specify objectives; (2) take an inventory of the physical, financial, and human resources available; (3) identify possible enterprises and their technical coefficients; (4) estimate gross margins for each enterprise; (5) choose a plan—the feasible enterprise combination that best meets the specified goals; and (6) develop a whole-farm budget that projects the profit potential and resource needs of the plan. Each step will be discussed and illustrated by an example.

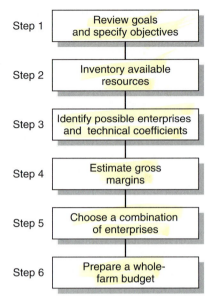

Figure 12-1 **Procedure for developing a whole-farm plan.**

Often the farm plan is developed for one year, and resources such as land, full-time labor, breeding livestock, machinery, and buildings are considered to be fixed in quantity. The usual objective for a one-year plan is to maximize the total gross margin for the farm. However, alternatives that assume an increase in the supply of one or more of these key resources can also be considered as can a complicated multi-year plan. Some more advanced techniques for analyzing the profitability of new capital investments are discussed in Chapter 17.

Review Goals and Specify Objectives

Most planning techniques assume that profit maximization is the primary management goal, but this objective is often subject to a number of both personal and societal restrictions. Maintaining the long-term productivity of the land, protecting the environment, guarding the health of the operator and workers, maintaining financial independence, and allowing time for leisure activities are all goals that affect the total farm plan. Certain enterprises may be included in the plan simply because of the satisfaction the operator receives from them regardless of their economic results, while others may be excluded for personal reasons even though they could be profitable.

After identifying goals, both business and personal, a manager should be able to specify a set of performance objectives. These can be defined in terms of crop yields, livestock production rates, costs of production, net income, or other measures.

Inventory Resources

The second step in the development of a whole-farm plan is to complete an accurate inventory of available resources. The type, quality, and quantity of resources available determine which enterprises can be considered in the whole-farm plan and which are not feasible.

Land

The land resource is generally the most valuable resource and one of the most difficult to alter. Land is also a complex resource with many characteristics that influence the type and number of enterprises to be considered. The following are some of the important items to be included in the land inventory.

1. Total number of acres available in cropland, pasture, orchards and vineyards, timber, and wasteland
2. Climatic factors, including temperature, annual rainfall, and length of the growing season
3. Soil types and factors such as slope, texture, and depth
4. Soil fertility levels and needs. A soil testing program may be needed as part of the inventory
5. The current water supply and irrigation system or the potential for developing it
6. Drainage canals and tile lines in existence, and any current or potential surface and subsurface drainage problems
7. Soil conservation practices and structures, including any current and future needs for improvement
8. The current soil conservation plan and any limitations it may place on land use or technology
9. Crop bases, established yields, long-term contracts, or other characteristics related to government programs or legal obligations
10. Existing and potential pest and weed problems that might affect enterprise selection and crop yields
11. Tenure arrangements and lease terms that may affect production decisions

This is also a good time to draw up a map of the farm showing field sizes, field layouts, fences, drainage ways and ditches, tile lines, and other physical features. A map can assist in planning changes or documenting past practices. If available, the cropping history of each field, including crops grown, yields obtained, fertilizer and lime applied, and pesticides used, can be recorded on a copy of the field map or in a computer database. This information is very useful for developing a crop program where a crop rotation is desirable or herbicide carryover may be a problem. Many counties now have extensive computerized databases of soil and land characteristics that can be accessed through state extension specialists or U.S. Department of Agriculture offices.

Buildings

The inventory of buildings should include a list of all structures, along with their condition, capacity, and potential uses. Livestock enterprises and crop storage may be severely limited in both number and size by the facilities available. Feed-handling equipment, forage and grain storage, water supply, manure handling, and arrangement and capacity of livestock facilities should all be noted in the inventory.

The potential for livestock production may also be affected by the location of the farmstead. Close proximity to streams, lakes, or nearby residents may restrict the type and volume of animals that can be raised or finished. In addition, there should be adequate land area available for environmentally sound manure disposal.

Labor

The labor resources should be analyzed for both quantity and quality. Quantity can be measured in months of labor currently available from the operator, family members, and hired labor, including its seasonal distribution. The availability and cost of additional full-time or part-time labor should also be noted, as the final farm plan might profitably use additional labor. Labor quality is more difficult to measure, but any special skills, training, and experience that would affect the possible success of certain enterprises should be noted.

Machinery

Machinery can be a fixed resource in the short run, and the number, size, and capacity of the available machinery should be included in the inventory. Particular attention should be given to any specialized, single-purpose machines. The capacity limit of a specialized machine, such as a cotton picker, will often determine the maximum size of the enterprise in which it is used.

Capital

Capital for both short-run and long-run purposes can be another limiting resource. The lack of ready cash or limited access to operating credit can affect the size and mix of enterprises that are chosen. Reluctance to tie up funds in fixed assets or to leverage the business through long-term borrowing may also limit expansion of the farming operation or the purchase of labor-saving technology.

Management

The last part of a resource inventory is an assessment of the management skills available for the business. What are the age and experience of the manager? What are the past performance of the manager and his or her capacity for making management decisions? What special skills or critical weaknesses are present? If a manager has no training, experience, or interest in a certain enterprise, that enterprise is likely to be inefficient and unprofitable. The quality of the management resource should be reflected in the technical coefficients incorporated into the farm budgets. Past success and records are the best indicators of future performance.

Other Resources

The availability of local markets, transportation, consultants, marketing quotas, or specialized inputs are also important resources to consider when developing the whole-farm plan.

Identify Enterprises and Technical Coefficients

For some producers, the decision about which enterprises to include in the farm plan has already been determined by personal experience and preferences, fixed investments in specialized equipment and facilities, or the regional comparative advantages for certain products. For them, the whole-farm planning process focuses on preparing the whole-farm budget for their plan. Other managers, though, will want to experiment with different enterprise combinations by developing a series of budgets and comparing them.

The resource inventory will show which new crop and livestock enterprises are feasible. Those requiring a resource that is not available can be eliminated from consideration, unless purchasing or renting this resource is possible. Custom and tradition should not be allowed to restrict the list of potential enterprises, only the resource limitations. Many farms have incorporated alternative or nontraditional enterprises into their farm plans, and some have been quite profitable.

The *technical coefficients* for an enterprise indicate how much of a resource is required to produce one unit of an enterprise. These technical coefficients, or resource requirements, are important in determining the maximum possible size of enterprises and the final enterprise combination.

The technical coefficients are developed to correspond to the budgeting unit for each enterprise, which would typically be 1 acre for crops and 1 head for livestock. For almost all enterprises, technical coefficients will be needed for land, labor, and capital. Technical coefficients for 1 beef cow, for example, might be 2 acres of pasture, 10 hours of labor, and $120 of operating capital. For some enterprises, machinery time or building use may also be critical resources that will need to be explicitly considered in the farm planning process. Thus, the technical coefficients for an acre of corn might be 1 acre of land, 4 hours of labor, $135 of operating capital, and 3.5 hours of tractor time.

Estimate the Gross Margin per Unit

Enterprise budgets, discussed in detail in Chapter 10, are important tools for farm planning. An enterprise budget is required for every enterprise that might be chosen for the farm plan, both those already being produced and any new alternatives being considered. Enterprise budgets provide the estimates of gross margin needed in the farm-planning process, and they also give information that can be used to develop the technical coefficients discussed above.

As shown in Chapter 10, gross margin is the difference between gross income and variable costs. It represents how much each unit of an enterprise contributes toward fixed costs and profit, after the variable costs of production have been paid. If maximization of profit is the goal, using gross margin, instead of net profit, may seem strange. However, because the plan is for the short run, fixed costs are constant regardless of the farm plan selected. Any positive gross margin represents a contribution toward paying those fixed costs. Therefore, in the short run, maximizing gross margin is equivalent to maximizing profit (or minimizing losses), because the fixed costs will not change.

Accurate enterprise budgets are of great importance in whole-farm planning. A good estimate of the gross margin for an enterprise requires the manager's best estimates of both gross revenue and variable costs. It is important that estimates of both selling prices and yields be accurate and that the yield estimates reflect the production practices employed. The variable cost estimates also need to reflect production practices, and these estimates require identifying the amount of each variable input used as well as the purchase price per unit. Using inaccurate enterprise budgets in the farm planning process will result in a less than optimal farm plan and a reduction in profit.

Choose the Enterprise Combination

Given the gross margins of the enterprises, the amount of each resource available on the farm, and the amount of each resource required per unit of each enterprise being considered, managers will attempt to find the combination of enterprises that provides the highest total profit for the farm. If there are more than a few enterprises to consider, or many resource restrictions, using paper and pencil to find the best combination would be difficult if not impossible. Fortunately, computer software programs exist to aid farm managers in their decision making. *Linear programming* (LP) is a mathematical technique that can be used to find the optimal combination of enterprises within the resource limits of the farm. Linear programming software is now widely available as an "add-on" to electronic spreadsheet programs. Linear programming will be discussed in detail later in this chapter.

Prepare the Whole-Farm Budget

The last step in the planning process is to prepare the *whole-farm budget*. A whole-farm budget can be used for the following purposes:

1. To estimate the expected income, expenses, and profit for a given farm plan
2. To estimate the cash inflows, cash outflows, and liquidity of a given farm plan

3. To compare the effects of alternative farm plans on profitability, liquidity, and other considerations

4. To evaluate the effects of expanding or otherwise changing the present farm plan

5. To estimate the need for, and availability of, resources such as land, capital, labor, livestock feed, or irrigation water

6. To communicate the farm plan to a lender, landowner, partner, or stockholder

The starting point for a whole-farm budget is the income and variable costs used for computing enterprise gross margins. As shown in Figure 12-2, these values are multiplied by the number of units of each enterprise in the plan to get a first estimate of total gross income and total variable costs. Other farm income that does not come directly from the budgeted enterprises, such as custom work income and fuel tax refunds, should then be added into the budget. Past records are a good source of information for estimating these additional sources of income.

Any costs not already included in the enterprise variable costs should now be added. In practice, expenses such as building repairs, car and truck expenses, interest, utilities, insurance, and other overhead costs are very difficult to allocate to specific enterprises and are affected very little by the final enterprise combination. These fixed costs are often called *indirect costs*. Other expenses, such as cash rent or property taxes, may apply to part of the acres in the plan but not to others. Although these and other fixed costs do not affect the selection of a short-term profit-maximizing plan, they may be a major portion of total expenses and should be included in the whole-farm budget. Budgets based on farm plans that include new investments in additional fixed assets should have their fixed costs revised accordingly.

Opportunity costs may or may not be included in the whole-farm budget. If they are included, the budgeted profit is economic profit. If not, it is accounting profit or net farm income.

Forms for organizing and recording whole-farm budgets are often available in farm record books. The Extension Service in each state may also have publications, forms, or computer software available. Use of such forms or computer programs will save time and improve the accuracy of the budget estimates.

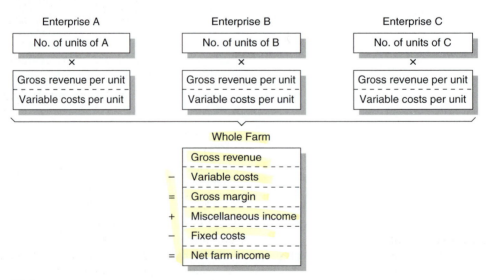

Figure 12-2   **Constructing the whole-farm budget.**

> ## Box 12-1 Whole-Farm Resource Budgets
>
> Although most whole-farm budgets are used to project income and expenses for a particular farm plan, a similar procedure can be used to estimate the quantities of key resources needed. Examples include the use of:
>
> - labor
> - feedstuffs
> - irrigation water
> - machinery time
>
> A detailed whole-farm resource budget can be used to estimate needs not only for the entire year but also for critical time periods within the year. These can be compared to the fixed quantities of resources available to see if potential bottlenecks exist. If shortages are projected, they can be resolved by:
>
> - changing the mix of enterprises to use less of the limited resource
>
> - changing the technical coefficients (for example, feeding a different ration or using herbicides to replace cultivation)
> - temporarily increasing the supply of the resource by actions such as hiring part-time labor, leasing or custom hiring machinery, or buying feed from outside sources.
>
> As more enterprises are included in the whole-farm plan that compete for the same resources, it becomes more important to create a budget that projects their requirements. This is especially true when the resource needs are quite variable throughout the year, such as labor for harvesting or irrigation water in mid-summer. By planning ahead, provisions can be made well in advance to have enough resources available at the right time. Table 21-3 in Chapter 21 shows how a detailed resource budget for labor can be developed for a whole farm.

EXAMPLE OF WHOLE-FARM PLANNING

The procedure used in whole-farm planning can be illustrated through the following example.

Goals and Objectives

The manager of the example farm wishes to choose a combination of crop and livestock activities that will maximize total gross margin for the farm for the coming year. For agronomic reasons, the operator wishes to carry out a crop mix that does not include more than 50 percent of the Class A cropland acres planted to cotton in any year.

Resource Inventory

Table 12-1 contains the resource inventory for the example farm. The land resource has been divided into three types: 400 acres of Class A

cropland, 200 acres of Class B cropland, and 600 acres of land that is in permanent pasture. Labor is the only other resource that is limited, with 2,400 hours available per year. Capital and machinery are both available in adequate amounts, and the few buildings available are sufficient for the livestock enterprises to be considered.

Enterprises and Technical Coefficients

Potential crop and livestock enterprises are identified and listed in Table 12-2. Any enterprise that is obviously unprofitable (e.g. has a negative gross margin) or that requires an unavailable resource should be eliminated. However, all feasible enterprises with a positive gross margin should be considered even if, at first glance, an enterprise appears less profitable than some others. The availability of surplus resources will sometimes cause one of the apparently less profitable enterprises to be included in the final plan.

TABLE 12-1	Resource Inventory for Example Farm

Resource	Amount and comments
Class A cropland	400 acres (not over 50% in cotton production)
Class B cropland	200 acres
Pasture	600 acres
Buildings	Only hay shed and cattle shed are available
Labor	2,400 hours available annually
Capital	Adequate for any farm plan
Machinery	Adequate for any potential crop plan, but all harvesting will be custom hired
Management	Manager appears capable and has experience with crops and beef cattle
Other limitations	Any hay produced must be fed on farm, not sold

As shown in Table 12-2, the manager of the example farm has identified three potential crop enterprises on Class A cropland (cotton, milo, and wheat) and two on Class B cropland (milo and wheat). Livestock enterprises are limited to beef cows or stocker steers because of the buildings available. The beef cow enterprise also requires some Class B land for hay production. Resource requirements per unit of each enterprise, or technical coefficients, are also shown in Table 12-2. For example, an acre of cotton requires 1 acre of Class A cropland, 4 hours of labor, and $250 of operating capital. Even though operating capital is not limited in this example, it is included for estimating cash flow needs.

Gross Margins

Table 12-3 contains the estimated income, variable costs, and gross margins for the seven potential enterprises to be considered in the whole-farm plan. Detailed breakdowns of the variable costs have been omitted to save space but would be included in the enterprise budgets,

which are done first. Careful attention should always be paid to estimating yields, output prices, input levels, and input prices. The accuracy of the whole-farm plan depends heavily on the estimated gross margins.

The Enterprise Combination

A procedure for choosing the most profitable combination of enterprises will be illustrated later in this chapter. For this example, assume that the farm manager chooses to produce 200 acres of cotton and 200 acres of wheat on Class A land, 150 acres of milo on Class B land, and 100 head of beef cows. The beef cows require 50 acres of Class B land for hay production.

The Complete Budget

The whole-farm budget for the chosen plan is shown in Table 12-4 under the Plan 1 section. Income and variable costs for individual enterprises were calculated by multiplying the income and expenses per unit by the number of units to be produced. Estimated net farm income is $50,150.

Note that an estimate of income from other sources, such as custom work done for neighbors, is included in the budget. Also shown are other expenses that do not vary directly with the number of crop acres produced or number of livestock raised. These items include property taxes, property and liability insurance, interest on fixed debt, depreciation, and miscellaneous expenses. Interest on borrowed operating capital may be shown here if a general operating loan or credit line for the entire farm is used. However, interest on an operating loan that is tied to an individual enterprise, such as for the purchase of feeder cattle, should be included in the enterprise budget data. Opportunity costs are not included in this example.

Alternative Plans

Table 12-4 also contains whole-farm budgets for two other farm plans. Plan 2 involves dropping

TABLE 12-2	**Potential Enterprises and Resource Requirements**

| | | Crops (per acre) | | | | | Livestock (per head) | |
| | | Class A Cropland | | | Class B Cropland | | | |
Resource	Quantity available	Cotton*	Milo	Wheat	Milo	Wheat	Beef cows	Stocker steers
Class A cropland (acres)	400	1	1	1	—	—	—	—
Class B cropland (acres)	200	—	—	—	1	1	0.5	—
Pasture (acres)	600	—	—	—	—	—	6	3
Labor (hours)	2,400	4	3	2.5	3	2.5	6	1
Operating capital ($)		250	100	80	80	65	470	550

*Limited to one-half of the Class A cropland for crop rotation needs.

TABLE 12-3	**Estimating Gross Margins Per Unit**

| | Class A Cropland | | | Class B Cropland | | Livestock | |
	Cotton (acre)	Milo (acre)	Wheat (acre)	Milo (acre)	Wheat (acre)	Beef cows (head)	Stocker steers (head)
Yield	500 lb.	80 cwt.	52 bu.	66 cwt.	36 bu.	—	—
Price ($)	0.80	2.50	3.75	2.50	3.75	—	—
Gross income ($)	400	200	195	165	135	675	620
Total variable costs ($)	250	100	80	80	65	470	550
Gross margin ($)	150	100	115	85	70	205	70

the beef cow enterprise and cash renting 300 more acres of Class A cropland to be equally divided between cotton and wheat. This plan has a projected net farm income of $54,650, an increase of $4,500 over Plan 1. There are additional fixed expenses for hired labor, cash rent, and miscellaneous items.

Plan 3 involves purchasing 600 acres of pasture land and adding 100 more beef cows to the plan. A loan of $150,000 at 8 percent interest to be repaid over 15 years is used to finance the purchase. Fifty acres of Class B land is shifted from wheat to produce more hay needed by the cows. Additional fixed expenses for some part-time hired labor and interest on the new loan are included under other expenses.

The profitability of this expansion should be analyzed over a period of more than one year. Therefore, the interest expense was calculated on the average principal amount that would be owed over the period of the loan—$75,000, rather than the initial debt of $150,000. For the first few years of the plan, the interest expense would be considerably higher, but it would also be lower in the later years. Plan 3 has a projected net farm income of $55,300, slightly higher than Plan 2.

TABLE 12-4	Example of a Whole-Farm Budget

	$/Unit	Plan 1 Units	Plan 1 Total	Plan 2 Units	Plan 2 Total	Plan 3 Units	Plan 3 Total
Gross income							
Cotton–A	$400	200	$ 80,000	350	$140,000	200	$ 80,000
Milo–A	200	0	0	0	0	0	0
Wheat–A	195	200	39,000	350	68,250	200	39,000
Milo–B	165	150	24,750	200	33,000	100	16,500
Wheat–B	135	0	0	0	0	0	0
Beef cows	675	100	67,500	0	0	200	135,000
Stocker steers	620	0	0	0	0	0	0
Total gross income			$211,250		$241,250		$270,500
Variable costs							
Cotton–A	$250	200	$ 50,000	350	$ 87,500	200	$ 50,000
Milo–A	100	0	0	0	0	0	0
Wheat–A	80	200	16,000	350	28,000	200	16,000
Milo–B	80	150	12,000	200	16,000	100	8,000
Wheat–B	65	0	0	0	0	0	0
Beef cows	470	100	47,000	0	0	200	94,000
Stocker steers	550	0	0	0	0	0	0
Total variable costs			$125,000		$131,500		$168,000
Total gross margin			$ 86,250		$109,750		$102,500
Other income			$ 5,000		$ 5,000		$ 5,000
Other expenses							
Property taxes			$ 5,600		$ 5,600		$ 6,200
Insurance			2,500		2,500		3,000
Interest on debt			17,000		17,000		23,000
Hired labor			0		3,500		3,500
Depreciation			10,500		10,500		10,500
Cash rent			0		15,000		0
Miscellaneous			5,500		6,000		6,000
Total other expenses			$ 41,100		$ 60,100		$ 52,200
Net farm income			$ 50,150		$ 54,650		$ 55,300
10% reduction in gross income			−21,125		−24,125		−27,050
Revised net farm income			$ 29,025		$ 30,525		$ 28,250

LINEAR PROGRAMMING

In Chapter 8, a method for finding the most profitable combination of two enterprises was presented. In reality, however, producers often choose between many different enterprises, and they usually do not have perfect knowledge of the production possibility curves for their farms. Linear programming (LP) is a mathematical procedure that uses a systematic technique to

---

Box 12-2 Whole-Farm Planning and Farming Systems

Most whole-farm planning and budgeting procedures treat the various crop and livestock enterprises under consideration as independent activities. At most, they may be considered to compete for some of the same fixed resources. However, there may be positive or negative interactions between enterprises that also need to be taken into account. Some examples include:

- fixation of nitrogen by legume crops, making nitrogen available for other crops that follow or are planted in conjunction with the legumes
- interruption of pest cycles by rotating dissimilar crops
- shading of low-growing crops by higher-growing crops in an intercropping system

- providing feed for livestock from crops on the same farm, and nutrients to crops through use of livestock manure
- passing young livestock from a breeding enterprise into a feeding enterprise on the same farm

The term *farming systems analysis* is given to the study of how various agricultural activities interact to achieve the overall goals of the farm or ranch. Although such interactions may at times be rather complex, ignoring them may lead to plans that fall considerably short of the farm's potential. Techniques such as input/output analysis and linear programming allow planners to model such interactions and create more realistic whole-farm plans and budgets.

---

find the "best" (e.g. most profitable) possible combination of enterprises. Linear programming models have a linear *objective function,* which is maximized or minimized subject to the resource restrictions.

Linear Programming Basics

In farm planning, normally the objective will be to maximize the total gross margin. In mathematical terms we can write the objective function (OBJ) this way

$$OBJ = GM(1) \times UNITS(1) + \\ GM(2) \times UNITS(2) + \\ GM(3) \times UNITS(3)$$

where GM(1) is the gross margin per unit (e.g. acre of crop or head of livestock) of the first enterprise and UNITS(1) is the number of units of this enterprise (acres or head) produced and so on for all the enterprises being considered in the plan. Each enterprise (corn, soybeans, stocker steers etc.) must be represented by a term in the equation. The order of the enterprises does not

matter, but it must be consistent throughout the linear programming model. Here, there are three enterprises being considered. For more enterprises, more terms must be added.

Developing the resource restrictions for the linear programming model requires the use of the technical coefficients, discussed previously in this chapter, and the overall resource limits. The general form of the restrictions is

$$X1 \times UNITS(1) + \\ X2 \times UNITS(2) + \\ X3 \times UNITS(3) \leq RESOURCE$$

where X1 represents the amount of a particular resource needed to produce one unit of the first enterprise, and so on. The total amount of the resource available for all uses, represented by RESOURCE, in the above equation, limits the number of units of the three possible enterprises that the farm can produce. Each enterprise being considered needs to be included in the equations. A restriction needs to be developed for every limited resource. The manager may also impose subjective constraints by arbitrarily setting a minimum or maximum level on some enterprises. A restriction to limit an enterprise, say enterprise 3, to no more than a certain amount would look like

$$UNITS(3) \leq LIMIT$$

where LIMIT is the maximum amount of enterprise 3 the manager will allow.

A linear programming model, with two enterprises, can be solved graphically, as shown in the Appendix to this chapter. Small linear programs can also be solved by hand using matrix algebra, but the process is tedious and, if there are more than a few enterprises or many restrictions, it is very easy to make mistakes. A computer program can solve quickly and accurately large linear programming problems, with many enterprises and complex restrictions.

The Linear Programming Tableau

Linear programming software is often available as an "add-on" to an electronic spreadsheet program. To solve an LP problem using these programs requires that the information be put in an organized form, often called the LP *tableau*. An example tableau is shown in Table 12-5. The first step in developing the tableau is to set up a column in the spreadsheet for each possible activity or enterprise. Two other columns are also needed, one for the type of restriction (less than or equal to, greater than or equal to, or equal to), and one for the limit on each resource. The limit column is sometimes called the "RHS" because, in the mathematical relation described previously, this value sits on the right-hand side of the inequality sign.

The first row in the LP tableau will hold the gross margin coefficients from the objective function (OBJ). The type is "MAX" for maximization. There will be no entry in the "Limit" columns in the OBJ row. The subsequent rows will represent the resource restrictions. The column for each enterprise contains the technical coefficients, that is, the amount of that resource required to produce one unit of enterprise. The "type" column shows the type of restriction (for example, "less than or equal to"), and the "limit" column shows the total amount of that resource available.

The example tableau in Table 12-5 uses the information provided in Tables 12-1, 12-2, and 12-3. The first row is the objective function, so the values are the gross margins for each enterprise, taken from Table 12-3. The next row is the first restriction, Class A cropland. As shown in Table 12-2, each Class A crop requires one acre of this cropland. A "1" is therefore entered in the

TABLE 12-5	**Linear Programming Tableau for the Farm Planning Example**

	Units	Class A cotton (acre)	Class A milo (acre)	Class A wheat (acre)	Class B milo (acre)	Class B wheat (acre)	Beef cows (head)	Stocker steers (head)	Type	Limit
Gross margin	$/unit	$150	$100	$115	$85	$70	$205	$70	MAX	
Class A land	acre	1	1	1	0	0	0	0	LE	400
Class B land	acre	0	0	0	1	1	0.5	0	LE	200
Pasture	acre	0	0	0	0	0	6	3	LE	600
Labor	hour	4	3	2.5	3	2.5	6	1	LE	2400
Rotation limit	acre	1	0	0	0	0	0	0	LE	200

columns for Cotton-Class A, Milo-Class A, and Wheat-Class A. Because the amount of Class A land used must be less than or equal to the total amount available, an "LE" for "less than or equal to" is entered in the "type" column. Finally, this row is completed by putting the total amount of Class A land available, 400 acres, in the "limit" column.

The next row, for the restriction on Class B cropland, is completed in the same manner. Each acre of Milo-Class B and Wheat-Class B will require one acre of Class B land. In addition, each cow-calf unit will require 0.5 acres of Class B land, as shown in Table 12-2. The total amount of Crop B land available is 200 acres, so 200 is the limit on this resource restriction.

For pasture, as shown in Table 12-2, each beef cow requires 6 acres and each stocker steer requires 3 acres. The limit on pasture is 600 acres. These technical coefficients and the limit appear in the corresponding columns of the pasture-restriction row. Labor requirements for each enterprise are found in Table 12-2, and the total amount available, 2400 hours, is found in Table 12-1. These values are used to fill in the labor-restriction row.

The final row of the tableau is a limit on cotton acreage, imposed by rotational considerations. Because capital, machinery, and management available are adequate for any farm plan,

no restrictions are developed for these resources in this case. In other situations, restrictions for these resources might be included.

Table 12-6 shows the solution to the example LP problem. The exact format of the computer output for the solution will depend on the computer package used to solve the problem; however, most packages will give information similar to that shown here. For the example, the optimal enterprise combination is: 200 acres of cotton on Class A land, 200 acres of wheat on Class A land, 150 acres of milo on Class B land, and 100 beef cows. Total gross margin (the objective value) for the farm is $86,250. The slack column indicates how much of a resource would be left over if this farm plan were followed. In this example, all resources except labor are completely used. For labor, 2,350 of the available 2,400 hours are used, leaving a slack value of 50 hours. The other entries in this table will be explained later in this chapter.

Additional Features of Linear Programming

Since computerized linear programming routines can quickly solve large problems, many different enterprises can be considered and detailed resource restrictions can be included. For example, the labor resources could be defined

TABLE 12-6	Linear Programming Solution to the Farm Planning Example					
Activity	Optimum level	Reduced Cost ($)	Rows	Level of use	Slack (unused)	Shadow price ($)
Cotton–Class A	200	0.00	Total Gross Margin	$86,250	.	.
Milo–Class A	0	−15.00	Class A Crop Land	400	0	115.00
Wheat–Class A	200	0.00	Class B Crop Land	200	0	85.00
Milo–Class B	150	0.00	Pasture	600	0	27.08
Wheat–Class B	0	−15.00	Labor	2350	50	0.00
Beef cows	100	0.00	Rotation limit	200	0	35.00
Stocker steers	0	−11.25				

on a monthly basis instead of annually, resulting in 12 labor restrictions instead of one. This refinement ensures that potential labor bottlenecks during peak seasons are not overlooked. Limits on operating capital can be defined in a similar manner, and the land resource can be divided into several classes. Additional restraints that reflect desirable land use practices, environmental considerations, farm program provisions, or subjective preferences of the manager can also be incorporated. Several enterprises can be defined for each crop or type of livestock, representing different types of technology, levels of production, or stages.

In addition to limitations imposed by fixed resources such as owned land and permanent labor, linear programming can incorporate activities that increase the supplies of certain resources. These activities could include renting land, hiring additional labor, or borrowing money. The optimal solution to the planning problem will then include the acquisition of additional resources, to the extent that total gross margin can be increased by doing so. For example, if we wanted to allow labor to be hired, we would include a column in our tableau labeled "hire labor." If labor costs $8.25 per hour to hire, our objective function value in this column would be −$8.25. The objective function value for hiring labor would be negative because hiring labor is a cost. In the labor restriction row, we would put a technical coefficient with the value "−1" in the "hire labor" column. This coefficient would be negative because hiring labor increases the amount of labor available rather than using up an hour. The negative number on the left side of the inequality would thus be equivalent to adding the hours on the right hand side.

It is also possible to transfer production from one enterprise to the resource supply needed by another enterprise and hence increase the amount of that resource available. For example, corn could be transferred from a corn enterprise to a feed resource, to increase the supply of feed available for livestock production. A transfer of this kind requires some careful thinking to be certain that the model does not also allow the sale of the same corn that is transferred to the feed resource.

Shadow Prices and Reduced Costs

Besides selecting the enterprises for the optimal farm plan, linear programming routines provide other useful information. Of particular use to many managers are values called *dual values*. There are two types of dual values, *shadow prices* and *reduced costs*. Shadow prices are the dual values on the resource restrictions, and reduced costs are dual values on the activities. Dual values—shadow prices and reduced costs—for our LP example can be found in Table 12-6.

For every resource restriction in the LP tableau, the LP software will calculate a shadow price, which represents the amount by which total gross margin would be increased if one more unit of that resource were available. For resources that are not completely used, this value would be zero. If a resource is completely used up by the enterprises selected in the plan, the shadow price tells the manager whether it may or may not be a good idea to acquire more of that resource. If the shadow price on a resource is higher than the cost of acquiring one more unit of that resource, the manager might consider obtaining more of that resource. Conversely, if the shadow price is less than the cost, it would not be profitable to acquire more of that resource.

The shadow prices shown in the last column of Table 12-6 indicate that having one more acre of Class B land, for example, would increase gross margin by $85, exactly the same value as the gross margin for one acre of milo, the most profitable crop on Class B land. On the other hand, the value of one more hour of labor is zero, because not all of the original labor supply is used. Similar interpretations can be placed on the other resource shadow prices. This information is useful for deciding how much the manager can afford to pay for additional resources.

For example, it would be profitable to rent more Class B land only if the cost were less than $85 per acre. It would not be profitable to hire extra labor given the current unused labor hours.

The reduced costs, found next to the enterprise levels in Table 12-6, tell the manager how much the total gross margin (e.g. the objective value) would be reduced by forcing into the plan one unit of an enterprise that was not included in the optimal plan. Only those enterprises not chosen in the optimal plan will have a non-zero reduced cost. The reduced cost of any enterprise that appears in the solution (at any level) is always zero. Hence, Class A cotton, Class A wheat, Class B milo, and beef cows have values of $0.00 for their reduced costs.

In our example, Class A milo, Class B wheat, and stockers all have non-zero reduced costs because these enterprises were not part of the optimal solution. Some software packages report all non-zero reduced costs as negative numbers (as seen in Table 12-6), while others may report them as positive numbers. The interpretation is the same, whether the number is reported with a negative sign or a positive sign.

The reduced cost for milo on Class A land indicates that forcing one acre of milo on Class A land into the solution would reduce total gross margin by $15.00. Similarly, the reduced costs for wheat on Class B land and for stocker steers indicate that forcing one acre of Class B wheat or one head of stocker steers into the solution would reduce total gross margin by $15.00 and $11.25, respectively. Before these enterprises could compete for a place in the farm plan, the gross margin per unit from these enterprises would have to increase by these amounts or more.

While linear programming is a good tool for finding the best enterprise combination, it does not substitute for human judgment. Most programming models do not consider goals other than maximization of gross margin, although some of the more sophisticated programming techniques can take into account such concerns as reduction of risk, conservation of resources,

interactions among enterprises, and optimal growth strategies over time. Model results are sometimes highly sensitive to the estimated gross margins, or the technical coefficients. Sometimes, a manager may accidentally omit an important restriction, such as management time or talent, leading to implausible results. The results of these computer models should therefore be interpreted cautiously. If the results do not make sense, the manager should check the gross margins, the technical coefficients, and the resource limits for accuracy and rerun the model as needed.

OTHER ISSUES

The whole-farm budget projects the most likely profit picture from following a particular farm plan. However, further analysis can provide some additional information about risk and liquidity.

Sensitivity Analysis

Although the whole-farm budget may project a positive net income, unexpected changes in prices or production levels can quickly turn that into a loss. Analyzing how changes in key budgeting assumptions affect income and cost projections is called *sensitivity analysis*.

At the end of the example in Table 12-4, a very simple sensitivity analysis was performed by reducing the anticipated gross farm income by 10 percent and recalculating the net farm income. This reduction could be caused by a decrease in production, selling prices, or both. Although Plan 3 has the highest projected net income in a typical year, it would have the lowest net farm income if there were a 10 percent drop in gross income. Additional sensitivity analysis could be carried out by constructing several entire budgets using different values for key prices and production rates. In Chapter 15, several more advanced approaches to evaluating the risk of a particular farm plan will be discussed.

Analyzing Liquidity

Liquidity refers to the ability of the business to meet cash flow obligations as they come due. A whole-farm budget can be used to analyze liquidity as well as profitability. This is especially important when major investments in fixed assets or major changes in noncurrent debt are being considered. Table 12-7 shows how net cash flow can be estimated from the whole-farm budget. Besides cash farm income, income from nonfarm work and investments can be added to total cash inflows. Cash outflows include cash farm expenses (but not noncash expenses such as depreciation), cash outlays to replace capital assets, principal payments on term debts (interest has already been included in cash farm expenses), and nonfarm cash expenditures for family living and income taxes.

Profitable farm plans will not always have a positive cash flow, particularly when they include a heavy debt load. Interest payments will be especially large during the first few years of the loan period, so it is wise to analyze liquidity for both the first year or two of a plan and for an average year. A negative cash flow projection indicates that some adjustments to the plan are needed if the business is going to be able to meet all of its obligations on time. All three plans in the example have a projected net cash flow that is positive, but Plan 3 would have a $17,750 negative cash flow if prices or yields fell 10 percent below expectations. In addition, the interest expense for the new loan projected in Plan 3 would be $12,000 the first year, instead of the average amount of $6,000 that was included in the budget. Therefore, severe liquidity problems could be encountered in the early years of Plan 3.

Long-Run versus Short-Run Budgeting

Short-run budgets that assume some resources are fixed should generally incorporate assumptions about prices, costs, and other factors that are expected to hold true over the next production period. However, when major changes in

TABLE 12-7	Example of Liquidity Analysis for a Whole-Farm Budget

	Plan 1	Plan 2	Plan 3
Cash inflows:			
Cash farm income	$216,250	$246,250	$275,500
Nonfarm income	20,000	20,000	20,000
	$236,250	$266,250	$295,500
Cash outflows:			
Cash farm expenses	$155,600	$181,100	$209,700
Term debt principal	10,500	10,500	20,500
Equipment replacement	17,000	17,000	17,000
Nonfarm expenses	38,500	38,500	38,500
	$221,600	$247,100	$285,700
Net cash flow	$14,650	$19,150	$ 9,800
10% reduction in cash farm income	−21,625	−24,625	−27,550
Revised net cash flow	−6,975	−5,475	−17,750

the supply of land, labor, or other assets, or in the way they are financed, are being contemplated, a longer-run perspective is needed. Very few farms or ranches are profitable every year, but a plan that involves a long-term investment or financing decision should project a positive net income in an average or typical year.

The following procedures should be used for developing a typical year budget:

1. Use average or long-term planning prices for products and inputs, not prices that are expected for the next production cycle. In particular, use prices that accurately reflect the long-term price relations among various products and inputs.
2. Use average or long-term crop yields and livestock production levels. Use past records as a guide. For new enterprises, use conservative performance rates.
3. Ignore carryover inventories of crops and livestock, accounts payable and receivable, or cash balances when estimating income,

expenses, and cash flows. In the long run, these will cancel out from year to year. Assume that units sold are equal to production within a typical year.

4. Assume that the borrowing and repayment of operating loans can be ignored when projecting cash flows in a typical year, because they will offset each other. If significant short-term borrowing is anticipated, however, the interest cost that results should be incorporated into the estimate of cash expenses.

5. Assume that enough capital investment is made each year to at least maintain depreciable assets at their current level; that is, to replace those that wear out.

6. Assume that the operation is neither increasing nor decreasing in size. This is especially critical when projecting liquidity for a typical year.

In some cases, the farm business may require several years to move from the current plan to a future plan. Profits and cash flows may be reduced temporarily because of inventory buildup, start-up costs, low production levels while learning new technology, and rapid debt repayment. Several transitional budgets may be needed to analyze conditions that will exist until the new farm plan is fully implemented.

The actual profitability of the operation may fall below levels projected by the whole-farm budget in some years. If this is due to unfavorable weather or low price cycles, however, the whole-farm plan chosen may still be the most profitable one for the long run.

SUMMARY

Whole-farm planning and the resulting whole-farm budget analyze the combined profitability of all enterprises in the farming operation. Planning starts with reviewing goals, setting objectives, and taking an inventory of the resources available. Feasible enterprises must be identified, and their gross income per unit, variable costs, and gross margins computed. The combination of enterprises chosen can then be used to prepare a whole-farm budget. Income and variable costs per unit are multiplied by the number of units to be produced and then combined with other farm income, fixed costs, and any additional variable costs to estimate net farm income. The completed whole-farm budget is an organized presentation of the sources and amounts of income and expenses. Whole-farm budgets can be based on either short-run or long-run planning assumptions, and can also be used to evaluate liquidity.

Linear programming can be used to select the combination of enterprises that maximizes gross margin without exceeding the supply of resources available. It can handle large, complex planning problems quickly and accurately, and also provides information such as the value of obtaining additional resources or the penalties for including certain enterprises.

QUESTIONS FOR REVIEW AND FURTHER THOUGHT

1. Why do goals have to be set before a whole-farm plan can be developed?

2. What should be included in a resource inventory needed for whole-farm planning?

3. How does a whole-farm budget differ from a partial budget or an enterprise budget?

4. Why can fixed costs be ignored when developing the whole-farm plan, but are included in the whole-farm budget?

5. Give some examples in which "fixed" costs would change when comparing whole-farm budgets for alternative farm plans.

6. Use linear programming software to find the profit-maximizing whole-farm plan using the following information:

| | | Resource requirements per acre | |
Resources	Resource limit	Corn	Soybeans
Land (acre)	800	1	1
Capital ($)	90,000	150	100
Labor (hour)	3,500	5	2.5
Gross margin per acre		$100	$80

7. What values could you change in a whole-farm budget to perform a sensitivity analysis?

8. What is the difference between analyzing profitability and analyzing liquidity?

9. What prices and yields would you use when developing a long-run whole-farm plan? A short-run plan? What other assumptions would change?

Appendix. Graphical Example of Linear Programming

The basic logic for solving a linear programming problem can be illustrated in graphical form for a small problem involving two enterprises, corn and soybeans, and three limited resources. The necessary information is shown in Table 12-8, with land, labor, and operating capital as the limiting resources. Gross margins and the technical coefficients are also shown in the table.

Resource limits and the technical coefficients are used to graph the possible enterprise combinations shown in Figure 12-3. The supply of land limits corn and soybeans to a maximum of 120 acres each. These points, A and A', are found on the axes and connected with a straight line. Any point on line AA' is a possible combination of corn and soybeans, given only the land restriction. Labor, however, restricts corn to a maximum of 100 acres (500 hours divided by 5 hours per acre) and soybeans to 166.7 acres (500 hours divided by 3 hours per acre). These points on the axes are connected by line BB'. Any point on line BB' is a possible combination of corn and soybeans permitted by the labor restriction. In a similar manner, line CC' connects the maximum corn acres permitted by the operating capital restriction

TABLE 12-8	Information for Linear Programming Example		

| | | Resource Requirements (per acre) | |
Resources	Resource limit	Corn	Soybeans
Land (acres)	120	1	1
Labor (hours)	500	5	3
Operating capital ($)	30,000	200	160
Gross margin ($)		120	96

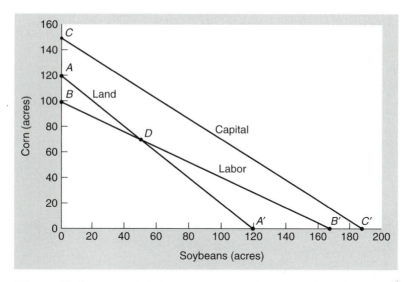

Figure 12-3   Graphical illustration of resource restrictions in a linear programming problem.

($30,000 ÷ $200 per acre = 150 acres) with the maximum soybean acres ($30,000 ÷ $160 per acre = 187.5 acres). Line CC′ identifies all the possible combinations based only on the operating capital restriction.

The sections of lines AA′, BB′, and CC′ that are closest to the origin of the graph, or line BDA′, represent the maximum possible combinations of corn and soybeans when all limited resources are considered together. Line BDA′ is actually a segmented production possibility curve similar to those for competitive enterprises discussed in Chapter 8 (see Figure 8-2). The graph reveals that operating capital is not actually a limiting resource. It is fixed in amount, but $30,000 is more than sufficient for any combination of corn and soybeans permitted by the land and labor resources.

The next step is to find which of the possible combinations of corn and soybeans will maximize total gross margin. Total gross margin from planting 100 acres of corn, the maximum possible amount if no soybeans are grown (represented by point B), is $12,000. Adding one acre of soybeans increases the gross margin by $96, but requires 3 hours of labor, which in turn causes 0.67 less acre of corn to be grown (remember, no extra labor is available). This subtracts (0.67 × $120 = $80) from total gross margin, for a net increase of $16.

This substitution can be continued until no more unused land is available (point D in Figure 12-3). At this point, increasing soybeans by one more acre requires one less acre of corn to be grown, resulting in a net decrease in gross margin of ($120 − $96) = $24. Thus, point D represents the combination of the two crops that maximizes gross margin. This combination is 70 acres of corn and 50 acres of soybeans, with a total gross margin of $13,200. Producing either more acres of corn or more acres of soybeans would only reduce the total gross margin.

Figure 12-4 shows the graphical solution to this example. Only the relevant production possibility curve, or line segment BDA′ is shown. The graphical solution to a profit-maximizing linear programming problem is the point where a line containing points of equal total gross margin just touches or is tangent to the production possibility curve on its upper side. This is point D in Figure 12-4, which

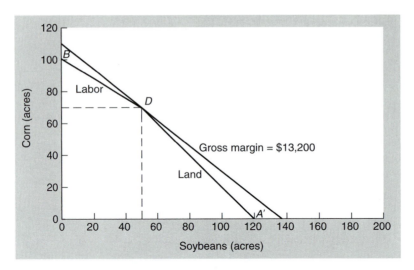

Figure 12-4   Graphical solution for finding the profit-maximizing plan using linear programming.

is just tangent to a line representing all possible combinations of corn and soybeans producing a total gross margin of $13,200. Higher gross margins are not possible, because they require combinations of enterprises or enterprise levels that are not permitted by the limited resources. Combinations other than 70 acres of corn and 50 acres of soybeans are possible, but they would have a total gross margin of less than $13,200.

The solution at point D was found in a manner similar to that used to find the profit-maximizing enterprise combination in Chapter 8, where the substitution ratio was equated with the price ratio. One basic difference is that linear programming generates a production possibility curve with linear segments, rather than a smooth, continuous curve, as in Figure 8-2. The solution will generally be at one of the corners or points on the production possibility curve, so the substitution ratio will usually not exactly equal the price ratio. In fact, because the variable costs change as the number of acres of each crop changes, the ratio of the gross margins per acre for each crop must be compared instead of the price ratio. The gross margin ratio will fall between the substitution ratios for the two most limiting resources. For the example, the substitution ratio of soybeans for corn is 0.67 along segment BD and 1.0 along segment DA'. The gross margin ratio is ($96 ÷ $120) = 0.80, which falls between the two substitution ratios.

CASH FLOW BUDGETING

CHAPTER OBJECTIVES

1. To identify cash flow budgeting as a tool for financial decision making and business analysis

2. To understand the structure and components of a cash flow budget

3. To illustrate the procedure for completing a cash flow budget

4. To describe both the similarities and differences between a cash flow budget and an income statement

5. To discuss the advantages and potential uses of a cash flow budget

6. To show how to use a cash flow budget when analyzing a possible new investment

Even the most profitable farms and ranches occasionally find themselves short of cash. Anticipating these shortages and having a plan to deal with them is an important management activity. A cash flow budget is a financial analysis tool with applications in both forward planning and the ongoing analysis of a farm or ranch business. Preparing one is the next logical step after finishing the whole-farm plan and budget. The cash flow budget will provide answers to some remaining questions. Is the plan financially feasible? Will there be sufficient capital available at the specific times it will be needed? If not, how much will need to be borrowed? Will the plan generate the cash needed to repay any new loans? These types of questions can be answered by preparing and analyzing a cash flow budget.

FEATURES OF A CASH FLOW BUDGET

A cash flow budget is a summary of the projected cash inflows and cash outflows for a business over a given period of time. This period of time is typically a future accounting period and is divided into quarters or months. As a forward planning tool, its primary purpose is to estimate the amount and timing of future borrowing needs and the ability of the business to repay these and other loans on time. Given the large amount of capital today's commercial farms and ranches require and often must borrow, cash flow budgeting is an important budgeting and financial management tool.

A discussion of cash flow budgeting must continually emphasize the word *cash*. All cash flows must be identified and included on the budget. Cash flows into a farm business from many sources throughout the year, and cash is used to pay business expenses and to meet other needs for cash. Identifying and measuring these sources and uses of cash is the important first step in constructing a cash flow budget. The concept of cash flows is shown graphically in Figure 13-1. It assumes that all cash moves through the business checking account, making it the central point for identifying and measuring the cash flows.

Two things make a cash flow budget substantially different from a whole-farm budget. First, a cash flow budget contains all cash flows, not just revenue and expenses, and it does not include any noncash items. For example, cash inflows would include cash from the sale of capital items and proceeds from new loans, but not inventory changes. Principal payments on debt and the full cost of new capital assets would be included as cash outflows, but depreciation would not. The emphasis is on cash flows, regardless of source or use and whether or not they are business revenue or expenses. For this reason, personal and nonfarm cash revenue and expenses are also included in a cash flow budget, because they affect the amount of cash available for farm business use.

The second major difference between a whole-farm budget and a cash flow budget is the latter's concern with the timing of revenue and expenses. A cash flow budget also includes "when" cash will be received and paid out as

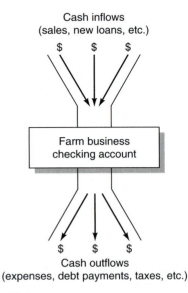

Figure 13-1 **Illustration of cash flows.**

well as "for what" and "how much." This timing is shown by preparing a cash flow budget on a quarterly or monthly basis. Because farming and ranching are seasonal businesses, most agricultural cash flow budgets are done on a monthly basis to permit a detailed analysis of the relation between time and cash flows.

The unique characteristics of a cash flow budget do not allow it to replace any other type of budget or any of the records discussed in Part II. It fills other needs and is used for different purposes. However, it will be shown shortly that much of the information needed for a cash flow budget can be found in the whole-farm budget and farm records.

Actual Versus Estimated Cash Flows

By definition, a cash flow budget contains estimates of cash flows for a future time period. However, it is possible to record and organize the actual cash flows for some past time period. The Farm Financial Standards Council recommends that a *Statement of Cash Flows* be developed as part of the standard end-of-year set of financial statements for the farm business. This statement is a condensed version of a typical cash flow budget, and it contains actual data for the accounting period.

A record of actual cash flows has several uses. First, if done monthly, it can be compared with monthly budgeted values at the end of each month. This comparison can provide an early warning of any substantial deviations while there is still time to determine causes and make corrections. Second, the Statement of Cash Flows can provide useful insight into the financial structure of the business and show how the operating, financing, and investing activities combine and interact as sources and uses of cash. Third, actual cash flows provide a good starting point for developing the next annual cash flow budget. With the totals and timing of past cash flows, it is relatively easy to make the adjustments needed to project these into the future. Reviewing past cash flows also prevents

some important items from being overlooked on the new budget.

Structure of a Cash Flow Budget

The structure and format of a cash flow budget are shown in Table 13-1 in condensed form. This condensed budget illustrates the sources and uses of cash, which need to be included on any cash flow form. Note that there are five potential sources of cash:

1. The beginning cash balance, or cash on hand at the beginning of the period
2. Farm product sales or cash revenue from the operation of the farm business
3. Capital sales, which is the cash received from the sale of capital assets such as land,

TABLE 13-1	Simplified Cash Flow Budget	
	Time period 1	**Time period 2**
1. Beginning cash balance	$ 1,000	$ 500
Cash inflow:		
2. Farm product sales	$ 2,000	$12,000
3. Capital sales	0	5,000
4. Miscellaneous cash income	0	500
5. Total cash inflow	$ 3,000	$18,000
Cash outflow:		
6. Farm operating expenses	$ 3,500	$ 1,800
7. Capital purchases	10,000	0
8. Miscellaneous expenses	500	200
9. Total cash outflow	$14,000	$ 2,000
10. Cash balance (line 5 − line 9)	−11,000	16,000
11. Borrowed funds needed	$11,500	0
12. Loan repayments (principal and interest)	0	11,700
13. Ending cash balance (line 10 + line 11 − line 12)	500	4,300
14. Debt outstanding	$11,500	$ 0

machinery, breeding livestock, and dairy cattle

4. Nonbusiness cash receipts, which would include nonfarm cash income, cash gifts, and other sources of cash
5. New borrowed capital or loans received (line 11)

The last source cannot be included in the cash inflow section, because borrowing requirements are not known until the cash outflows are matched against the cash inflows.

Table 13-1 also shows the four general uses of cash. They are:

1. Farm operating expenses, which are the normal and usual cash expenses incurred in producing the farm revenue
2. Capital purchases, which would be the full purchase price of any new capital assets such as land, machinery, and dairy or breeding livestock
3. Nonbusiness and other expenses, which would include cash used for living expenses, income and social security taxes, and any uses of cash not covered elsewhere
4. Principal payments on debt. Interest payments should also be included here, unless they were included as part of the operating expenses.

The difference between total cash inflows and total cash outflows for any time period is shown as the ending cash balance for that period.

Only two time periods are shown in Table 13-1, but once the basic procedure is understood, it can be extended to any number of time periods. In the first time period, the total cash inflow of $3,000 includes the beginning cash balance. The total cash outflow is $14,000, leaving a projected cash balance of ($11,000). This deficit will require borrowing $11,500 to provide a $500 minimum ending cash balance.

Total cash inflow for the second time period is estimated to be $18,000, resulting in a cash balance of $16,000 after subtracting a total cash outflow of $2,000. This large cash balance permits paying off the debt incurred in the first time period, which is estimated at $11,700 when interest of $200 is included. The final result is an estimated $4,300 cash balance at the end of the second time period. Following this same procedure for all subsequent time periods traces out the projected level and timing of borrowing and debt repayment potential.

CONSTRUCTING A CASH FLOW BUDGET

Considerable information is needed to construct a cash flow budget. For simplicity and to ensure accuracy, a logical, systematic approach should be followed. The following steps summarize the process and information needs.

1. Develop a whole-farm plan. It is impossible to estimate cash revenues and expenses without knowing what crops and livestock will be produced.
2. Take inventory. This should include existing crops and livestock available for sale during the budget period.
3. Estimate crop production and, for combination crop and livestock farms, estimate livestock feed requirements. This step projects crops available for sale after meeting livestock feed requirements. It may show a need to purchase feed if production plus beginning inventory is less than what is needed for livestock.
4. Estimate cash receipts from livestock enterprises. Include both livestock included in the beginning inventory and those to be produced and sold within the year. Sales of livestock products such as milk and wool should also be included.
5. Estimate cash crop sales. First, determine a desired ending inventory for feed use and sale next year. Next, compute the amount available for sale as beginning inventory *plus* production *less* livestock feed requirements *less* desired ending inventory.
6. Estimate other cash income such as revenue from custom work or government farm program payments. If the budget

includes both business and personal cash flows, include rent, interest, or dividends on nonfarm investments plus any other nonfarm sources of cash revenue.

7. Estimate cash farm operating expenses. Reviewing actual cash flows for the last year will prevent overlooking items such as property taxes, insurance, repairs, and other cash expenses not directly related to crop and livestock production.

8. Estimate personal and nonfarm cash expenses. Included here would be cash needed for living expenses, income and social security taxes, and any other nonfarm cash expenses. If the budget is for the farm business only, simply estimate the cash withdrawals that will be needed.

9. Estimate purchases and sales of capital assets. Include the full purchase price of any planned purchases of machinery, buildings, breeding livestock, and land, as well as the total cash to be received from the sale of any capital assets.

10. Find and record the scheduled principal and interest payments on existing debt. This will be primarily noncurrent debt, where the amounts and dates of each payment are shown on a repayment schedule. Any carryover current debt from the previous year should also be included here.

The data estimated and organized in these steps can now be entered on a cash flow form. With a few more calculations, the end result will be an estimate of borrowing needs for the year, the ability of the business to repay these loans, and the timing of each. Key assumptions about selling prices, input costs, and production levels should be well documented before the plan is presented to a lender.

A Cash Flow Budgeting Form

Printed forms for completing a cash flow budget are available from many sources, including lending agencies and the agricultural extension service in most states. These sources, as well as commercial software firms, may have computer programs that can be used. A cash flow budget can also be constructed by anyone familiar with any of the spreadsheet software programs. Table 13-2 illustrates one type of form for a cash flow budget. To save space, only the headings for the first three columns are shown. Other forms may differ in organization, headings, and details, but all provide the same basic information.

The first 13 lines of the form are for recording cash inflows projected from both farm and nonfarm sources. Since both nonfarm cash income and expenses will affect the cash available for farm business use, these items are included in the budget, even though they are not directly related to the farm business. The total annual amount expected for each cash inflow is recorded in the Total column. This amount is then allocated to the month or months when it will be received.

Cash farm operating expenses are listed on lines 14 through 34, with the total on line 35. As with all entries on a cash flow budget, the projected total for each expense item is put in the Total column, and this amount is then allocated to the month or months when the cash will actually be needed. The sum of total expenses for the individual months should always be compared with the sum in the Total column on line 35. This cross check will show whether any errors were made when allocating individual cash expenses to specific months.

Several other possible cash outflows are shown on lines 36 through 47. Capital expenditures for replacement or expansion of machinery and equipment, breeding livestock, land, and buildings require cash. The full purchase price for any capital expenditures should be recorded, even when borrowing will be used to make the purchase. One purpose of the budget is to estimate the necessary amount of such borrowing. Family living expenses, income and social security taxes, and other nonfarm cash expenses should be entered on lines 39 through 43. All personal expenses, such as automobile expenses and medical and life insurance premiums, should be included in family living expenses.

TABLE 13-2	Form for a Cash Flow Budget

Cash Flow Budget

Name: I.M. Farmer	Total	Jan	Feb	March
1 Beginning cash balance				
Operating receipts:				
2 Grain and feed				
3 Feeder livestock				
4 Livestock products				
5 Other				
6				
Capital receipts:				
7 Breeding livestock				
8 Machinery and equipment				
9				
Nonfarm income:				
10 Wages and salary				
11 Investments				
12				
Total cash inflow				
13 (add lines 1–12)				
Operating expenses:				
14 Seed				
15 Fertilizer and lime				
16 Chemicals				
17 Other crop expenses				
18 Gas, oil, lubricants				
19 Hired labor				
20 Machine hire				
21 Feed and grain				
22 Feeder livestock				
23 Livestock expenses				
24 Repairs—machinery				
25 Repairs—buildings				
26 Cash rent				
27 Supplies				
28 Property taxes				
29 Insurance				
30 Utilities				
31 Auto and pickup (farm share)				
32 Other farm expenses				
33				
34				
35 Total cash operating expenses				

| TABLE 13-2 | *(CONTINUED)* | | | |

Cash Flow Budget

Name: I.M. Farmer	Total	Jan	Feb	March
Capital expenditures:				
36	Machinery and equipment				
37	Breeding livestock				
38					
Other expenditures:				
39	Family living expenses				
40	Income tax and social security				
41	Other nonfarm expenses				
42					
43					
Scheduled debt payments:				
44	Current debt—principal				
45	Current debt—interest				
46	Noncurrent debt—principal				
47	Noncurrent debt—interest				
48	Total cash outflow				
	(add lines 35–47)				
49	Cash available (line 13 − line 48)				
New borrowing:				
50	Current				
51	Noncurrent				
52	Total new borrowing				
Payments on new current debt				
53	Principal				
54	Interest				
55	Total debt payments				
	(line 53 + line 54)				
56	Ending cash balance				
	(lines 49 + 52 − 55)				
Summary of debt outstanding				
57	Current				
58	Noncurrent				
59	Total debt outstanding				

Lines 44 through 47 are used to enter the scheduled principal and interest payments on debt incurred in past years. Entries here would be payments on noncurrent debt, as well as on any current debt carried over from the past year. Only scheduled payments on old debt are entered in this section, because payments on any new debt incurred during the coming year will be computed and entered in a later section.

Lines 35 through 47 are summed to get total annual cash outflows and totals for each month, with the amount entered on line 48. Next, the total estimated cash available at the end of the month is calculated by subtracting line 48 from line 13, with the result entered on line 49. If the total cash outflow is greater than the total cash inflow, the cash available will be negative, and new borrowing or other adjustments will be needed to obtain a positive ending cash balance. Any new borrowing is entered on lines 50 or 51, depending on the type of loan. The total new borrowing is added to the cash available on line 49 to find the ending cash balance for the month. No cash is available to pay down new current debt, so line 55 will be zero in this case.

If the cash available on line 49 is greater than zero, the total cash inflow for the month was greater than the total cash outflow. This amount can be used to pay off part or all of any new current debt incurred earlier in the year or carried over to a future time period. The principal and interest are entered on lines 53 and 54, and these amounts are adjusted to permit a positive ending cash balance. After finding the ending cash balance by subtracting line 55 from line 49 (line 52 is zero in this case), the result is entered on line 56. The same amount is transferred to line 1 of the next month as its beginning cash balance.

Lines 57 through 59 are not a necessary part of a cash flow budget. However, they summarize the debt situation for the business and provide some useful information. For each type of debt, the amount outstanding at the end of any month will equal the amount at the end of the previous month, plus any new debt, minus any principal payments made during the month. Lines 57 and 58 should be summed for each month, and the total debt outstanding should be entered on line 59. This figure shows the pattern of total debt and its changes throughout the annual production cycle. Because of the seasonal

nature of agricultural production, income, and expenses, this pattern will often repeat itself each year. Some lenders specify a maximum amount of debt the operator can have at any given time during the year. This part of the budget helps project whether this limit will allow sufficient capital to be borrowed.

Example of a Cash Flow Budget

A completed cash flow budget is shown in Table 13-3. It will be used to review the budgeting steps and to point out several special calculations. The first estimates needed are the beginning cash balance on January 1 and all sources and amounts of cash inflows for the year. These amounts are entered in the Total column and then allocated to the month or months when the cash will actually be received. This example shows an estimated total annual cash inflow of $267,100, including the beginning cash balance of $2,500. Notice that the total cash inflow for any month other than January cannot be determined until its beginning cash balance is known. The ending cash balance for each month becomes the beginning cash balance for the next month. This requires completing the budget for one month before beginning the next month.

The next step is to estimate total cash operating expenses by type, placing each amount in the Total column. Each expense estimate is then allocated to the appropriate month(s), and the total expenses for each month are entered on line 35. In the example, the total annual cash operating expenses are projected to be $108,500, with the largest expenditures in November and March. The same procedure is followed for capital expenditures, family living expenses, income taxes, and social security payments. Another important requirement for cash is the scheduled principal and interest payments on debt outstanding at the beginning of each year. These amounts are shown on lines 44 through 47. Total cash outflow, found by summing lines 35 through 47, is $272,700 in the example. All the above steps should be completed before any of the calculations on lines 50 to 59 are attempted.

Beginning with the month of January, note that total cash inflow is $23,300, and total cash outflow is $4,650, leaving $18,650 of cash available at the end of January (line 49). There is no need to borrow any money in January, and there are no new loans that can be paid off. Therefore, lines 50 through 55 are all zero, and the $18,650 becomes the ending cash balance, as shown on line 56. The first column of lines 57 and 58 indicate there was no current debt and $340,000 of noncurrent debt at the beginning of the year. Because there are no changes in the debt balances during the month, these figures are transferred to the January column, showing $340,000 of total debt outstanding at the end of January.

The next step is to transfer the January ending cash balance of $18,650 to the beginning cash balance for February. Calculations for February are similar to those for January, with one exception. There is a $6,000 principal payment on noncurrent debt, which reduces its balance to $334,000, as shown on line 58, and the total debt outstanding to $334,000.

February's ending cash balance of $15,250 is transferred to the beginning balance for March, where the first new borrowing occurs. Cash outflow exceeds cash inflow by $76,300 in March, requiring the borrowing of $76,800 to cover this amount and leave a small projected ending balance of $500. New machinery will be purchased in March at a cost of $60,000, so a new noncurrent loan of $40,000 is included as part of the new borrowing. The remaining $36,800 needed is shown as a new current loan. After the appropriate adjustments in the loan balances on lines 57 and 58, March will end with an estimated $410,800 of debt outstanding.

This example shows that additional new borrowing will also be needed in April, May, June, and July. Payments can be made in August when the feeder livestock are sold, and in October when some crop income is received. Repayments are handled in the following manner, using August as the example. A total of $70,050

in cash (line 49) is available. All of this cannot be used for principal payments, as there will be some interest due, and a positive ending balance is needed. A principal payment of $67,400 plus $2,137 interest on this amount will leave an ending balance of $513. The current debt outstanding must be reduced by the $67,400 principal payment, and the noncurrent debt by the $6,000 principal payment (line 46). Total debt outstanding at the end of August is $373,300.

The principal repaid in August assumes that the oldest of the new current borrowing was repaid first. It included the $36,800 borrowed in March, the $16,400 borrowed in April, the $6,900 borrowed in May, and $7,300 of the June borrowing, for a total of $67,400. The interest rate was assumed to be 9 percent, and the interest due was calculated as below, using the equation

$$interest = principal \times rate \times time$$

	Principal	Interest
March borrowing:	$36,800 \times 9\% \times 5/12$ of a year =	$1,380
April borrowing:	$16,400 \times 9\% \times 4/12$ of a year =	492
May borrowing:	$\ 6,900 \times 9\% \times 3/12$ of a year =	155
June borrowing:	$\ 7,300 \times 9\% \times 2/12$ of a year =	110
Totals	$67,400	$2,137

For convenience, principal payments were assumed to be made in $100 increments.

Sufficient cash was available in October to repay all remaining new current debt, which consisted of $22,300 left from the June borrowing and the $3,000 borrowed in July. Interest due was calculated as above and amounted to $737. Large cash balances are projected for October, November, and December after all new current debt has been repaid. The budget shows that new borrowing is projected to peak in July, with $92,700 of new current debt and $40,000

TABLE 13-3 | Example of a Cash Flow Budget

Name: I.M. Farmer

Cash Flow Budget

Year 20XX

		Total	Jan	Feb	March	April	May	June	July	Aug	Sept	Oct	Nov	Dec
1	Beginning cash balance	2,500	2,500	18,650	15,250	500	450	450	500	450	513	663	26,176	37,376
	Operating receipts:													
2	Grain and feed	160,000	20,000	20,000								60,000	60,000	
3	Feeder livestock	88,000								88,000				
4	Livestock products													
5	Other	5,400									5,400			
6														
	Capital receipts:													
7	Breeding livestock	3,800								3,800				
8	Machinery and equipment													
9														
	Nonfarm income:													
10	Wages and salary	7,200	600	600	600	600	600	600	600	600	600	600	600	600
11	Investments	200	200											
12														
13	Total cash inflow (add lines 1–12)	267,100	23,300	39,250	15,850	1,100	1,050	1,050	1,100	92,850	6,513	61,263	86,776	37,976
	Operating expenses:													
14	Seed	5,800			5,800									
15	Fertilizer and lime	26,000			8,000	10,000							8,000	
16	Chemicals	4,000			4,000									
17	Other crop expenses	2,600										2,600		
18	Gas, oil, lubricants	3,000					1,500							1,500
19	Hired labor	6,000			500	1,500	1,500	500				1,000	1,000	
20	Machine hire	800						400			400			
21	Feed and grain	4,000		800		800		800		800				800
22	Feeder livestock	36,000											36,000	
23	Livestock expenses	7,500	1,000	500	500	1,500	500	1,000	500				1,000	1,000
24	Repairs—machinery	3,600	500	500	500		600			500	500	500		
25	Repairs—buildings	1,800									1,800			
26	Cash rent													
27	Supplies	1,000		250			250			250			250	
28	Property taxes	3,400			1,700							1,700		
29	Insurance	700		700										
30	Utilities	600	50	50	50	50	50	50	50	50	50	50	50	50
31	Auto and pickup (farm share)	1,200	100	100	100	100	100	100	100	100	100	100	100	100
32	Other farm expenses	500		100		100		100		100		100		
33														
34														
35	Total cash operating expenses	108,500	1,650	3,000	21,150	14,050	4,500	2,950	650	1,800	2,850	6,050	46,400	3,450

TABLE 13-3 (CONTINUED)

Name: I.M. Farmer

Cash Flow Budget

Year 20XX

		Total	Jan	Feb	March	April	May	June	July	Aug	Sept	Oct	Nov	Dec
	Capital expenditures:													
36	Machinery and equipment	60,000			60,000									
37	Breeding livestock	1,000						1,000						
38														
	Other expenditures:													
39	Family living expenses	36,000	3,000	3,000	3,000	3,000	3,000	3,000	3,000	3,000	3,000	3,000	3,000	3,000
40	Income tax and social security	8,000			8,000									
41	Other nonfarm expenses													
42														
43														
	Scheduled debt payments:													
44	Current debt—principal	0												
45	Current debt—interest	0												
46	Noncurrent debt—principal	32,000		6,000				20,000		6,000				
47	Noncurrent debt—interest	27,200		12,000				3,200		12,000				
48	Total cash outflow (add lines 35–47)	272,700	4,650	24,000	92,150	17,050	7,500	30,150	3,650	22,800	5,850	9,050	49,400	6,450
49	Cash available (line 13 − line 48)	(5600)	18,650	15,250	(76,300)	(15,950)	(6,450)	(29,100)	(2,550)	70,050	663	52,213	37,376	31,526
	New borrowing:													
50	Current	92,700			36,800	16,400	6,900	29,600	3,000					
51	Noncurrent	40,000			40,000									
52	Total new borrowing	132,700			76,800	16,400	6,900	29,600	3,000					
	Payments on new current debt													
53	Principal	92,700								67,400		25,300		
54	Interest	2,874								2,137		737		
55	Total debt payments (line 53 + line 54)	95,574								69,537		26,037		
56	Ending cash balance (lines 49 + 52 − 55)	31,526	18,650	15,250	500	450	450	500	450	513	663	26,176	37,376	31,526
	Summary of debt outstanding													
57	Current (beginning of $0)	0	0	0	500		60,100	89,700	92,700	25,300	25,300	0	0	0
58	Noncurrent (beg. of $340,000)	340,000	340,000	334,000	374,000	374,000	374,000	354,000	354,000	348,000	348,000	348,000	348,000	348,000
59	Total debt outstanding	340,000	340,000	334,000	410,800	427,200	434,100	443,700	446,700	373,300	373,300	348,000	348,000	348,000

of new noncurrent debt for the new machinery. With the repayment capacity exhibited in this cash flow budget, a borrower should have no trouble obtaining necessary funds from a lender.

The large cash balances toward the end of the year suggest several alternatives that should be considered. First, arrangements might be made to invest this money for several months rather than have it sit idle in a checking account. Any interest earned would be additional income. Second, the excess cash could be used to make early payments on the noncurrent loans to reduce the interest charges. A third alternative relates to the $40,000 new noncurrent loan used to purchase machinery in March. Up to $29,000 of this amount could have been included as new current debt and been paid off by November. This alternative would reduce the ending noncurrent debt to $319,000, and the ending cash balance to slightly less than $1,000 after paying interest on the additional $29,000 of current debt.

This last alternative would require paying for all the new machinery in the year of purchase and would add considerable risk to the operation. If prices and yields were below expectations, there might not be sufficient cash available to repay all the new current debt. For this reason, using current debt to finance the purchase of long-lived items such as machinery and buildings is not recommended. A better practice would be to obtain a noncurrent loan and then make additional loan payments in years when extra cash is available.

This example assumed that all monthly cash shortages would be met by new current borrowing. However, other potential solutions to monthly cash shortages should be investigated first. For example, could some products be sold earlier than projected to cover negative cash balances? Could some expenditures, particularly large capital expenditures, be delayed until later in the year? Would it be possible to change the due date of some expenditures such as loan payments, insurance premiums, and family expenditures to a later date? These adjustments would reduce or eliminate the amount of new current borrowing needed for some months and reduce interest expense.

USES FOR A CASH FLOW BUDGET

The primary use of a cash flow budget is to project the timing and amount of new borrowing that the business will need during the year and the timing and amount of loan repayments. Other uses and advantages are:

1. A borrowing and debt repayment plan can be developed to fit an individual farm business. The budget can prevent excessive borrowing, and it shows how repaying debts as quickly as possible will save interest.
2. A cash flow budget may suggest ways to rearrange purchases and scheduled debt repayments to minimize borrowing. For example, capital expenditures and insurance premiums might be moved to months where a large cash inflow is expected.
3. A cash flow budget can combine both business and personal financial affairs into one complete plan.
4. A bank or other lending agency is better able to offer financial advice and spot potential weaknesses or strengths in the business based on a completed cash flow budget. A realistic line of credit can be established.
5. By planning ahead and knowing when cash will be available, managers can obtain discounts on input purchases by making a prompt cash payment.
6. A cash flow budget can also have a payoff in tax planning by pointing out the income tax effects of the timing of purchases, sales, and capital expenditures.
7. A cash flow budget can help spot an imbalance between current and noncurrent debt and suggest ways to improve the situation. For example, too much current debt relative to noncurrent debt can create cash flow problems.

> ### Box 13-1   Does Positive Cash Flow Mean a Profit?
>
> Not necessarily! While a positive net cash flow is certainly preferred to a negative one, it doesn't always mean a positive profit. There are too many differences between a cash flow budget and an income statement. A cash flow budget does not include noncash items such as inventory changes and depreciation which do affect profit. It does include cash items such as loan payments, new loans, and withdrawals for personal use which do not affect profit.
>
> A large positive net cash flow and a low profit can occur whenever there are receipts from sales of capital assets such as land and breeding livestock, cash gifts or inheritances, large nonfarm revenue, new loans, large depreciation, or inventory decreases. Negative cash flow and a good profit could result from substantial inventory increases, cash purchases of capital assets, and larger than usual debt payments and personal withdrawals.

Other uses and advantages of a cash flow budget could be listed, depending on the individual financial situation. The alert manager will find many ways to improve the planning and financial management of a business through the use of a cash flow budget.

MONITORING ACTUAL CASH FLOWS

A cash flow budget can also be used as part of a system for monitoring and controlling the cash flows during the year. The control function of management was discussed earlier, and one control method using a cash flow budget will be illustrated here. A form such as the one shown in Table 13-4 provides an organized way to compare budgeted cash flows with actual amounts.

Budgeted total annual cash flow for each item can be entered in the first column. The total budgeted cash flow to date is entered in the second column, and the actual cash flow to date in the third column. If the figures are updated monthly, or at least quarterly, it is easy to make a quick comparison of budgeted and actual cash flows.

This comparison is a means of monitoring and controlling cash flows, particularly cash outflows, throughout the year. Outflows that are exceeding budgeted amounts are quickly identified, and action can be taken to find and correct the causes. Estimates for the rest of the year can also be revised. Actual results at the end of the year can be used to make adjustments in budgeted values for next year, which will improve the accuracy of future budgeting efforts.

INVESTMENT ANALYSIS USING A CASH FLOW BUDGET

The discussion to this point has been on using a cash flow budget that covers the entire business for one year. There is another important use for a cash flow budget. Any major new capital investment, such as the purchase of land, machinery, or buildings, can have a large effect on cash flows, particularly if additional capital is borrowed to finance the purchase.

Borrowed capital requires principal and interest payments, which represent new cash outflows. The question to answer before making the new investment is: Will the new investment generate enough additional cash income to meet its additional cash requirements? In other words, is the investment financially feasible, as opposed to economically profitable? An example illustrates this use of a cash flow budget. Suppose a farmer is considering the purchase of an irrigation system to irrigate 120 acres. The information on page 211 has been gathered to use in completing the cash flow analysis:

TABLE 13-4	A Form for Monitoring Cash Flows

Name _____   Year _____

		Annual Total	Budget to Date	Actual to Date
1	Beginning cash balance			
2	Operating receipts: Grain and feed			
3	Feeder livestock			
4	Livestock products			
5	Other			
6				
7	Capital receipts: Breeding livestock			
8	Machinery and equipment			
9	Other			
10	Nonfarm income: Wages and salary			
11	Investments			
12	Other			
13	Total cash inflow (add lines 1–12)			
14	Operating expenses: Seed			
15	Fertilizer and lime			
16	Chemicals			
17	Other crop expenses			
18	Gas, oil, lubricants			
19	Hired labor			
20	Machine hire			
21	Feed and grain			
22	Feeder livestock			
23	Livestock expenses			
24	Repairs—machinery			
25	Repairs—buildings and improvements			
26	Cash rent			
27	Supplies			
28	Property taxes			
29	Insurance			
30	Utilities			
31	Auto and pickup (farm share)			
32	Other farm expenses			
33				
34				
35	Total cash operating expenses			

Cost of irrigation system	$60,000
Down payment in cash	18,000
Capital borrowed (@ 12% for 3 years)	42,000
Additional crop income from yield increase ($140 per acre)	16,800
Additional crop expenses from higher input levels ($30 per acre)	3,600
Irrigation expenses ($25 per acre)	3,000

Table 13-5 summarizes this information in a cash flow format. Because this is a long-lived capital investment, it is important to look at the cash flow over a number of years, rather than month by month for one year as was done for the whole-farm cash flow budget.

The new loan requires a principal payment of $14,000 each year of the three-year loan, plus interest on the unpaid balance. This obligation generates a large cash outflow requirement during the first three years, causing a negative net cash flow for these years. Once the loan is paid off in the third year, there is a positive net cash flow in the following years. This result is quite common when a large part of the purchase price is borrowed and the loan must be paid off in a relatively short time.

This investment is obviously going to cause a cash flow problem the first years. Does this mean the investment is a bad one? Not necessarily. The irrigation system should last for more than the five years shown in the table and will continue to generate a positive cash flow in later years. Over the total life of the irrigation system, there would be a positive net cash flow, perhaps a substantial one. The problem is how to get by the first three years.

At this point, the purchase of the irrigation system should be incorporated into a cash flow budget for the entire farm. This budget may show that other parts of the farm business are generating enough excess cash to meet the negative cash flow that would result from the purchase of the irrigation system. If not, one possibility would be to negotiate with the lender for a longer loan with smaller annual payments. This solution would help reduce the cash flow problem but would extend principal and interest payments over a longer time period and increase the total amount of interest paid.

This cash flow analysis did not include any effects the investment would have on income taxes. However, a new investment can have a large impact on income taxes to be paid, which

TABLE 13-5	Cash Flow Analysis for an Irrigation Investment

			Year		
	1	**2**	**3**	**4**	**5**
Cash inflow:					
Increase in crop income	$16,800	$16,800	$16,800	$16,800	$16,800
Cash outflow:					
Additional crop expenses	3,600	3,600	3,600	3,600	3,600
Irrigation expenses	3,000	3,000	3,000	3,000	3,000
Principal payments	14,000	14,000	14,000	0	0
Interest payments	5,040	3,360	1,680	0	0
Total cash outflow	25,640	23,960	22,280	6,600	6,600
Net cash flow	(−8,840)	(−7,160)	(−5,480)	10,200	10,200

in turn affects the net cash flows. Accuracy is improved if *after-tax* cash flows are used in this type of analysis. (See Chapter 16 for a discussion of income taxes.)

Land purchases often generate negative cash flows for a number of years, unless a substantial down payment is made. New livestock facilities may require significant cash for construction, breeding stock, and feed before any additional cash income is generated. Orchards and vineyards may take a number of years before they become productive and generate cash revenue. These examples illustrate the importance of analyzing a new investment with a cash flow budget. It is always better to know about and solve a cash flow problem ahead of time than to be faced with an unpleasant surprise. In some cases, there may be no way to solve a projected negative cash flow associated with a new investment. In this situation, the investment would be financially infeasible, because the cash flow requirements cannot be met.

SUMMARY

A cash flow budget is a summary of all cash inflows and outflows for a given future time period. The emphasis is on cash and all cash flows, regardless of type, source, or use. Both farm and personal cash needs are included on a cash flow budget. No noncash entries are included. A cash flow budget also includes the timing of cash inflows and outflows to show how well they match up. The final result provides an estimate of the borrowing needs of the business, its debt repayment capacity, and the timing of both.

A cash flow budget can also be used to do a financial feasibility analysis of a proposed investment. The investment may cause major changes in cash income and cash expenses, particularly if new borrowing is used to finance the purchase. A cash flow analysis concentrating on just the cash flows resulting from the purchase will indicate whether the investment will generate the cash needed to meet the cash outflows it causes. If not, the purchase needs to be reconsidered, or further analysis done, to be sure the cash needed is available from other parts of the business.

QUESTIONS FOR REVIEW AND FURTHER THOUGHT

1. Why is machinery depreciation not included on a cash flow budget?

2. Identify four sources of cash inflows that would not be included on an income statement but which would be on a cash flow budget. Why are they on the cash flow budget?

3. Identify four types of cash outflows that would not be included on an income statement but would be on a cash flow budget. Why are they on the cash flow budget?

4. Identify four noncash entries found on an income statement but not on a cash flow budget.

5. Discuss the truth or falsity of the following statement: A cash flow budget is used primarily to show the profit from the business.

6. Discuss how you would use a cash flow budget when applying for a farm business loan.

7. Assume you would like to make an investment in agricultural land in your area. Determine local land prices, cash rental rates, and the cash expenses you would have as the land owner. Construct a 5-year cash flow budget for the investment, assuming that 60 percent of the purchase price is borrowed with a 20-year loan at 10 percent interest. Would the investment be financially feasible without some additional source of cash inflow?

8. A feasible cash flow budget should project a positive cash balance for each month of the year, as well as for the entire year. When making adjustments to the budget to achieve this, should you begin with the annual cash flow or the monthly values? Why?

IMPROVING MANAGEMENT SKILLS

The basic management skills presented in Parts I, II, and III will now be extended and applied to specific areas of the farm or ranch business. In addition, some of the simplifying assumptions, such as knowing prices and production in advance, will be relaxed, and decision making in a risky environment will be discussed.

A beginning farmer or rancher must decide on what type of business organization to use, and this decision should be reviewed throughout the life of the business. Business organization refers to the legal and operational framework in which management decisions are made and carried out. The basic choices include sole proprietorship, partnership, corporation, and others. They are discussed in Chapter 14. Over time, changes in family size and involvement, goals, tax rules, and financial conditions will require adjustments to the organizational structure of the business.

Agriculture is a risky business. Few decisions can be made with complete information about the future. Usually, there is uncertainty about prices, yields, and other production and financial conditions. In Chapter 15, some methods are presented to improve decision making under uncertainty, and some techniques are discussed for reducing the risk inherent in agricultural production, marketing and finance.

A profitable farm or ranch business eventually will have to use part of its profits to meet its income tax obligations. This affects the cash flow of the business and slows the growth in equity. Chapter 16 discusses how management decisions affect income taxes and why a manager should consider the tax effects of all decisions made throughout the year. It also discusses some basic tax management strategies that may be useful to meet a goal of maximizing after-tax profit.

Many resources used in agriculture are long-term investments that tie up large amounts of capital. These investment decisions can affect the financial condition of the business for many years. Chapter 17 explains various investment analysis techniques, such as capital budgeting, net present value, and internal rate of return. Using these tools will help farm managers make better decisions about long-term capital investments and financing methods.

Many of the analysis tools presented in Part II can be used in the control function of management. These are consolidated in Chapter 18 and combined with some additional analytical measures to perform a whole-farm business analysis. Measures of solvency, liquidity, profitability, and efficiency can be compared against certain goals and standards. Areas of strength and weakness can be identified, and specific problems addressed to improve overall business performance and meet the operator's goals.

FORMS OF FARM BUSINESS ORGANIZATION

CHAPTER OBJECTIVES

1. To describe the sole proprietorship, partnership, corporation, limited liability company, and cooperative as primary forms of business organization available to farmers

2. To discuss the organization and characteristics of each form of business organization

3. To compare the advantages and disadvantages of each form of business organization

4. To show how income taxes are affected by the form of business organization

5. To summarize the important factors to be considered when selecting a form of business organization

6. To compare alternative arrangements for transferring the income, ownership, and management of a farm business from one generation to the next

Any business, including a farm or ranch, can be organized in a number of different legal and business formats. Many managers change the type of organization during the life of their business to better meet changing goals and objectives.

The three most common forms of business organization for farms and ranches are: (1) *sole (or individual) proprietorship,* (2) *partnership,* and (3) *corporation.* Each one has different legal and organizational characteristics, and is subject to different income tax regulations. According to data from the 1997 U.S. Census of Agriculture, nearly 86 percent of U.S. farms and ranches are organized as sole proprietorships, approximately 9 percent are partnerships, and 5 percent are corporations. In addition, some farms and ranches are organized as limited liability companies or cooperatives, or have operating agreements with other businesses.

The proper choice of organization depends on factors such as the size of the business, the number of people involved in it, the career stage and age of the primary operators, and the owners' desires for passing on their assets to their heirs. A final choice of business organization should be made only after analyzing all the possible long-run effects on the business and on the individuals involved.

LIFE CYCLE

Each farm business has a life cycle, with four stages: entry, growth, consolidation, and exit. These stages are depicted in Figure 14-1. The *entry* stage includes choosing farming as a career, selecting enterprises, acquiring and organizing the necessary resources, and establishing a financial base.

The *growth* stage involves expansion in the size of the business, typically through purchasing or leasing additional land or by increasing the scale of the livestock enterprises. This stage often uses debt capital to finance the expansion and requires good financial planning and risk management. It may even include merging with another operator. Following the growth stage, the emphasis usually turns toward *consolidation* of the operation. Debt reduction becomes a priority, and increased efficiency is preferred to increased size. Early planning and merging of the next generation into the business, however, may allow it to continue in the growth or consolidation stage for some time without showing a decline in size.

As the farm operator nears retirement, attention turns toward reducing risk, liquidating the business, or transferring property to the next

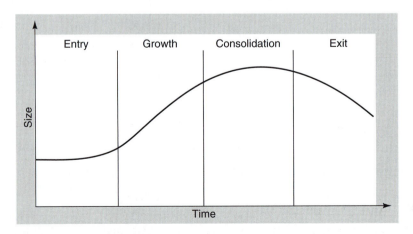

Figure 14-1   Illustration of the life cycle of a farm business.

Box 14-1   The Life Cycle of a Farm Business

Arthur and Beatrice Campbell entered farming by renting 240 acres from Arthur's uncle when they were in their early twenties. They also milked several cows that Beatrice had acquired during her 4-H years. Ten years later, they bought the land from Arthur's uncle and rented another 160 acres. They kept adding cows until the herd numbered 60 head.

When they reached their early fifties, they decided that their farming operation had grown large enough. They concentrated on paying off all the debts they had incurred, making their milking system more efficient, and improving the genetics of their herd.

None of their children had an interest in farming as a career, so they sold off the dairy herd when they reached the age of 60. A few years later, they sold their machinery and rented their land to a younger farmer who lives nearby. However, they continue to live on the farm and take an active role in their community.

---

generation. In this *exit* stage, business size may actually decline. The income tax consequences of liquidation or transfer must be considered, along with the need for adequate retirement income. Another concern is to provide equitable treatment to all the children who may choose farming as a career, as well as to those who pursue other interests.

The life-cycle stage in which the farm is currently operating is an important consideration in selecting the form of business organization. Total capital invested, size of debt, income tax considerations, the goals of the owners and operators, and other factors are likely to be different in each stage. Farmers and ranchers may find that a different form of business organization is appropriate for different stages in the life cycle. However, careful and early thought must be given to how the change from one type of organization to another will be handled.

SOLE PROPRIETORSHIP

A sole proprietorship is easy to form and easy to operate, which accounts for much of its popularity.

Organization and Characteristics

In a sole proprietorship, the owner both owns and manages the business, assumes all the risks, and receives all profits and losses. Its distinguishing characteristic is the single owner, who acquires and organizes the necessary resources, provides the management, and is solely responsible for the success or failure of the business, as well as for servicing all business debts. Often the sole proprietorship could be more accurately called a family proprietorship, because on many farms and ranches, the husband, wife, and children are all involved in ownership, labor, and management.

A sole proprietorship is established simply by starting to operate the business. No special legal procedures, permits, or licenses are required. A sole proprietorship is not limited in size by either the amount of inputs that can be used or the amount of commodities produced. The business can be as large or small as the owner desires. There can be any number of employees. Additional management may be hired, and property may even be co-owned with others. A sole proprietorship does not necessarily need to own any assets; one can exist even when all land and machinery are leased.

Advantages

The advantages of a sole proprietorship are its simplicity and the freedom the owner has in operating the business. No other operator or owner needs to be consulted when a management decision is made. The owner is free to organize and operate the business in any legal manner, and all business profits and losses belong to the sole proprietor.

A sole proprietorship is also flexible. The manager can quickly make decisions regarding investments, purchases, sales, enterprise combinations, and input levels, based solely on his or her best judgment. Assets can be quickly purchased or sold, money borrowed, or the business even liquidated if necessary, although the latter may require the concurrence of a landlord or lender.

Disadvantages

The management freedom inherent in a sole proprietorship also implies several responsibilities. Owners of sole proprietorships are personally liable for any legal difficulties and debts related to the business. Creditors have the legal right to attach not only the assets of the business but the personal assets of the owner to fulfill any unpaid financial obligations. This feature of a sole proprietorship can be an important disadvantage for a large, heavily financed business where the owner has substantial personal and nonfarm assets. Business failure can result in these assets being acquired by creditors to pay the debts of the sole proprietorship.

The size of a sole proprietorship is limited by the capital available to the single owner. If only a limited amount is available, the business may be too small to realize many economies of size, making it difficult to compete with larger and more efficient farms. At the other extreme, the management abilities and time of the single owner may be insufficient for a very large business, making it difficult to become an expert in any one area. Thus, a large sole proprietorship may need to hire additional management expertise or consulting time.

Another disadvantage of a sole proprietorship is a lack of business continuity. It is difficult to bring children into the business on any basis other than as employees or tenants. Death of the owner also means the business may have to be liquidated or reorganized under new ownership. This can be time consuming and costly, resulting in a smaller inheritance and less income for the heirs during the transition period.

Income Taxes

The owner of a business organized as a sole proprietorship pays income taxes on any business profit at the tax rates in effect for individual or joint returns. Business profits and capital gains are added to any other taxable income earned to determine the individual's total taxable income. Earned income is subject to self-employment and Medicare tax.

JOINT VENTURES

While a sole proprietorship offers the manager maximum flexibility and independence, other forms of business organization are possible. They allow two or more operators to combine their respective abilities and assets to achieve levels of efficiency or other goals that might be unattainable on their own.

Such businesses are called *joint ventures*. Of the several types of joint ventures, those most often used in agriculture include:

1. Operating agreements
2. Partnerships
3. Corporations
4. Limited liability companies
5. Cooperatives

Each type of joint venture has its own unique characteristics related to property ownership, distribution of income, taxation, continuity, liability, and formal organization. In many cases, only part of the total farm business is included in the joint venture. Sometimes the joint venture affects only one enterprise. In other cases, some assets, such as land, are excluded to protect them from the unlimited liability of the joint venture, or to make ownership shares more equal.

Particular care should be taken to distribute the income earned by a joint venture in an equitable manner. Businesses with an operating agreement may simply divide gross income, as will be illustrated later. In more complex businesses, owners of assets that are used but not owned by the joint venture may be paid a fair market rental rate. People who lend money to the business receive interest payments. People who contribute labor should be paid a fair wage, even if they are also owners of the business. And, finally, profits earned by the business are distributed on the basis of ownership share, patronage rate, or some other mutually agreeable criterion.

It is important to distinguish among payments made as returns to rented assets, borrowed capital, labor, and equity capital, even though the recipients may be some or all of the same people. Figure 14-2 illustrates this relation.

The primary advantages of a joint venture over a sole proprietorship involve combining capital and management. Pooling the capital of the members of the joint venture allows a larger business to be formed, which can be more efficient than two or more smaller businesses. It can also increase the amount of credit available, allowing further increases in business size. The total supply of management and labor is also increased by pooling the capabilities of all the members. Management efforts can be divided, with each person specializing in one area of the business, such as crops, livestock, marketing, or accounting. It is also easier for one operator to be absent from the business when another operator is available to fill in.

OPERATING AGREEMENTS

Sometimes two or more sole proprietors carry on some farming activities jointly while maintaining individual ownership of their own resources. Such an activity is often called an *operating agreement*. They tend to be informal, limited arrangements.

In most operating agreements, all parties pay the costs related to ownership of their own assets, such as property taxes, insurance, maintenance, and interest on loans. Operating expenses such as the cost of seed, fertilizer, veterinarian fees, or utilities may be shared among the parties in a fixed proportion, often in the same proportion as the fixed costs. In other cases, one party or another may pay for all of certain operating costs, such as fuel or feed, as a matter of convenience. In either case, the general principle for an operating agreement is to share income in the same proportion as total resources are contributed, including both fixed assets and operating costs.

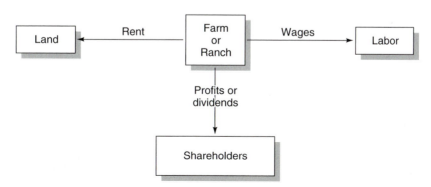

Figure 14-2   Distribution of income from a joint venture.

An enterprise budget can be a useful tool for comparing the resource contributions by each party in an operating agreement. The example in Table 14-1 is based on the cow/calf enterprise budget found in Table 10-3. In this situation, party A owns all the pasture, buildings, and fences for the cow herd, as well as all the livestock. Party A also pays the cost of maintaining the pastures and repairing fences and other livestock facilities.

Party B will provide all the feed and labor for the cow/calf enterprise and pay all other variable costs. The question, then, is how should the income from the cattle sold be divided? Table 14-1 shows that party A is contributing 52 percent of the total costs, and party B is contributing 48 percent. Income can be divided in the same proportion, or for convenience, equally to each person, since the results are close to a 50/50 division of costs.

An alternative would be to sum the value of the unpaid resources contributed by each party (land, facilities, livestock, machinery, and labor, in this example) and share all cash operating expenses and income in the same proportion. In the example, party A's annual contributions of livestock, facilities, machinery, and land are valued at a total of $196.50 per cow unit, while the labor contribution of Party B is valued at $30.00. Party A is providing 87 percent of the fixed resources, and party B, 13 percent. They could agree to divide all the remaining costs, as well as gross income, in the same proportion. This may not be convenient however, unless a

TABLE 14-1	Example Budget for a Cow/Calf Joint Enterprise (One Head)		
Item	Value	Party A	Party B
Operating expenses			
Hay	$ 90.00		$ 90.00
Grain and supplements	$ 56.00		$ 56.00
Salt and minerals	$ 2.40		$ 2.40
Pasture maintenance	$ 22.50	$ 22.50	
Veterinary and health	$ 10.00		$ 10.00
Livestock facilities	$ 8.00	$ 8.00	
Machinery and equipment	$ 5.00		$ 5.00
Breeding expense	$ 5.00	$ 5.00	
Labor	$ 30.00		$ 30.00
Miscellaneous	$ 10.00		$ 10.00
Interest on variable costs	$ 11.95	$ 1.78	$ 10.17
Ownership expenses			
Interest on breeding herd	$ 75.00	$ 75.00	
Livestock facilities			
Depreciation and interest	$ 10.00	$ 10.00	
Machinery and equipment			
Depreciation and interest	$ 6.50	$ 6.50	
Land charge	$105.00	$105.00	
Total cost	$447.35	$233.78	$213.57
Percent contribution	100%	52%	48%

joint operating bank account is established. Moreover, the arrangement then begins to look like a partnership, which may be neither intended nor desirable.

Two cautions should be observed when valuing resource contributions. The opportunity costs for all unpaid resources, such as labor and capital assets, must be included along with cash costs, and all costs must be calculated for the same time period, usually one year. It would not be correct, for example, to value land at its full, current market value when labor and operating costs are included for only one year. An annual interest charge or rental value for the land should be used instead.

PARTNERSHIPS

A *partnership* is an association of two or more persons who share the ownership of a business to be conducted for profit. Partnerships can be organized to last for a brief time or for a long duration.

Two types of partnerships are recognized in most states. The *general partnership* is the most common, but the *limited partnership* is used for some situations. Both types have the same characteristics, with one exception. The limited partnership must have at least one general partner, but can have any number of limited partners. Limited partners cannot participate in the management of the partnership business, and in exchange, their financial liability for partnership debts and obligations is limited to their actual investment in the partnership. The liability of general partners, however, can extend even to their personal assets.

Limited partnerships are most often used for businesses such as real estate development and cattle feeding, where investors want to limit their financial liability and do not wish to be involved in management. Most farm and ranch operating partnerships are general partnerships. For this reason, the discussion in the remainder of this section will apply only to general partnerships.

Organization and Characteristics

There are many possible patterns and variations in partnership arrangements. However, the partnership form of business organization has three basic characteristics:

1. A sharing of business profits and losses
2. Shared control of property, with possible shared ownership of some property
3. Shared management of the business

The exact sharing arrangement for each of these characteristics is flexible and should be outlined in the partnership agreement.

Oral partnership agreements are legal in most states but are not recommended. Important points can be overlooked, and fading memories over time on the details of an oral agreement can create friction between the partners. Problems can also be encountered when filing the partnership income tax return if arrangements are not well documented.

A written partnership agreement is of particular importance if the profits will not be divided equally. The courts and the Uniform Partnership Act, which is a set of regulations enacted by most states to govern the operation of partnerships, assume an equal share in everything, including management decisions, unless there is a written agreement specifying a different arrangement. For any situation not covered in the written agreement, the regulations in the state partnership act apply.

The written agreement should cover at least the following points:

1. *Management* State who is responsible for which management decisions, and how they will be made.
2. *Property ownership and contribution* List the property that each partner will contribute to the partnership and describe how it will be owned.
3. *Share of profit and losses* Carefully describe the method for calculating profits and losses, and the share going to each partner, particularly if there is an unequal division.

Box 14-2   **An Example of a Partnership**

McDonald and Garcia formed a partnership for the purpose of finishing feeder pigs. They agreed to each put in $30,000 of operating capital and the feed they had on hand. McDonald will put up the finishing building and pay for the feeders, waterers, fans, and other equipment. Garcia will contribute a tractor and machinery to handle manure. The table below summarizes the value of their contributions.

Assets contributed	Total	McDonald	Garcia
Operating capital	$60,000	$30,000	$30,000
Current feed inventory	25,600	9,500	16,100
Livestock equipment	21,500	20,200	1,300
Tractor and machinery	27,000	0	27,000
Building and manure storage	85,000	85,000	0
Total	$219,100	$144,700	$74,400
Ownership share	100%	66%	34%

After looking at the values, they agreed to divide net income and losses from the partnership in the ratio of two-thirds to McDonald and one-third to Garcia. However, Garcia will spend about 25 hours per week operating the finishing facility, and McDonald will work only 5 hours per week, so they agreed that the partnership would pay each of them a fair wage for their labor before net income is calculated. They also agreed to leave one-half of the net income in the partnership the first year, to increase their operating capital.

---

4. *Records*   Records are important for the division of profits and for maintaining an inventory of assets and their ownership. Designate who will keep what records as part of the agreement.
5. *Taxation*   Include a detailed account of the tax basis of property owned and controlled by the partnership, and copies of the partnership information tax returns.
6. *Termination*   State the date that the partnership will be terminated, if one is known or can be determined.
7. *Dissolution*   The termination of the partnership on either a voluntary or involuntary basis requires a division of partnership assets. The method for making this division should be described to prevent disagreements and an unfair division.

Partners may contribute land, capital, labor, management, and other assets to the partnership capital account. Compensation for labor contributed or assets rented to the partnership should be paid first, because these may not be in the same proportions as the ownership. The remaining profits generally are divided in proportion to the ownership of the partnership, which is usually based on the initial value of the assets contributed to the business. Some partners prefer to share partnership profits equally, however, regardless of contributions.

Property may be owned by the partnership, or the partners may retain ownership of their individual property and rent it to the partnership. When the partnership itself owns property, any partner may sell or dispose of any asset without the consent and permission of the other partners.

This aspect of a partnership suggests that retaining individual ownership of some assets and renting them to the partnership may be desirable.

A partnership can be terminated in a number of ways. The partnership agreement may specify a termination date. If not, a partnership will terminate upon the incapacitation or death of a partner, bankruptcy, or by mutual agreement between the partners. Termination upon the death of a partner can be prevented by placing provisions in the written agreement that allow the deceased partner's share to pass to the estate and hence to the legal heirs.

In addition to the three basic characteristics mentioned earlier, factors that indicate a particular business arrangement is legally a partnership include:

1. Joint ownership of assets in the partnership capital account
2. Operation under a firm name
3. A joint bank account
4. A single set of business records
5. Management participation by all parties
6. Sharing of profits and losses

Joint ownership of property does not by itself imply a partnership. However, a business arrangement with all or most of these characteristics may be a legal partnership, even though a partnership was not intended. These characteristics should be kept in mind for landlord-tenant arrangements, for example. A partnership can be created by a leasing arrangement based on shares, unless steps are taken to avoid most of the conditions mentioned above.

Advantages

A partnership is easier and cheaper to form than a corporation. It may require more records than a sole proprietorship, but not as many as a corporation. While each partner may lose some individual freedom in making management decisions, a carefully written agreement can maintain most of this freedom.

A partnership is a flexible form of business organization, in which many types of arrangements can be accommodated and included in a written agreement. It fits situations such as when parents desire to bring children and their spouses into the business. The children may contribute only labor to the partnership initially, but the partnership agreement can be modified over time to allow for their increasing contributions of management and capital.

Disadvantages

The unlimited liability of each general partner is an important disadvantage of a partnership. A partner cannot be held personally responsible for the *personal* debts of the other partners. However, each partner can be held personally and individually responsible for any lawsuits and financial obligations arising from the operation of the partnership. If the partnership does not have sufficient assets to cover its legal and financial obligations, creditors can bring suit against all partners individually to collect any money due them. In other words, a partner's personal assets can be claimed by a creditor to pay partnership debts.

This disadvantage takes on a special significance, considering that any partner, acting individually, can act for the partnership in legal and financial transactions. For this reason, if for no other, it is important to know and trust your partners and to have the procedures for making management decisions included in the partnership agreement. Too many partners or an unstructured management system can easily create problems.

Like a sole proprietorship, a partnership has the disadvantage of poor business continuity. It can be unexpectedly terminated by the death of one partner or a disagreement among the partners. Dissolution of any business is generally time consuming and costly, particularly when it is caused by the death of a close friend and partner or by a disagreement that results in bad feelings among the partners. The required sharing of management decision making and the loss of some personal freedom in a partnership are always potential sources of conflict between the partners.

Laws governing the formation and taxation of partnerships are less detailed than for corporations. Unfortunately, this allows many farm partnerships to operate with minimal records and little documentation of how resources were contributed and how income has been divided. This can make a fair dissolution of the partnership very difficult.

Income Taxes

A partnership does not pay any income taxes directly. Instead it files an information income tax return reporting the income and expenses for the partnership. Income from the partnership is then reported by the individual partners on their personal tax returns, based on their respective shares of the partnership income as stated in the partnership agreement.

In a 50/50 partnership, for example, each partner reports one-half of the partnership income, capital gains, expenses, depreciation, losses, and so forth. These items are included with any other income the partner has, to determine the total income tax liability. Partnership income, therefore, is taxed at individual tax rates, with the exact rate depending on the amount of partnership income and other income earned by each partner/taxpayer.

Partnership Operation

Because a partnership is usually formed for a relatively long period, contributions to the partnership can be valued on the basis of current market values for the assets contributed by each party at the time the partnership was started. All operating expenses should be paid by the partnership, usually from a separate partnership account. If either or both partners contribute labor, they should be paid a representative and fair wage by the partnership.

In some partnerships, major assets such as land and buildings are not included in partnership property. Instead, the partnership pays rent to the owners of these assets (who may also be the partners) for their use. In such cases, the rented assets are not included when calculating the relative ownership shares of the partnership.

Net income may either be paid out or left in the partnership account to increase business equity. As long as all withdrawals (not including wages or rent) are made in the same proportion as the ownership of the partnership, the ownership shares will not change. However, if one partner leaves more profits in the business, such as by retaining more breeding stock, or contributes additional assets, the ownership percentages will change. Sometimes this is done as a way to move toward more equal ownership of the partnership.

When a partnership is liquidated, the proceeds should be distributed in the same proportion as the ownership. For this reason, it is important to keep detailed and accurate records of the property utilized by a partnership, and of how wages, rent, and net income are distributed to the partners.

CORPORATIONS

A *corporation* is a separate legal entity that must be formed and operated in accordance with the laws of the state in which it is organized. It is a legal "person," separate and apart from its owners, managers, and employees. This separation of the business entity from its owners distinguishes a corporation from the other forms of business organization. As a separate business entity, a corporation can own property, borrow money, enter into contracts, sue, and be sued. It has most of the basic legal rights and duties of an individual.

The number of farm and ranch corporations, though still small, has been increasing. The majority are classified as family farm corporations with a relatively small number of related shareholders.

Organization and Characteristics

The laws affecting the formation and operation of a corporation vary somewhat from state to state. In addition, several states have laws that prevent a corporation from engaging in farming or ranching, or that place special restrictions on farm and ranch corporations. For this and other reasons, competent legal advice should be sought before attempting to form a corporation.

While state laws do vary, there are some basic steps that will generally apply to forming a farm corporation. These are:

1. The incorporators file a preliminary application with the appropriate state official. This step may include reserving a name for the corporation.
2. The incorporators draft a pre-incorporation agreement outlining the major rights and duties of the parties after the corporation is formed.
3. The articles of incorporation are prepared and filed with the proper state office.
4. The incorporators turn property and/or cash over to the corporation in exchange for shares of stock representing their ownership share of the corporation.
5. The shareholders meet to organize the business and elect directors.
6. The directors meet to elect officers, adopt bylaws, and begin business in the name of the corporation.

Three groups of individuals are involved in a farming corporation: shareholders, directors, and officers. The shareholders own the corporation. Stock certificates are issued to them in exchange for property or cash transferred to the corporation. The number of shareholders may be as few as one in some states, while three is the minimum number in several other states. As owners, the shareholders have the right to direct the affairs of the corporation, which is done through the elected directors and at annual meetings. Each shareholder has one vote for each share of voting stock owned. Therefore, any shareholder with 51 percent or more of the outstanding voting stock has majority control over the business affairs of the corporation. In some corporations, contributions of capital are exchanged for bonds instead of stock. Bonds carry a fixed return but no voting privileges.

The directors are elected by the shareholders at each annual meeting, and they hold office for the following year. They are responsible to the shareholders for the management of the business. The number of directors is normally fixed by the articles of incorporation. Meetings of the directors are held to conduct the business affairs of the corporation and to set broad management policy to be carried out by the officers.

The officers of a corporation are elected by the board of directors and may be removed by them. They are responsible for the day-to-day operation of the business within the guidelines established by the board. The officers' authority flows from the board of directors, to whom they are ultimately responsible. A corporation president may sign certain contracts, borrow money, and perform other duties without board approval but will normally need board approval before committing the corporation to large financial transactions or performing certain other acts.

In many small family farm corporations, the shareholders, directors, and officers are all the same individuals. To an outsider, the business may appear to be operated like a sole proprietorship or partnership. Even the directors' meetings may be held informally around a kitchen table, but a set of minutes must be kept for each official meeting.

Two types of corporations are recognized for federal income tax purposes. The first is the regular corporation, which is also called a C corporation. A second type of corporation is the S corporation, sometimes referred to as a tax-option corporation. Both types have many of the same characteristics. However, there are certain restrictions on the formation of S corporations. Some of the more important restrictions include:

1. There must be no more than 75 shareholders. A husband and wife are considered one shareholder, even if both own stock.

2. Shareholders are limited to individuals (foreign citizens are excluded), estates, and certain types of trusts. Other corporations cannot own stock in an S corporation.
3. There may be only one class of stock, but the voting rights may be different for some shareholders.
4. All shareholders must initially consent to operating as an S corporation.

There are also certain limitations on the types and amounts of income that a corporation can receive from specified sources and still maintain the S corporation status. The special taxation of an S corporation will be discussed in a later section.

Advantages

Corporations provide limited liability for all the shareholder/owners. They are legally responsible only to the extent of the capital they have invested in the corporation. Personal assets of the shareholders cannot be attached by creditors to meet the financial obligations of the corporation. This advantage may be negated if a corporation officer is required to personally co-sign a note for corporation borrowing. In this case, the officer can be held personally responsible for the debt if the corporation cannot meet its responsibilities.

A corporation, like a partnership, provides a means for several individuals to pool their resources and management. The resulting business, with a larger size and the possibility of specialized management, can provide greater efficiency than two or more smaller businesses. Credit may also be easier to obtain because of the business continuity advantage of a corporation. The business is not terminated by the death of a shareholder, because the shares simply pass to the heirs and the business continues. However, a plan for management continuity should exist by having more than one person involved in management and capable of taking over responsibility.

A corporation provides a convenient way to divide and transfer business ownership. Shares of stock can be easily purchased, sold, or given as gifts without actually transferring title to specific parcels of land or other assets. Transferring shares does not disrupt or reduce the size of the business and is a convenient way for a retiring farmer to gradually transfer part of an ongoing business to the next generation.

There can be income tax advantages to incorporation, depending on the size of the business, how it is organized, and the income level of the shareholders. Any C corporation is a separate taxpaying entity, and as such, is subject to different tax rates than individuals. Current personal and corporate tax rates are shown in Table 14-2.

One advantage of a C corporation is that the owners are usually employees as well. This allows the tax deductibility of certain fringe benefits provided to the shareholder/employees, such as premiums for health, accident, and life insurance. It also is easier to allocate income among individuals by setting salaries, rents, and dividends.

Disadvantages

Corporations are more costly to form and maintain than sole proprietorships and partnerships. Certain legal fees are necessary when organizing a corporation, and legal advice will be needed on a continuing basis to ensure compliance with state regulations. An accountant may also be needed during the formation period and throughout the life of the corporation to handle financial records and tax-related matters. Most states require various fees when filing the articles of incorporation, and some type of annual operating fee or tax on corporations, which are not assessed on the other forms of business organization.

Doing business as a corporation requires that shareholders and directors meetings are held, minutes are kept of directors meetings, and annual reports are filed with the state. However, if forming a corporation results in better business and financial records being kept, this might be viewed as an advantage rather than a disadvantage, because

TABLE 14-2	Personal and Corporate Income Tax Rates (2003)

Personal			Corporate	
Taxable Income*		Marginal tax rate (%)	Taxable Income	Marginal tax rate (%)
Single	Married filing joint return			
$0 to $6,000	$0 to $12,000	10	Up to $50,000	15
$6,000 to $28,400	$12,000 to $47,450	15	$50,000 to $75,000	25
$28,400 to $68,800	$47,450 to $114,650	27	$75,000 to $100,000	34
$68,800 to $143,500	$114,650 to $174,700	30	$100,000 to $335,000	39
$143,500 to $311,950	$174,700 to $311,950	35	Over $335,000	34 to 38
Over $311,950	Over $311,950	38.6		

*The personal tax brackets will be increased in future years to adjust for the effects of inflation.

better information for making management decisions will be available.

Income Taxes

Another disadvantage of a C corporation is the potential double taxation of income. After the corporation pays tax on its taxable income, any after-tax income distributed to the shareholders as dividends is considered taxable income to the shareholders. It is taxed at their applicable individual rates. Many small farm corporations avoid some of this double taxation by distributing most of the corporation income to the shareholders as rent, wages, salaries, and bonuses. These items are tax-deductible expenses for the corporation but taxable income for the shareholders/employees. However, any wages and salaries paid must be reasonable for bona fide work performed for the corporation, and not a means to avoid double taxation.

The S corporations do not pay income tax themselves but are taxed like a partnership; hence, the name "tax-option" corporation. The corporation files an information tax return, but the shareholders report their share of the corporation income, expenses, and capital gains according to the proportion of the total outstanding stock they own. This income (or loss) is in-

cluded with the shareholders' other income, and tax is paid on the total based on the applicable individual rates. The income tax treatment of an S corporation avoids the dividend double-taxation problem of a C corporation.

Only a partial treatment of the income tax regulations that are applicable to both types of corporations is included here. There are many rules pertaining to special situations and special types of income and expenses. All applicable tax regulations should be reviewed with a qualified tax consultant before the corporate form of business organization is selected.

Corporation Operation

A corporation can be formed and operated in a manner similar to a partnership. The shares of the corporation are divided in the same proportion as the original contributions of equity capital. It should be noted that both partnerships and corporations can receive contributions of debt from the partners, as well as assets.

The number of shares to be issued is an arbitrary decision by the stockholders. The initial value of each share is found by dividing the beginning equity of the corporation by the number of shares to be issued. As with a partnership, the corporation can pay salaries for labor contributed

Box 14-3   An Example of a Corporation

Suppose McDonald and Garcia had decided to form a corporation for finishing feeder pigs instead of a partnership. The assets they contributed and their value would be exactly the same as in the partnership example. These are summarized in the table below.

Assets contributed	Total	McDonald	Garcia
Operating capital	$ 60,000	$ 30,000	$30,000
Current feed inventory	25,600	9,500	16,100
Livestock equipment	21,500	20,200	1,300
Tractor and machinery	27,000	0	27,000
Building and manure storage	85,000	85,000	0
Total	$219,100	$144,700	$74,400
Liabilities contributed			
Term loan on buildings	56,000	56,000	0
Equity contributed	$163,100	$ 88,700	$74,400
Number of shares @ $100	1,631	887	744

McDonald had to borrow $56,000 to construct the finishing building. In this case, the shareholders agreed to have the corporation take over this debt and make the principal and interest payments as they come due.

Including the $56,000 loan in the capitalization of the corporation lowered the initial equity of the business to $163,100. McDonald and Garcia agreed to issue 1,631 shares of stock with a book value of $100 each. These shares are divided in proportion to the initial equity contributions: 887 shares to McDonald, and 744 shares to Garcia.

The corporation will pay each of them wages based on their contribution of labor. Any dividends paid out will be divided in proportion to the number of shares owned.

by shareholders, as well as rent for the use of assets not owned by the corporation. Distributions of net income are in the form of dividends.

Table 14-3 summarizes the important features of each of the three main forms of business organization. The advantages and disadvantages of each feature should be evaluated carefully before one is selected.

LIMITED LIABILITY COMPANIES

A *limited liability company* (LLC) is a relatively new type of business organization that closely resembles a partnership, but offers its members the advantage of limited liability. This means that creditors or other claimants can pursue the assets of the LLC to satisfy debts or other obligations, but cannot pursue personal or business assets owned individually by the members of the LLC. This is a significant advantage to potential investors.

Limited liability companies can include any number of members, all of whom can participate in management. Ownership is distributed according to the fair market value of assets contributed, just as in a partnership. Likewise, net farm income from an LLC is passed to the individual

TABLE 14-3	Comparison of Forms of Farm Business Organization		
Category	**Sole proprietor**	**Partnership**	**Corporation**
Ownership	Single individual	Two or more individuals	A separate legal entity that is owned by its shareholders
Life of business	Terminates on death	Agreed-on term, or terminates at death of a partner	Forever unless fixed by agreement; in case of death, stock passes to heirs
Liability	Proprietor is liable	Each partner is liable for all partnership obligations, even to personal assets (except limited partners)	Shareholders are not liable for corporate obligations; in some cases, individual stockholders may be asked to co-sign corporation notes
Source of capital	Personal investments, loans	Partnership contributions, loans	Shareholder's contributions, sale of stock, sale of bonds, and loans
Management decisions	Proprietor	Agreement of partners	Shareholders elect directors who manage the business or hire a manager
Income taxes	Business income is combined with other income on individual tax return	Partnership files IRS information report; each partner's share of partnership income is added to her or his individual taxable income	Regular C corporation: Corporation files a tax return and pays income tax; salaries to shareholder employees are deductible; shareholders pay tax on dividends received
			Tax-option (S) corporation: Shareholders report their share of income, operating loss, and long term capital gain on individual returns; IRS information report filed by corporation

tax returns of the members in proportion to their shares of ownership. An LLC is chartered by state law, so organizational requirements vary. However, it is still a good idea to keep a careful record of contributions and distributions of assets and income, and to write down the agreed-on rules of operation.

Unlike a corporation, the LLC cannot deduct the cost of fringe benefits, such as insurance plans or use of vehicles, that are provided to employees who are also members of the LLC. Neither does it automatically continue in the event of the death of a member.

Limited liability companies are relatively new, so some questions about their legal nature are still untested. However, they offer an attractive alternative for farm families who desire the simplicity and flexibility of a partnership combined with the limited financial liability offered by a corporation.

COOPERATIVES

Farmers have used *cooperatives* for many years as a means of obtaining inputs and services or for marketing products jointly. In some countries, cooperatives for the purpose of farm production have been formed by groups of small-scale landholders or farm workers to gain efficiencies in production.

In the United States, farm cooperatives that actually produce grain or livestock have been rare, but some states have passed laws that encourage the formation of small-scale cooperatives for this purpose. Often they are made up of independent farmers who wish to carry out one particular operation jointly. Examples include sow cooperatives that produce feeder pigs for their members to finish, or a cooperative that grows high-quality seed for its members.

Cooperatives are actually a special type of corporation. They require articles, by-laws, and detailed records. Members who contribute capital enjoy limited liability on those contributions. Net income is passed on to members before it is subject to income tax. Cooperatives can also provide tax deductible benefits to owner/members.

Unlike other corporations, cooperatives can pay a maximum return of 8 percent annually to shareholders. Remaining net income is distributed to members as "patronage refunds," based on how much business each member does with the cooperative, rather than on shares of ownership. Control of a cooperative also differs from the other forms of business organizations, in that all members have one vote each when it comes to making decisions, regardless of how much of the cooperative they own. Most agricultural cooperatives limit membership to active farmers.

In recent years a new form of farmer cooperative, known as a "closed" or "new generation" cooperative has become popular. Members must contribute a significant amount of equity capital to join, and agree to sell a certain volume of production to the cooperative on a fixed schedule. Memberships can be bought and sold. These new cooperatives have been formed mostly for value-added processing, but also include livestock finishing and egg-laying operations.

Perhaps the key factor in making a farming cooperative successful is a true desire to cooperate. Each member must perceive that working together will help obtain benefits that outweigh the necessity of giving up some degree of managerial independence.

TRANSFERRING THE FARM BUSINESS

At the beginning of this chapter, the life cycle of a farm business was illustrated (Fig. 14-1). If the exit stage of one operator coincides with the entry stage of the next operator, the transfer process may be relatively simple. Livestock, equipment, and machinery may be sold or leased to the new operator or dispersed by an auction. Land may be sold outright, sold on an installment contract, or rented to the new operator. The outward size and structure of the business may change very little in the process.

In many family farming situations, however, the next generation is ready to enter the business while the current generation is still in the growth or consolidation stage. This brings up several important questions.

1. Is the business large enough to productively employ another person or family? If the labor supply will be decreased through retirement of the present operators or a reduction in the amount of labor hired, then a new operator can be utilized efficiently. If not, the business may have to expand to provide more employment.

2. Is the business profitable enough to support another operator? Simply adding more labor does not necessarily produce more net income. If an additional source of income such as social security payments will become available to the older generation, then income from the farm can be more easily diverted to the new operator.

If adding another operator will require expanding the business, a detailed whole-farm budget should be completed. Liquidity, as well as profitability, should be analyzed, especially if additional debt will be used to finance the expansion. The net cash flow that will be available in a typical year should be projected for each person, as well as for the business as a whole, to avoid unexpected financial problems.

3. Can management responsibilities be shared? If the people involved do not have compatible personalities and mutual goals, then even a profitable business may not provide a satisfying career. Farmers and ranchers who are accustomed to working alone and making decisions independently may find it very difficult to accommodate a partner. Parents and children may find it especially difficult to work as equal partners in management.

Key Areas to Transfer

Three key areas of a farm business must be transferred: income, ownership, and management. Typically they are not all transferred at the same time, however.

Income can be transferred initially by simply paying the new operator a wage, which could also include some type of bonus or profit sharing. As the younger individual acquires more assets, part of the income can be based on a return on investment as well.

Ownership can be transferred by allowing the younger generation to gradually acquire property, such as by saving breeding stock or investing in machinery. Longer-life assets can be sold on contract or by outright sale. Property can also be gifted to the eventual heirs, but if annual giving exceeds certain limits, gift tax may have to be paid. Gifting of assets also involves finding a way to treat nonfarm heirs fairly. If the farm business is organized as a corporation, individual shares of stock can be sold or given away over time. This is much easier than transferring specific assets or ownership shares of assets. Finally, careful estate planning will assure an orderly transfer of property upon the death of the owner.

Management may be the most difficult part of a farming operation to transfer. The new operator may be given responsibility for one particular enterprise or for a certain management area, such as feeding, breeding, or record keeping. Allowing younger operators to rent additional land or produce a group of livestock on their own while using the farm's machinery or facilities helps them learn management skills without putting the entire operation at risk.

The older generation must understand that there are many different ways to carry out farming enterprises. Knowing when to offer advice and when to keep silent will help make the management transfer occur smoothly.

Stages in Transferring the Business

The specific arrangements chosen for transferring income, ownership, and management depend on the type of organization that is ultimately desired and how quickly the family wants to complete the transfer. A testing stage of one to five years is recommended, during which the entering operator is employed and receives a salary and bonus or incentive, and begins to own or rent some additional assets. Forms of business organization and acquisition of assets that would be difficult to liquidate should be avoided until both parties determine their ultimate goals, and whether or not they can work together. Following the testing stage, at least three alternative types of arrangements are possible, as illustrated in Figure 14-3.

1. The *spin-off* option involves the separation of the operators into their own individual operations. A spin-off works best when both operators wish to be financially and operationally independent, or when expansion possibilities for the original business are limited. The separate operations may still trade labor and the use of equipment, or even own some assets together, in order to realize some economies of size.

2. The *takeover* option occurs when the older generation phases out of active labor and management, usually to retire or enter another occupation. Expansion of the business may not be necessary. Transfer can take place by renting or selling the farm or ranch to the younger generation. A detailed budget should be developed to

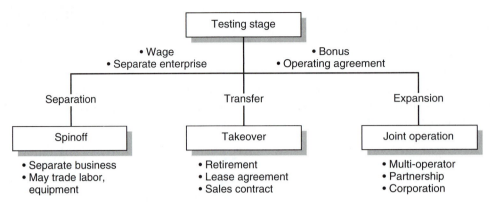

Figure 14-3   **Alternatives for farm business transfer.**

make sure that enough income will be available for living expenses after lease or mortgage payments are made.

3. A *joint operation* may be developed when both generations wish to continue farming together. This arrangement often involves an expansion of the business to adequately employ and support everyone. The farm may be organized with a joint operating agreement, as a partnership, or as a corporation. These multi-operator family farming operations include some of the most profitable and efficient farm businesses to be found. However, it is crucial that effective personal working and management relationships are developed.

Of course, the testing stage may also end in a decision by the younger generation to not enter farming as a career. This should not be considered a failure but as a realistic assessment of a person's values and goals.

SUMMARY

This chapter is an introduction to Part IV, which discusses acquiring and managing the resources necessary to implement a farm or ranch production plan. Before the resources to implement the plan are actually acquired, a form of business organization must be selected. The selection should be reviewed from time to time as the business grows and moves through the stages of the life cycle.

A farm or ranch business can be organized as a sole proprietorship, a partnership, a corporation, a limited liability company, or a cooperative. There are advantages and disadvantages to each form of business organization, depending on the size of the business, stage of its life cycle, and the desires of the owners. The right business organization can make transferring farm income, ownership, and management to the next generation easier. A testing stage in which members of the younger generation work as employees can lead to a separate spin-off, a takeover of the existing business, or a joint operation.

QUESTIONS FOR REVIEW AND FURTHER THOUGHT

1. What are the differences among the four stages in the life cycle of a farm business? Think about a farm or ranch with which you are familiar. In which stage is it?

2. Why do you think the sole proprietorship is the most common form of farm business organization?

3. What general advantages does a joint venture have over a sole proprietorship? Disadvantages?

4. How does an operating agreement differ from a partnership?

5. For the joint operating agreement example shown in Table 14-1, how would you divide gross income if party A and party B each owned one-half of the livestock?

6. Explain the importance of putting a partnership agreement in writing. What should be included in a partnership agreement?

7. Explain the difference between a general partnership and a limited partnership.

8. Does a two-person partnership have to be 50/50? Can it be a 30/70 or a 70/30 partnership? How should the division of income be determined?

9. Explain the differences between C and S corporations. When would each be advantageous?

10. Why might a partnership or corporation keep more and better records than a sole proprietorship?

11. Compare the personal and corporate tax rates shown in Table 14-2 in this chapter. Would a sole proprietor (joint return) or a C corporation pay more income taxes on a taxable income of $25,000? On $50,000? On $150,000?

12. What special characteristics do limited liability companies and cooperatives have?

13. What form of business organization would you choose if you were just beginning a small farming operation on your own? What advantages would this form have for you? What disadvantages?

14. What form of business organization might be preferable if you had just graduated from college and were joining your parents or another established operator in an existing farm? What advantages and disadvantages would there be to you? To your parents?

MANAGING RISK AND UNCERTAINTY

CHAPTER OBJECTIVES

1. To identify the sources of risk and uncertainty that affect farmers and ranchers

2. To show how risk and uncertainty affect decision making

3. To discuss how attitude and financial stability affect risk bearing

4. To illustrate several ways to measure the degree of risk attached to alternative management actions

5. To demonstrate several methods that can be used to help make decisions under risky conditions

6. To discuss tools that can be used to reduce risk or control its effects

Decision making was discussed in Chapter 2 as a principal activity of management, and Chapters 7 through 13 introduced some useful principles and techniques for making good management decisions. These earlier chapters implicitly assumed that all the necessary information about input prices, output prices, yields, and other technical data was available, accurate, and known with certainty. This assumption of perfect knowledge simplifies the understanding of a new principle or concept, but it seldom holds true in the real world of agriculture.

We live in a world of uncertainty. There is an old saying that "Nothing is certain except death and taxes." Many agricultural decisions have outcomes that take place months or years after the initial decision is made. Managers find that their best decisions often turn out to be less than perfect because of changes that take place between the time the decision is made and the time the outcome of that decision is finalized.

Crop farmers must make decisions about what crops to plant and what seeding rates, fertilizer levels, and other input levels to use early in the cropping season. The ultimate yields and prices obtained will not be known with certainty until several months later, or even several years later in the case of perennial crops. A rancher who has decided to expand a beef cow herd by raising replacement heifers must wait several years before the first income is received from the calves of the heifers kept for the herd expansion. Unfortunately, farmers and ranchers can do little to speed up the biological processes in crop and livestock production, or to make them more predictable.

When an outcome is more favorable than expected, a manager may regret not having implemented the decision more aggressively or on a larger scale. However, in this case, the financial health of the operation has been enhanced, not threatened. The real risk comes from unexpected outcomes with adverse results, such as low prices, drought, or disease. Risk management is mostly concerned with reducing the possibility of unfavorable outcomes, or at least softening their effects.

SOURCES OF RISK AND UNCERTAINTY

The sources of risk and uncertainty in agriculture are many. What are the risks associated with selecting enterprises, determining the proper levels of feed and fertilizer to be used, hiring a new employee, or borrowing money? What makes the outcomes of these decisions difficult to predict? Although the sources of risk in agriculture are quite diverse, they can be summarized into five broad management areas: production and technical, price and market, financial, legal, and personal.

Production and Technical Risk

Manufacturing firms know that the use of a certain collection of inputs will almost always result in the same quantity and quality of output, with only minor deviations. This is not the case with most agricultural production processes. Crop and livestock performances depend on biological processes that are affected by weather, diseases, insects, weeds, feed conversion rates, and genetics. These factors cannot be predicted with certainty.

The relative importance of various causes of insured crop losses is shown in Figure 15-1. Nearly all of them are related to weather. Cropping programs must be chosen, fertilizer applied, and money borrowed, all before the weather and its effects on production are known. Horticultural crops in particular are susceptible to unexpected freezes and frost, and a whole array of pests and diseases. Livestock producers also face important production risks. Cold, wet weather in the spring or dry weather in the summer can be devastating to range production of sheep and cattle. Disease infestations may force a producer to liquidate an entire flock or herd.

Another source of production risk is new technology. There is always some risk involved when changing from tested and reliable production methods to something that is new. Will the new technology perform as expected? Has it

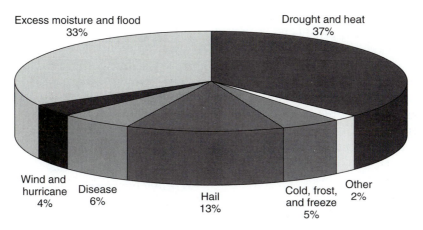

Excess moisture and flood
33%

Drought and heat
37%

Wind and
hurricane
4%

Disease
6%

Hail
13%

Cold, frost,
and freeze
5%

Other
2%

Figure 15-1   **Causes of insured crop losses.**
(Source: National Crop Insurance Services).

been thoroughly tested? What if it costs more? These and other questions must be considered before a new technology is adopted. However, not adopting a successful new technology means the operator may miss out on additional profits and become less competitive. The risk associated with adopting new technology is captured in an old saying: "Be not the first upon whom the new is tried nor the last to lay the old aside."

New crops may be touted as having high profit potential. But their reaction to dry weather or insects may be untested.

Price and Market Risk

Price variability is another major uncertainty in agriculture. Commodity prices vary from year to year, as well as day to day, for reasons beyond the control of an individual producer. The supply of a commodity is affected by farmers' production decisions, in many states and countries, government policies, and the weather. Demand for a commodity is a result of consumer preferences and incomes, exchange rates, export policies, the strength of the general economy, and the price of competing products. Some price movements follow seasonal or cyclical trends, which can be predicted, but even these movements exhibit a great deal of volatility.

Sometimes market access is a source of risk. A processor or packer may go out of business, and leave no viable marketing channel. Or buyers may impose quality standards or quantity restrictions that a producer cannot meet easily.

Box 15-1   Corn Borer Resistant Hybrids: A Risk Reducer?

New seed corn varieties have been introduced that have built-in resistance to the European corn borer. Does this technology decrease or increase production risk? Of course, the risk of yield losses due to corn borer damage is diminished. But will yields be as high in a normal year as for traditional hybrids? Will grain quality be as good? Will benefits outweigh the higher cost of the seed? Will consumers accept genetically engineered crop varieties? Until the new varieties are widely tested, the answers to these questions represent new sources of risk.

Costs represent another source of price risk. Input prices tend to be less variable than output prices but still add to uncertainty. Several times in recent decades shortages of petroleum have caused sudden increases in energy costs, which carried through to fuel, fertilizer, and pesticide prices. Likewise, livestock producers who purchase feeder animals and/or feed are especially susceptible to volatile input prices.

Financial Risk

Financial risk is incurred when money is borrowed to finance the operation of the business. This risk is caused by uncertainty about future interest rates, a lender's willingness to continue lending at the levels needed now and in the future, changes in market values of loan collateral, and the ability of the business to generate the cash flows necessary for debt payments. In Chapter 19, the financial risk that is created by using borrowed capital is discussed in detail.

Production, marketing, and financial risk exist on most farms and ranches, and are interrelated. The ability to repay debt depends on production levels and prices received for the production. Financing the production and storage of commodities depends on the ability to borrow the necessary capital. Therefore, all three types of risk need to be considered together, particularly when developing a whole-farm risk management plan.

Legal Risk

As farmers and nonfarmers come into closer contact in rural areas, more regulation of agricultural production can be expected. Moreover, increased awareness of food safety is affecting how products are being grown and processed. Livestock producers need to be aware of withdrawal periods for antibiotics, as well as rules about locating production sites and handling manure. Violations can bring expensive fines and lawsuits. Losses also occur when milk must be thrown out due to high levels of residue, or when livestock carcasses are condemned.

Farmers can also be subject to legal actions or liability suits for accidents caused by machinery or livestock, or for violating laws regarding health and safety or treatment of hired workers. Ignorance of the law is not an acceptable excuse for not following it, so good managers need to be informed about current rules and regulations.

Personal Risk

No matter how much capital is invested in land, livestock, or machinery, the most irreplaceable assets on a ranch or farm are the manager and key employees. The risk that one of them could suffer a sudden injury or illness is very real— farming is traditionally a hazardous occupation. Some health problems appear only after prolonged exposure to dust, odors, or sun.

Key employees also can be lost due to retirement, a career change, or relocation. If no one else is informed about or skilled in the employee's area of responsibility, significant production losses can occur before a replacement is hired.

Finally, family disputes or divorce settlements can divert property, financial assets, or cash from a farm business and bring about economic losses, as well as personal stress. Mental health problems among farmers often go untreated and unreported, but can eventually lead to severe financial and personal losses.

RISK-BEARING ABILITY AND ATTITUDE

Ranchers and farmers vary greatly in their willingness to take risks and in their abilities to survive any unfavorable outcomes of risky actions. Therefore, the level of risk that a farm business should accept is very much an individual decision. Certainly, good risk management does not mean eliminating all risk. Rather, it means limiting risk to a level that the operators are willing and able to bear.

Ability to Bear Risk

Financial reserves play a big part in determining an operation's risk-bearing ability. Farms with a large amount of equity capital obviously can stand larger losses before they become insolvent. Highly leveraged farms, with a high value of debt relative to assets, can lose equity quickly, because their volume of production is high relative to their own capital. These farms are also more vulnerable to financial risks such as rising interest rates.

Cash flow commitments also affect risk-bearing ability. Families with high fixed living expenses, educational expenses, or health care costs are less able to withstand a low-income year and should not expose themselves to as much risk as other operations. Farmers who have more of their assets in a liquid form, such as a savings account or marketable grain and livestock, have secure off-farm employment, or can depend on relatives or friends to assist them in a financial emergency also have greater risk-bearing ability.

Willingness to Bear Risk

Some farmers and ranchers refuse to take risks even though they have no debt and a strong cash flow. They may have experienced financial setbacks in the past or be concerned about having enough income for retirement. Most farm and ranch operators are risk avoiders. They are willing to take some risks, but only if they can reasonably expect to increase their long-run returns by doing so. Age, equity, financial commitments, past financial experiences, the size of the gains or losses involved, family responsibilities, familiarity with the risky proposition, emotional health, cultural values, and community attitudes all influence the amount of risk producers are willing to take.

EXPECTATIONS AND VARIABILITY

The existence of risk adds complexity to many decisions. When managers are uncertain about the future, they often use some type of average or "expected" values for yields, costs, or prices. There is no assurance that this value will be the actual outcome each time, of course, but decisions must be made based on the best information available. To analyze risky decisions, a manager needs to understand how to form expectations, how to use probabilities, and how to analyze the whole distribution of potential outcomes.

Probabilities are very useful when forming expectations. The true probabilities for various outcomes are seldom known, but subjective probabilities can be derived from whatever information is available plus the experience and judgment of the individual. The probability of rain in a weather forecast or odds of a futures contract exceeding a certain price are examples of subjective probabilities. Because each individual has had different experiences and may interpret the available information differently, subjective probabilities will vary from person to person. This is one reason why different individuals make different decisions when they are faced with the same risky alternatives.

Forming Expectations

Several methods can be used to form expectations about future events. Once a "best estimate" is chosen, it can be used for planning and decision making, until more information allows a better estimate to be made.

Most Likely

One way to form an expectation is to choose the value that is most likely to occur. This procedure requires a knowledge of the probabilities associated with each possible outcome, either actual or subjective. This can be based on past occurrences or on analysis of current conditions. The outcome with the highest probability is selected as the most likely to occur. An example is contained in Table 15-1, where six possible ranges of wheat yields are shown along with the estimated probabilities of the actual yield falling into each one. Using the "most likely" method to form an expectation, a yield of 29 to 35 bushels per acre would be selected. For budgeting purposes, we could use the midpoint of the range, or 32 bushels. There is no assurance the actual yield will be between 29 and 35 bushels per acre in any given year, but if the probabilities are correct, it will occur about 35 percent of the time over the long run. The "most likely" method is especially useful when there is only a small number of possible outcomes to consider.

Averages

Two types of averages can be used to form an expectation. A *simple average,* or *mean,* can be calculated from a series of actual past results. The primary problem is selecting the length of the data series to use in calculating the simple average. Should the average be for the past 3, 5, or 10 years? As long as the fundamental conditions that affect the observed outcomes have not changed, as many observations should be used as are available.

In some cases, the fundamental conditions have changed. New technology may have increased potential crop yields, and long-term changes in supply and demand may have affected market prices. In these cases, a method that gives recent values more importance than the older ones can be used to calculate a *weighted average.* A weighted average can also be used when true or subjective probabilities of the expected outcomes are available, but are not all equal. A weighted average that uses probabilities as weights is also called an *expected value.* The expected value is an estimate of what the simple average would be if the event were repeated many times. Of course, the accuracy of the expected value depends on the accuracy of the probabilities used.

Table 15-2 shows an example of using both simple and weighted average methods. Price information from 5 previous years is used to predict the average selling price for beef cattle during the coming year. One projection is the simple average of the past five years, or $67.98. Alternatively, the most recent values can be assigned greater weights than the observations

TABLE 15-1	Using Probabilities to Form Expectations

Possible wheat yields (bushels/acre)	Number of years actual yield was in this range	Probability
0–14	1	5%
15–21	2	10%
22–28	5	25%
29–35	7	35%
36–42	4	20%
43–51	1	5%
Total	20	100%

that occurred longer ago, as shown in the last column of Table 15-2. The assigned weights should always add to 1.00. Each price is multiplied by its assigned weight, and the results are summed. The expected price is $66.64 using the weighted average method. This method assumes that recent prices (which have been lower) more accurately reflect current supply and demand conditions, while the simple average treats each year's result with equal importance.

Expert Opinions

Many types of newsletters, weather reports, and electronically delivered predictions are available to help producers project supply and demand conditions. Some even offer probabilities that certain outcomes will occur. Professional forecasters usually have access to more information and more sophisticated analysis tools than individual producers. However, their recommendations may not fit a particular farm's production situation or risk-bearing ability.

Futures Markets

Many agricultural commodities are bought and sold for future delivery at a central location called a *futures market.* Several contracts for each product are traded, each with a different delivery date. The futures prices represent the approximate prices that all the people buying

TABLE 15-2	Using Averages to Form an Expected Value for the Price of Beef Cattle		

		Weighted average	
Year	**Average annual price**	**Weight**	**Price × weight**
5 years ago	$72.30	.10	7.23
4 years ago	69.60	.15	10.44
3 years ago	71.60	.20	14.32
2 years ago	65.30	.25	16.32
Last year	61.10	.30	18.33
Summation	$339.90	1.00	$66.64
Simple average: $\dfrac{\$339.90}{5} = \$67.98$		Weighted average = $66.64	

and selling contracts collectively think will exist at a future date. Later, the role that futures markets play in helping producers reduce price risk will be explained.

Variability

A manager who must select from two or more alternatives should consider another factor in addition to the expected values. The *variability* of the possible outcomes around the expected value is also important. For example, if two alternatives have the same expected value, most managers will choose the one whose potential outcomes have the least variability, because there will be smaller deviations with which to deal.

Variability can be measured in several ways.

Range

One simple measure of variability is the difference between the lowest and highest possible outcomes, or the *range*. Alternatives with a smaller range are usually preferred over those with a wider range if their expected values are the same. Range is not the best measure of variability, however, because it does not consider the probabilities associated with the highest and lowest values, nor the other outcomes within the range and their probabilities.

Standard Deviation

A common statistical measure of variability is the *standard deviation*.[1] It can be estimated from a sample of past actual outcomes for a particular event, such as historical price data for a certain week of the year. A larger standard deviation indicates a greater variability of possible outcomes, and therefore, a greater likelihood that the actual outcome will be quite different from the expected value.

Coefficient of Variation

The standard deviation is difficult to interpret, however, when comparing two types of occurrences that have quite different means. The occurrence with the higher mean value often has a larger standard deviation, but is not necessarily more risky. In this situation, it is more useful to look at the relative variability. The *coefficient of variation* measures variability relative to the mean and is found by dividing the standard deviation by the mean. Smaller coefficients of

---

[1]Standard deviation is equal to the square root of the variance. The equation for variance is

$$\text{Variance} = \frac{\sum\limits_{i=1}^{n}(X_i - \overline{X})^2}{n - 1}$$

where X_i is each of the observed values, $\overline{X}$ is the average of the observed values, and n is the number of observations.

variation indicate that the distribution has less variability compared to its mean than other distributions.

$$\text{Coefficient of variation} = \frac{\text{standard deviation}}{\text{mean}}$$

Table 15-3 shows historical data for corn and soybean yields on an individual farm. If we assume that production potential has not changed over time, we can use the simple means as estimates of the expected yields for next year, and the variations from the mean to calculate the standard deviations. Notice that corn had a greater standard deviation than soybeans. However, calculating the coefficients of variation shows that soybean yields were actually more variable than corn yields when compared to the mean. Thus, an operator who wished to reduce yield risk might prefer corn to soybeans.

Cumulative Distribution Function

Many risky events in agriculture have an almost unlimited number of possible outcomes, and the probability of any one of them occurring

TABLE 15-3	Historical Corn and Soybean Yields for an Individual Farm

Year	Corn (bushels/acre)	Soybeans (bushels/acre)
1	125	35
2	145	45
3	141	38
4	88	28
5	105	33
6	129	44
7	118	40
8	75	21
9	132	37
10	127	48
Mean (expected value)	118.5	36.9
Standard deviation	22.7	8.2
Coefficient of variation	0.19	0.22

TABLE 15-4	Cumulative Probability Distributions for Corn and Soybean Yields

Corn (bushels/acre)	Soybeans (bushels/acre)	Cumulative probability (%)
75	21	5
88	28	15
105	33	25
118	35	35
125	37	45
127	38	55
129	40	65
132	44	75
141	45	85
145	48	95

becomes very small. A useful format for portraying a large number of possible outcomes is a *cumulative distribution function,* or CDF. The CDF is a graph of the values for all possible outcomes for an event and the probability that the actual outcome will be equal to or less than each value. The outcome with the smallest possible value has a cumulative probability of nearly zero, while the largest possible value has a cumulative probability of 100 percent.

The steps in creating a CDF are as follows:

1. List a set of possible values for the outcome of an event or strategy, and estimate their probabilities. For example, the data from Table 15-3 can be used as a set of possible values for corn and soybean yields. If it is assumed that each of the 10 historical observations has an equal chance of occurring again, each one represents 10 percent of the total possible outcomes or distribution.
2. List the possible values in order from lowest to highest, as shown in Table 15-4.
3. Assign a *cumulative probability* to the lowest value equal to one-half of the range it represents. Each observation actually represents one segment or range out of the

total distribution, so it can be assumed that the observation falls in the middle of the range. For the example, the lowest yield observed represents the first 10 percent of the distribution, so it can be assigned a cumulative probability of 5 percent.

4. Calculate the cumulative probabilities (probability of obtaining that value or a smaller one) for each of the other values by adding the probabilities represented by all the smaller values to one-half of that value's own probability. In the example, the remaining observed yields would have cumulative probabilities of 15 percent, 25 percent, and so on.

5. Graph each pair of values and connect the points, as shown in Figure 15-2.

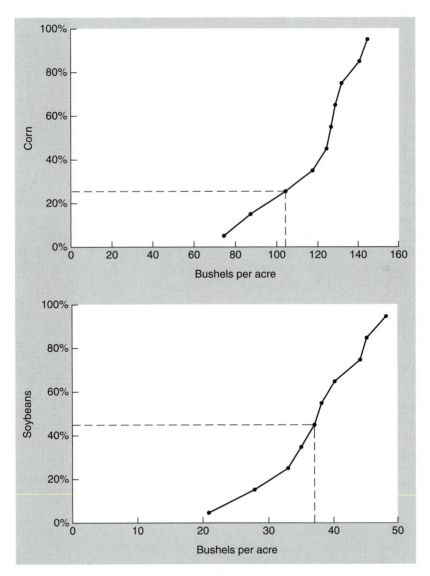

Figure 15-2   Cumulative distribution functions for corn and soybean yields.

The cumulative distribution function permits a view of all possible results for a certain event. The more vertical the graph, the less variability there is among the possible outcomes. Note that the upper portions of the graphs in Figure 15-2 are steeper than the lower portions, indicating that the positive yield responses to good weather are not as significant as the negative responses to poor growing conditions.

DECISION MAKING UNDER RISK

Making risky decisions requires careful consideration of the various strategies available and the possible outcomes of each. The process can be broken down into several steps:

1. Identify the possible sources of risk.
2. Identify the possible *outcomes* that can occur from an *event*, such as various weather conditions or prices, and their probabilities.
3. List the alternative *strategies* available.
4. Quantify the consequences or *results* of each possible outcome for each strategy.
5. Estimate the risk and expected returns for each strategy, and evaluate the trade-offs among them.

An example can be used to illustrate these steps. Assume that a wheat farmer plants a given number of acres of wheat in the fall. Stocker steers can be purchased in the fall and grazed on the wheat over the winter and sold at a known contract price in the spring. The farmer's main source of risk (*step 1*) is the weather, because it affects how much grazing will be available. Assume that there are three possible outcomes for this event—good, average, or poor weather (*step 2*)—and their probabilities are estimated to be 20, 50, and 30 percent, respectively. The selection of probabilities is important. They can be estimated by studying past weather events as well as near-term forecasts.

If too few steers are purchased and the weather is *good*, there will be excess grazing available, and an opportunity for additional profit will be lost. If too many steers are purchased and

the weather is *poor*, there will not be enough grazing, extra feed will have to be purchased, and profit will be reduced or a loss incurred. The third possibility, of course, is for *average* weather to occur with a typical supply of feed available.

The farmer is considering three alternative actions: purchase 300, 400, or 500 steers (*step 3*). These choices are the *decision strategies*. The same three weather outcomes are possible for each strategy, which creates nine potential combinations of results to consider. Once the elements of the problem are defined, it is helpful to organize the information in some way. Two ways of doing this are with a decision tree or with a payoff matrix.

Decision Tree

A *decision tree* is a diagram that traces out several possible management strategies, the potential outcomes from an event, and their results. Figure 15-3 is a decision tree for the previous example. Note that it shows three potential weather outcomes for each of three strategies, the probability for each outcome (which is the same regardless of the strategy chosen), and the estimated net returns for each of the nine potential consequences. For example, if 300 steers are purchased, the net return is $20,000 with good weather, $10,000 with average weather, and only $6,000 with poor weather (*step 4*).

The expected value for each strategy is the weighted average of the possible outcomes, found by multiplying each outcome by its probability and summing the results. Based on these values alone, it might be expected that the farmer would select the "Buy 400" strategy, as it has the highest expected value, $12,200. However, this overlooks the possibility of experiencing poor weather and only breaking even. Ways of making a decision taking this risk into account will be shown later (*step 5*).

Payoff Matrix

A *payoff matrix* contains the same information as a decision tree but is organized in the form of a contingency table. The top part of Table 15-5

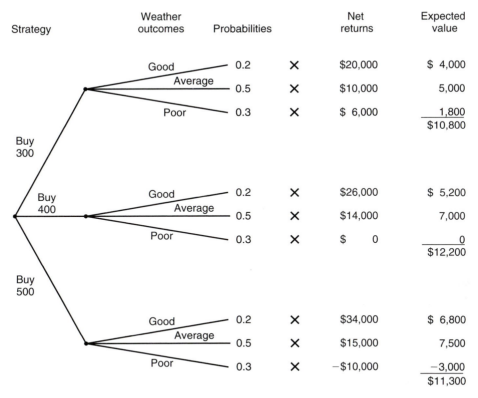

Strategy	Weather outcomes	Probabilities		Net returns	Expected value
	Good	0.2	×	$20,000	$ 4,000
	Average	0.5	×	$10,000	5,000
	Poor	0.3	×	$ 6,000	1,800
Buy 300					$10,800
Buy 400	Good	0.2	×	$26,000	$ 5,200
	Average	0.5	×	$14,000	7,000
	Poor	0.3	×	$ 0	0
					$12,200
Buy 500	Good	0.2	×	$34,000	$ 6,800
	Average	0.5	×	$15,000	7,500
	Poor	0.3	×	−$10,000	−3,000
					$11,300

Figure 15-3   Decision tree for stocker steer example.

shows the consequences of each strategy for each of the potential weather outcomes for the stocker steer example. The expected values, as well as the minimum and maximum values and range of outcomes, are shown in the bottom part of the payoff matrix. Both decision trees and payoff matrices can be used to organize the consequences of one or more related events. However, if each event has many possible outcomes, the diagram or table can become quite large.

Decision Rules

Several decision rules have been developed to help choose the appropriate strategy when faced with a decision involving risk. Using different rules may result in the selection of different strategies. The appropriate rule to use will depend on the decision maker's attitude toward risk, the financial condition of the business, cash flow requirements, and other individual factors.

Because these factors vary widely among decision makers, it is impossible to say that a certain rule is best for everyone.

Most Likely Outcome

This decision rule identifies the outcome that is most likely to occur (has the highest probability) and chooses the strategy with the best consequence for that outcome. Table 15-5 indicates that average weather has the highest probability, at 0.5, and the "Buy 500" strategy has the greatest net return for that outcome, $15,000. This decision rule is easy to apply, but does not consider the variability of the consequences nor the probabilities of the other possible outcomes.

Maximum Expected Value

This decision rule says to select the strategy with the highest expected value. The expected value represents the weighted average result for

| TABLE 15-5 | Payoff Matrix for Stocker Steer Problem |

| Weather outcomes | Probability | Purchase strategy | | |
		Buy 300	Buy 400	Buy 500
Good	0.2	$20,000	$26,000	$34,000
Average	0.5	10,000	14,000	15,000
Poor	0.3	6,000	0	−10,000
Expected value		10,800	12,200	11,300
Minimum value		6,000	0	−10,000
Maximum value		20,000	26,000	34,000
Range		14,000	26,000	44,000

a particular strategy based on the estimated probabilities of each possible result occurring.

Both Figure 15-3 and Table 15-5 show that the "Buy 400" strategy has the highest expected value, $12,200, so it would be selected using this rule. This rule will result in the highest average net return over time, but it ignores the variability of the outcomes. In the example, poor weather has only a 30 percent chance of occurring. However, there is no guarantee that it will not occur two or three years in a row, with the resulting $0 net return each year. The rule of choosing the maximum expected value disregards variability and should be used only by managers who have good risk-bearing ability and are not strong risk avoiders.

Risk and Returns Comparison

Managers who do not have high risk-bearing ability must consider the risk associated with various strategies, as well as the expected returns. Any strategy that has both a lower expected return and higher risk than another strategy should be rejected. Such is the case with the "Buy 500" strategy in Table 15-5. It has a lower expected value than the "Buy 400" strategy, $11,300, and also more risk, because it is the only alternative that could incur a loss. The "Buy 300" strategy also has a lower expected return than the "Buy 400" strategy but is less

risky. Therefore, managers who are very risk conscious might prefer this strategy.

Safety First

The safety first rule concentrates on the worst possible outcome for each strategy and ignores the other possible outcomes. The decision maker assumes that better than expected outcomes pose no serious problems, whereas the unfavorable outcomes are of real concern. Therefore, the strategy with the best possible result among the worst-case outcomes is selected; that is, the one with the "least bad" value. Referring to Table 15-5, application of the safety first rule would result in selection of the "Buy 300" strategy, because its minimum consequence of a $6,000 profit is higher than the other minimums, which are a $0 profit for buying 400 steers and a loss of $10,000 for buying 500 steers. The safety first rule is appropriate for a business that is in such poor financial condition that it could not survive the consequences of even one bad year.

Break-Even Probability

Knowing the probability that each strategy will result in a financial loss can also help the decision maker choose among them. Suppose, for example, that the farmer with the corn and soybean yield histories shown in Table 15-3 calculated that a yield of 105 bushels per acre for corn or 37 bushels per acre for soybeans was needed to just pay all production costs. From the CDF graphs in Figure 15-2, the probability of realizing less than the break-even level of production can be estimated by drawing a vertical line from the break-even yield to the CDF line, then a horizontal line to the cumulative probability scale. In this example, the probability of suffering an economic loss (producing a below break-even yield) is approximately 25 percent for corn and 45 percent for soybeans. Thus, soybeans carry more financial risk. However, the risk of a loss also must be weighed against the expected average return from each crop.

TOOLS FOR MANAGING RISK

Fortunately, farmers and ranchers have a variety of management tools available to them for softening the consequences of taking risky actions. Some of the tools are used to reduce the amount of risk the manager faces, while others help soften the impact of an undesirable result. They all follow one of four general approaches, however.

1. Reduce the variability of possible outcomes. The probability of a bad result is decreased, but the probability of a good result is often reduced, as well.
2. Set a minimum income or price level, usually for a fixed charge. Most insurance programs operate this way. The cost of the risk reduction is known, and the probability of achieving a better than average result is not affected.
3. Maintain flexibility of decision making. Managers do not "lock in" decisions for long periods of time, in case price or production prospects change.
4. Improve the risk-bearing ability of the business, so that an adverse result is less likely to affect the survival of the farming operation.

Farmers and ranchers use many examples of all four types of risk management tools.

Production Risk Tools

Variable crop yields, uncertain livestock production rates, and uneven product quality are the evidence of production risk. Several strategies can be used to reduce production risk.

Stable Enterprises

Some agricultural enterprises have a history of producing more stable income than others. Modern technology may be able to control the effects of weather on production, or government programs and marketing orders may control prices or the quantity of a commodity that can be sold. For example, irrigation will produce more stable crop yields than dryland farming in areas where rainfall is marginal or highly variable during the growing season. At the other extreme, enterprises such as livestock finishing, where both buying and selling prices can vary, and growing perishable products such as flowers, fruits, and vegetables, tend to have highly variable incomes.

Diversification

Many farms produce more than one product to avoid having their income depend totally on the production and price of a single commodity. If profit from one product is poor, profit from producing and selling other products may prevent total profit from falling below acceptable levels. In agricultural production, diversifying by producing two or more commodities will reduce income variability if all prices and yields are not low or high at the same time.

Table 15-6 shows an example of how diversification can work to reduce income variability. Based on the average net farm income for a 10-year period for a sample of Kansas farms, specialization in hog production provided the greatest average annual income per farm, $78,285. However, the diversified crop and livestock farms had the least variable net incomes, as measured by the coefficient of variation. The lack of a strong income correlation among the enterprises "smooths out" the annual income under diversification.

TABLE 15-6	Comparison of Specialized and Diversified Farms

| | Farm Type | | | |
	Hog	Beef Cow	Beef Backgrounding	Crop and Livestock
Net Farm Income (average, 1992–2001)	$78,285	$16,737	$19,795	$36,907
Standard Deviation	59,855	20,670	28,037	24,243
Coefficient of Variation	.76	1.24	1.42	.66

Source: *Kansas Farm Management Association,* 2001.

How much will diversification reduce income variability in an actual farm situation? The answer depends on the price and yield correlations among the enterprises selected. If prices or yields for all of the enterprises tend to move up and down together, little is gained by diversifying. The less these values tend to move together, or the more they move in opposite directions, the more income variability will be reduced by diversification. Likewise, adding a highly variable enterprise to a stable one may actually increase whole farm risk.

Since weather is the primary factor influencing crop yields, crops with the same growing season tend to have a strong positive yield correlation. The yield correlations among crops that have different growing seasons and are susceptible to different insects and diseases will be somewhat lower. Production rates among different types of livestock are less closely correlated, and there is little correlation between crop yields and livestock performance.

Most studies on the price correlations for major agricultural commodities show that pairs of commodities with a strong yield correlation often have a positive price correlation as well, because year-to-year production changes have a major impact on prices. Some specialty crops such as fruits and vegetables, however, may show a weak or even negative price correlation with some of the major field crops, and crop and livestock prices are mostly independent of each other.

Diversification plans can include nonfarm activities as well. Investing in stocks and bonds, carrying out a part-time business unrelated to agriculture, or holding a nonfarm job can all improve the stability of family income. Diversification may also mean giving up the benefits of specializing in one enterprise in order to gain the benefits from less variability in net income.

Insurance

There are several types of insurance that will help reduce both production and financial risk. Formal insurance can be obtained from an insurance company to cover events that could threaten business equity and survival. An alternative is for the business to be self-insured, that is, to maintain some type of readily available or liquid financial reserves in case a loss does occur. Without these financial reserves, a crop failure, major windstorm, or fire might cause such a large financial setback that the business could not continue.

Of course, financial reserves held in a form that can be easily liquidated, such as a savings account, often earn a lower rate of return than the same capital would if it were invested in the farm business or some other long-term investment. This sacrifice in earnings is the opportunity cost of being self-insured and should be compared to the premium cost for an insurance policy that would provide the same protection.

Farmers and ranchers use several common types of insurance against production risks:

1. *Property insurance* Property insurance protects against the loss of buildings, machinery, livestock, and stored grain from fire, lightning, windstorm, theft, and other perils. Property insurance is relatively inexpensive, while the loss from a serious fire or windstorm can be devastating. Therefore, most farmers and ranchers choose to carry at least a minimum level of property insurance on their more valuable assets.

2. *Multiple peril crop insurance* Multiple peril crop insurance (MPCI) is an insurance program backed by the U.S. Department of Agriculture, with policies sold through private insurers. Production guarantees can be purchased for up to 85 percent of the "proven yield" for the insured farm. When actual yields fall short of the guarantee, the producer is paid a fixed price for each bushel of loss. Examples of covered losses would be those due to drought, floods, hail, an early freeze, and insect damage. Added coverage for specific perils such as hail and fire are also available. A number of private companies provide this type of coverage, and the cost will depend on the amount of coverage desired and the past frequency and intensity of hailstorms in the area.

3. *Revenue insurance* A relatively new type of insurance policy allows crop producers to purchase a guarantee for a certain level of gross income per acre. If the producer's actual yield multiplied by the actual market price at harvest is below the guarantee, the insurance company pays the farmer the difference. Thus, the producer can count on at least a minimum amount of gross income per acre. Some crop revenue insurance policies increase the level of the revenue guarantee if market prices rise

from the time the insurance is purchased until harvest. This feature is especially useful for producers who forward price much of their crop before harvest, or who need a guaranteed supply of feed for their livestock. The concept of revenue insurance is beginning to be extended to some livestock species as well.

Another program called Adjusted Gross Revenue insurance allows a farmer to purchase a guaranteed minimum gross income for the entire farm. It is especially useful for farms growing non-traditional crops and livestock.

Extra Production Capacity

When poor weather delays planting or harvesting of crops, many farmers depend on having excess machinery or labor capacity to help them catch up. They incur higher than necessary machinery ownership costs or wages in some years as insurance against crop losses that could occur because of late planting or harvesting in other years. Some operators also prefer to own newer machinery in order to reduce the risk of breakdowns at crucial times or unexpected repair bills.

Share Leases

In many states, *crop share* or *livestock share* leases are common. The landowner usually pays part of the operating expenses and receives a portion of the crops or livestock produced instead of a cash rental payment. In this way, the risk of poor production, low selling prices, or high input costs is shared between the tenant and the owner. The tenant also needs less operating capital under a share lease than a cash lease. Some tenants use a *variable cash lease* to achieve similar risk reduction. Both types of leases are described in Chapter 20.

Custom Farming and Feeding

Rather than risk uncertain prices and yields, some operators prefer to custom farm cropland. They perform all machinery field operations for

---

Box 15-2   Managing Risk with Crop Insurance

Paul Edmundson raises 1,200 acres of soybeans each year in the Delta. One year, to protect his investment, he purchased multiple peril crop insurance. His proven average yield based on past production records was 40 bushels per acre. He chose to insure at the 75 percent production level, so he received a yield guarantee of (40 × 75%) = 30 bushels. He also chose the maximum price guarantee available that year, $5.50 per bushel. His premium cost for this coverage was $7.50 per acre.

That year, a late-summer flood cut his average yield to only 21 bushels per acre. He received a loss payment from his insurance company for (30 − 21) = 9 bushels, at $5.50 per bushel, or $49.50 per acre.

The next year, he decided to switch to crop revenue insurance. His insurance company offered him a policy for 75 percent of his projected gross income, based on his proven yield of 40 bushels per acre and that year's insurance price of $6.00 per bushel. His revenue guarantee was (40 × $6.00 × 75%) = $180 per acre.

In the fall, the Edmundson soybeans yielded 35 bushels per acre, but the market price dropped to $4.60, so his actual gross income was only (35 × $4.60) = $161 per acre. He received an insurance payment for his revenue deficit equal to ($180 − 161) = $19 per acre.

---

a landowner in exchange for a fixed payment. The landowner takes all the price and yield risk. Custom feeding is a similar arrangement. Livestock producers feed animals or poultry owned by investors in their own facilities for a fixed price per head or per space, or a fixed rate per day. Contracts for raising or caring for breeding livestock are also available. Although custom livestock production contracts pass on most production risk to the livestock owner, some do contain penalties for excessive death loss. All custom agreements should be analyzed carefully to compare their potential risks and returns to those of being an independent producer and marketer.

Input Procurement

Some livestock feeders depend on a reliable source of feed or feeder livestock to keep their facilities full. A long-term contract with a supplier reduces the risk of having to operate below capacity. The price may be determined by a set formula based on quality factors and current market prices. Other key inputs may be secured by advance contracts, as well, such as labor crews for harvesting fruits and vegetables.

Market Risk Tools

Market risk exists because of the variability of commodity prices and because the manager does not know what future prices will be when making the decision to produce a commodity. Several methods can be used to reduce price variability or to set a satisfactory price in advance of when crops or livestock are ready for sale.

Spreading Sales

Instead of selling all of a crop at one time, many farmers prefer to sell part of it at several times during the year. For example, 25 percent of the crop may be sold every 3 months, or one-sixth every 2 months. Spreading sales avoids selling all the crop at the lowest price of the year but also prevents selling all of it at the highest price. The result of spreading sales should be an average price received that is near the average annual price for the commodity.

Livestock sales can also be spread throughout the year. This can be accomplished by feeding several groups during the year or by calving or farrowing several times per year. Spreading sales of dairy and egg products occurs naturally, due to the continuous nature of their production.

Contract Sales

Producers of some specialty crops, such as seed, nursery stock, and fruits and vegetables, often sign a contract with a buyer or processor before planting the crop. This contract will usually specify certain management practices that are to be followed, as well as the price to be received for the crop and possibly the amount to be delivered. A contract of this type removes the price risk at planting time, and guarantees the producer will have a market. However, quality standards may be strict, creating added production risk.

It is also possible to obtain a *forward price contract* for many field crops and some types of livestock. Many grain and livestock buyers will contract to purchase a given amount of these commodities at a set price for delivery at a later date or at regular intervals. These contracts are often available during the growing season as well as after the crop has been harvested and stored. Contract sales remove price uncertainty but do not allow selling at a higher price if markets rise later in the year. An exception is a *minimum price contract,* which guarantees the seller a certain price, usually slightly below expected levels, but allows the commodity to be sold at the actual market price if it is above the minimum. The contract may impose a penalty if the producer cannot deliver the agreed on quantity or quality of commodity due to production problems.

Hedging

A market price can be established in advance by hedging on a commodity futures market. Hedging is possible before the crop is planted as well as during the growing season or while grain is stored. Livestock can also be hedged at the time of purchase, or at any time during the feeding period.

Hedging involves selling a commodity futures contract instead of the actual commodity, usually because the manager is currently unable or unwilling to deliver the commodity at that time. The contract is purchased by a buyer at a futures market exchange somewhere, who may represent a processor who wants to lock in the price of the commodity for future use, or a speculator who is hoping to sell the contract later for a higher price. Although futures contracts for some commodities allow for actual delivery when the contract expires, the contract usually is repurchased and the commodity is sold on the local cash market. Cash and futures prices usually move up or down together. Thus, any gain or loss that occurred because the cash market went up or down is offset by a corresponding loss or gain on the futures contract that was being held.

Before beginning a hedging program, a manager should carefully study the hedging process and the futures market and have a good understanding of *basis*, or the normal difference between the futures contract price and the local cash market price. The basis can become wider or narrower while the futures contract is being held, meaning that gains and losses in the cash and futures markets do not exactly offset each other. This variation is called *basis risk*. Basis risk should be taken into account, but basis is less variable and more predictable than cash prices.

Hedging can also be used to lock in the price of inputs to be purchased in the future, such as feedstuffs or feeder livestock. In this case, a futures contract is purchased (rather than sold) in advance, before it is feasible to take delivery of the input, then resold when the actual input is purchased.

Commodity Options

Many marketers dislike the fact that although forward contracting or hedging protects them against falling prices, it also prevents them from benefiting from rising prices. They prefer to use commodity options that set a minimum price in exchange for paying a set fee or premium, but still allow them to sell at a higher price should it occur.

A manager who wants to protect against a price decline will generally buy the right to sell a futures contract at a specified price, called a *put option*. If the market goes down (both cash and futures), the value of the put goes up and offsets the loss in value of the commodity the farmer is

---

Box 15-3   Reducing Price Risk by Hedging

Plainview Feeders, Inc. has just filled one of their feedlots with a new set of cattle that have an expected cost of production as follows:

average purchase cost of cattle per head	$450
expected feed cost (500 pounds of gain @ $.60)	300
nonfeed costs (200 days @ $.08)	16
	$766

At an average selling weight of 1,150 pounds, their break-even selling price is $67 per hundredweight (cwt.). The finished cattle market is about $70 per cwt. now, and June cattle futures contracts are selling at $74 per cwt. They decide to *hedge* the cattle by selling a futures contract.

By June, when the cattle are ready to sell, the market has trended downward somewhat. Plainview sells the cattle for $65 per cwt. and buys back the futures contract for $69 per cwt. Their net price is:

sold futures contract	+$74 per cwt.
bought back futures contract	− 69
sold for cash price	+ 65
	=$70 per cwt.

Although the market declined by $5, they still netted $70 due to the $5 gained on the futures contract.

What if the market had trended upward instead? Suppose by June the futures market price was $80 per cwt. and the cash price was $75. Their net price would have been:

sold futures contract	+$74 per cwt.
bought back futures contract	− 80
sold for cash price	+ 75
	=$69 per cwt.

Their net selling price would have been nearly the same regardless of whether the cattle market went up or down. The only thing affecting it was the *basis,* or the difference between the futures price and the cash price at the time the cattle were sold, which was $4 in the first example and $5 in the second one.

---

holding. If the market goes up, the value of the put goes down and may eventually reach zero. If the market rises even further, the value of the commodity being held continues to rise, but there is no further decline in the value of the put (because it is already at zero), and the farmer gains. When it is time to sell the commodity, the put is also sold, if it still has some value. Thus, for the cost of buying the put (the premium), the farmer is protected against falling prices but can still benefit from rising prices. The commodity option provides a type of price insurance.

Flexibility

Some management strategies allow the operator to change a decision if price trends or weather conditions change. Planting annual crops instead of perennial or permanent crops is one example.

Investing in buildings and equipment that can be used for more than one enterprise is another. Many grain producers build storage bins so they can delay marketing until prices are more favorable. In the case of livestock, animals may be sold as feeder livestock or finished to slaughter weight. Renting certain assets such as land or machinery instead of buying them is another example of maintaining management flexibility.

Financial Risk Tools

Reducing financial risk requires strategies to maintain liquidity and solvency. Liquidity is needed to provide the cash to pay debt obligations and to meet unexpected financial needs in the short run. Solvency is related to long-run business survival, or having enough assets to adequately secure the debts of the business.

Fixed Interest Rates

Many lenders offer loans at either fixed or variable interest rates. The fixed interest rate may be higher when the loan is made but prevents the cost of the loan from increasing if interest rates trend upward.

Self-Liquidating Loans

Self-liquidating loans are those that can be repaid from the sale of the loan collateral. Examples are loans for the purchase of feeder livestock and crop production inputs. Personal loans and loans for land or machinery are examples of loans that are not self-liquidating. The advantage of self-liquidating loans is that the source of the cash to be used for repayment is known and relatively dependable.

Liquid Reserves

Holding a reserve of cash or other assets that are easily converted to cash will help the farm weather the adverse results of a risky strategy. However, there may be an opportunity cost for keeping funds in reserve rather than investing them in the business or other long-term assets.

Credit Reserve

Many farmers do not borrow up to the limit imposed on them by their lender. This unused credit or credit reserve means additional loan funds can be obtained in the event of some unfavorable outcome. This technique does not directly reduce risk but does provide a safety margin. However, it also has an opportunity cost, equal to the additional profit this unused capital might have earned in the business.

Owner Equity

In the final analysis, it is the equity of the business that provides its solvency and much of its liquidity. Therefore, equity should be increased steadily, particularly during the early years of the business, by retaining profits in the business or attracting outside capital.

Legal Risk Tools

Business Organization

Farms and ranches can be organized under several different legal forms. Some of them, such as corporations, limited liability companies, and cooperatives, offer more protection from legal liabilities and damages than others. Chapter 14 contains more details.

Estate Planning

Having a will and an estate plan that provide for the orderly transition of a ranch or farm business to heirs will often save thousands of dollars in estate and income taxes or revenue lost due to interrupted management and division of an efficiently sized business. Having all farm leases and contracts in writing will also reduce legal problems down the road.

Liability Insurance

Liability insurance protects against lawsuits by third parties for personal injury or property damage for which the insured or employees may be liable. Liability claims on a farm may occur when livestock wander onto a road and cause an accident, or when someone is injured on the farm property. The risk of a liability claim may be small, but some of the damages awarded in recent years have been very large. Most people find liability insurance an inexpensive method to provide some peace of mind and protection against unexpected events.

Personal Risk Tools

Health Insurance

Self-employed farmers and ranchers often have difficulty obtaining health insurance coverage at a reasonable cost. The after-tax cost has been reduced somewhat by making more of the premiums tax deductible. Raising the levels at which coverage begins, and obtaining policies through farm organizations also can reduce costs. Given the high cost of medical treatments and unpredictability of health problems, no prudent manager should be without some type of health insurance. This also applies to workers' compensation insurance for employees.

Life Insurance

Life insurance is available to provide protection against losses that might result from the untimely death of the farm operator or a member of the family. The insurance proceeds can be used to meet family living expenses, pay off existing debts, pay inheritance taxes, and meet other expenses related to transferring management and ownership of the business. Care should be taken to select the most suitable type of life insurance for each individual's needs and interests.

Safety Precautions

Common sense, attention to the job at hand, and not getting in a hurry will go a long way toward avoiding accidents and injuries. Common safety measures include keeping machinery guards in place, shutting off machinery before making repairs or adjustments, not moving equipment on public roads after dark, following prescribed procedures for applying pesticides and fertilizers, and avoiding proximity to large, unrestrained livestock.

Backup Management

When only one person is informed about key aspects of the farm business, a sudden accident or illness could seriously disrupt both day-to-day and long-term operations. Key employees, spouses and attorneys should know the location of tax, financial and legal records, and be able to step in when the main operator is unable to continue.

Which of these many risk management tools are employed by a particular farm or ranch will depend on the type of risks being faced, the financial stability of the business, and the risk-bearing attitude of the manager.

SUMMARY

We live in a world of uncertainty. Rarely do we know the exact what, when, where, how, and how much of any decision and its possible outcomes. Decisions must still be made, however, using whatever information and techniques are available. No one will make a correct decision every time, but decision making under uncertainty can be improved by knowing how to identify possible events and strategies, estimate the value of possible outcomes, and analyze their variability.

Decision trees, payoff matrices, and cumulative distribution functions can be used to organize the outcomes of different strategies. Several decision rules can be used to choose among risky alternatives. Some consider only expected returns, some take into account the variability of outcomes both above and below the mean, and some look only at adverse results.

Production, marketing, financial, legal, and personal risk can be reduced or controlled using a number of techniques. Some reduce the range of possible outcomes, while others guarantee a minimum result in exchange for a fixed cost, provide more flexibility for making decisions, or increase the risk-bearing ability of the business.

QUESTIONS FOR REVIEW AND FURTHER THOUGHT

1. List at least five sources of risk and uncertainty for farmers in your area. Classify them as production, price, financial, legal, or personal risk. Which are the most important? Why?

2. How might a young farmer with heavy debt view risk, compared with an older, established farmer with little debt?

3. Might these same two farmers have different ideas about the amount of insurance they need? Why?

4. How do subjective probabilities differ from true probabilities? What sources of information are available to help form them?

5. Select five prices that you feel might be the average price for lean market hogs next year, being sure to include both the lowest and highest price you would expect. Ask a classmate to do the same.
   - a. Who has the greatest range of expected prices?
   - b. Calculate the simple average for each set of prices. How do they compare?
   - c. Assign your own subjective probabilities to each price on your list, remembering that the sum of the probabilities must equal 1.0. Compare them.
   - d. Calculate the expected value for each set of prices, and compare them.
   - e. Would you expect everyone in the class to have the same set of prices and subjective probabilities? Why or why not?

6. Assume that average annual wheat prices in your region for the past 10 years have been as follows:

Year 1	$3.45	Year 6	$2.56
Year 2	3.51	Year 7	3.28
Year 3	3.39	Year 8	3.87
Year 4	3.08	Year 9	2.64
Year 5	2.42	Year 10	2.80

   - a. Calculate the simple average or mean. Calculate the coefficient of variation given that the standard deviation is 0.478.
   - b. Draw a *cumulative distribution function* (CDF) for the price of wheat using the data above.
   - c. From the CDF, estimate the probability that the annual average price of wheat will be $3.00 or less in a given year.

7. Identify the steps in making a risky decision, and give an example.

8. Suppose that a crop consultant estimates that there is a 20 percent chance of a major insect infestation, which could reduce your per-acre gross margin by $60. If no insect damage occurs, you expect to receive a gross margin of $120 per acre. Treatment for the insect is effective but costs $15 per acre. Show the possible results of treating or not treating in a decision tree and in a payoff matrix. What is the expected gross margin for each choice? What is the range between the high and low outcomes for each choice?

9. Would a manager who did not need to consider risk choose to treat or not treat for insects in Question 8? What about a manager who could not afford to earn a gross margin of less than $90 per acre?

10. Give two examples of risk management strategies that fall into each of the following general categories:
   - a. Reduce the range of possible outcomes.
   - b. Guarantee a minimum result for a fixed cost.
   - c. Increase flexibility of decision making.
   - d. Improve risk-bearing ability.

11. Describe one important risk management tool or strategy that will help cope with each of the following types of risk on a ranch or farm.
   - a. Production
   - b. Market
   - c. Financial
   - d. Legal
   - e. Human

16

MANAGING INCOME TAXES

CHAPTER OBJECTIVES

1. To show the importance of income tax management on farms and ranches

2. To identify the objectives of tax management

3. To discuss the differences between the cash and accrual methods of computing taxable income, and the advantages and disadvantages of each

4. To explain how the marginal tax system and social security taxes are applied to taxable income

5. To review a number of tax management strategies that can be used by farmers and ranchers

6. To illustrate how depreciation is computed for tax purposes and how it can be used in tax management

7. To note the difference between ordinary and capital gain income and how they are taxed

Federal income taxes have been imposed since 1913 and affect all types of businesses, including farms and ranches. Income taxes are an unavoidable result of operating a profitable business. However, management decisions made over time can have a large impact on the timing and amount of income tax due. For this reason, a farm manager needs a basic understanding of the tax regulations in order to analyze possible tax consequences of management decisions. Since these decisions are made throughout the year, tax management is a year-round problem, not something to be done only when the tax return is completed.

Farm managers do not need and cannot be expected to have a complete knowledge of all the tax regulations in addition to all the other knowledge they need. However, a basic understanding and awareness of the tax topics to be discussed in this chapter enable a manager to recognize the possible tax consequences of some common business decisions. This basic information also has another function. It helps a manager identify decisions that may have large and complex tax consequences. This should be the signal to obtain the advice of a tax accountant or attorney who is experienced in farm tax matters. Expert advice *before* the management decision is implemented can be well worth the cost and save time, trouble, and taxes at a later date.

The income tax regulations are numerous, complex, and subject to change as new tax legislation is passed by Congress. Only some of the more basic principles and regulations can be covered in this chapter. They will be discussed from the standpoint of a farm business operated as a sole proprietorship. Farm partnerships and corporations are subject to different regulations in some cases, and anyone contemplating using these forms of business organization should obtain competent tax advice before making the decision. The reader is also advised to check the current tax regulations for the latest rules. Tax rules and regulations can and do change frequently.

Objectives of Tax Management

Any manager with a goal of profit maximization needs to modify that goal when considering income taxes. This goal should now become "the maximization of long-run, after-tax profit." Income tax payments represent a cash outflow for the business, leaving less cash available for other purposes. The cash remaining after paying income taxes is the only cash available for family living expenses, debt payments, and new business investment. Therefore, the goal should be the maximization of after-tax profit, because it is this amount that is finally available for use at the manager's or owner's discretion. Notice that the goal is not to minimize income taxes. This goal could be met by arranging to have little or no taxable income.

A very short-run objective of tax management is to minimize the taxes due on a given year's income at the time the tax return is completed and filed. However, this income is the result of decisions made throughout the year and perhaps in previous years. It is too late to change these decisions. Effective tax management requires a continuous evaluation of how each decision will affect income taxes, not only in the current year but in future years. It should be a long-run rather than a short-run objective, and certainly not something to be thought of only once a year at tax time. Therefore, tax management consists of managing income, expenses, and the purchases and sales of capital assets in a manner that will maximize long-run after-tax profits.

Tax management is not tax evasion but is avoiding the payment of any taxes that are not legally due and postponing the payment of taxes whenever possible. Several tax management strategies are available that tend to postpone or delay tax payments but not necessarily reduce the amount paid over time. However, any tax payment that can be put off until later represents additional cash available for business use for one or more years.

> ### Box 16-1   Tax Management or Tax Loophole?
>
> Tax management, or minimizing taxes, is sometimes equated with identifying and using "tax loopholes." This is often an unfair assessment. While some inequities undoubtedly exist in the tax regulations, what some call tax loopholes were often intentionally legislated by Congress for a specific purpose.
>
> This purpose may be to encourage investment and production in certain areas or to increase investment in general as a means of promoting an expanding economy and full employment. Taxpayers should not feel reluctant or guilty about taking full advantage of all legal ways to reduce income taxes.

TAX ACCOUNTING METHODS

A unique feature of farm business income taxes is the choice of tax accounting methods allowed. Farmers and ranchers are permitted to report their taxable income using either the cash or accrual method of accounting. All other taxpayers engaged in the production and sale of goods where inventories may exist are required to use the accrual method. Even though farmers and ranchers often have inventories, they may elect to use either one. The choice is made when the individual or business files the *first* farm tax return. Whichever method is used on this first tax return must be used in the following years.

It is possible to change tax accounting methods by obtaining permission from the Internal Revenue Service and paying a fee. The request must be filed on a special form and in a timely manner. It must also include a reason(s) for the desired change, along with other information. A request may be denied and, if granted, may add complexity to the accounting procedures during the change-over period. Therefore, it is advisable to study the advantages and disadvantages of each method carefully and make the better choice when the first farm tax return is filed.

The Cash Method

Under the cash accounting method, income is taxable in the year it is received as cash, or "constructively received." Income is constructively received when it is credited to an account or was

available for use before the end of the taxable year. An example of the latter situation would be a check for the sale of grain that the elevator was holding for the customer to pick up. If the check was available on December 31 but was not picked up until several days later, it would still be income constructively received in December. The check was available in December, and the fact it was not picked up does not make it taxable income in the following year.

Expenses under cash accounting are tax deductible in the year they are actually paid, regardless of when the item was purchased or used. An example would be a December feed purchase that was charged on account and not paid for until the following January. This would be a tax-deductible expense in the following year and not in the year the feed was actually purchased and used. An exception to this rule is the cost of items purchased for resale, which includes feeder livestock. These expenses can be deducted only in the taxable year in which they are sold. This means that the expense of purchasing feeder cattle or other items purchased for resale needs to be held over into the next year if the purchase and sale do not occur in the same year.

Inventories are not used to determine taxable income with the cash method of tax accounting. Income is taxed when received as cash and not as it accumulates in an inventory of crops and livestock. The cash method has several advantages and disadvantages.

Advantages

A large majority of farmers and ranchers use the cash method for calculating taxable income. There are a number of advantages that make this the most popular method.

1. *Simplicity*   The use of cash accounting requires a minimum of records, primarily because there is no need to maintain inventory records for tax purposes.
2. *Flexibility*   Cash accounting provides maximum flexibility for tax planning at the end of the year. Taxable income for any year can be adjusted by the timing of sales and payment of expenses. A difference of a few days can put the income or expense into one of two years.
3. *Capital gain from the sale of raised breeding stock*   More of the income from the sale of raised breeding stock will often qualify as long-term capital gain income under cash accounting. This income is not subject to social security taxes and is taxed at a lower rate than ordinary income. This subject will be discussed in detail in a later section.
4. *Delaying tax on growing inventory*   A growing business with an increasing inventory can postpone paying tax on the inventory until it is actually sold and converted into cash.

Disadvantages

There are several disadvantages to cash accounting that may make it less desirable in certain situations.

1. *Poor measure of income*   As discussed in Chapter 6, changes in inventory values are needed to measure net farm income accurately for a given accounting period. Because inventory changes are not included when determining taxable income under cash accounting, the resulting value does not accurately reflect net farm income for the year.
2. *Potential for income variation*   Market conditions may make it desirable to store one year's crop for sale the following year and to sell that year's crop at harvest. The result is no crop income in one year and income from two crops in the following year. Because of the progressive nature of the income tax rates, more total tax may be paid than if the income had been distributed more evenly between the years.
3. *Declining inventory*   During years with declining inventory values, more tax will be paid because there is no inventory decrease to offset the cash sales from inventory.

The Accrual Method

Income under the accrual method of tax accounting is taxable in the year it is earned or produced. Accrual income includes any change, from the beginning to the end of the taxable year, in the value of crops and livestock in inventory as well as any cash income received. Therefore, any inventory increase during the tax year is included in taxable income, and inventory decreases reduce taxable income. An account receivable, such as income earned for custom work done for a neighbor, is "earned" income. It is therefore taxable income even though a cash payment has not been received.

Expenses under the accrual method are deductible in the tax year in which they are incurred, whether or not they are paid. In essence, this means any accounts payable at the end of the tax year are deductible as expenses, because they represent expenses that have been incurred but not yet paid. Another difference from cash accounting is the cost of items purchased for resale, which can be deducted in the year of purchase even though the items have not been resold. Their cost is offset by the increase in ending inventory value for the same items.

Advantages

Accrual tax accounting has several advantages over cash accounting, which should be considered when choosing a tax accounting method.

1. *Better measure of income*   Taxable income under accrual accounting is calculated much like net farm income was in Chapter 6. For this reason, it is a better measure of net farm income than cash accounting and keeps income taxes paid up on a current basis.

2. *Reduces income fluctuations*   The inclusion of inventory changes prevents wide fluctuations in taxable income if the marketing pattern results in the production from two years being sold in one year.

3. *Declining inventory*   During years when the inventory value is declining, as might happen when a farmer is slowly retiring from the business, less tax will be paid under accrual accounting. The decrease in inventory value offsets some of the cash receipts.

Disadvantages

The disadvantages of accrual accounting offset many of the advantages for many farm businesses. This makes cash accounting the more popular method.

1. *Increased record requirements*   Complete inventory records must be kept for proper determination of accrual income. They should also be kept for use in calculating net farm income, but the additional complexity of a tax inventory and valuation of that inventory may discourage farmers from using accrual accounting.

2. *Increasing inventory*   When the inventory is increasing, accrual accounting results in tax being paid on the inventory increase before the actual sale of the products. Because no cash has yet been received for these products, this can cause a cash flow problem.

3. *Loss of capital gain on raised breeding stock*   Accrual accounting will result in losing much, if not all, of the capital gain income from the sale of raised breeding stock. This factor alone generally makes

accrual accounting less desirable for taxpayers with a breeding herd who raise most of their herd replacements. (See later section on "Capital Gain and Livestock.")

4. *Less flexibility*   Accrual tax accounting is less flexible than cash accounting. It is more difficult to adjust taxable income at year-end through the timing of sales and expenses.

Tax Record Requirements

Regardless of the accounting method chosen, complete and accurate records are essential for both good tax management and the proper reporting of taxable income. Both agricultural production and the tax regulations have become too complex for records to be a collection of receipts and canceled checks tossed in a shoe box. Poor records often result in two related and undesirable outcomes: the inability to verify receipts and expenses in case of a tax audit, and paying more taxes than may be legally due. In either case, good records do not cost, they pay in the form of lower income taxes.

Complete tax records include a listing of receipts and expenses for the year. A tax depreciation schedule is also needed for all depreciable property to determine annual tax depreciation and to calculate any gain, loss, or depreciation recapture when the item is sold. Permanent records should be kept on real estate and other capital items, including purchase date and cost, depreciation taken, cost of any improvements, and selling price. These items are important to determine any gain or loss from sale or for gift and inheritance tax purposes.

A computerized record system can be very helpful in keeping tax records. Accuracy, speed, and convenience are some of the advantages. In addition to keeping a record of income and expenses, some computer accounting programs also have the ability to maintain a depreciation schedule and calculate each year's tax depreciation. Some print out the year's income and expenses in the same format as a farm tax return, and others can print a copy of the completed farm tax return.

Box 16-2 **How Long Should Tax Records Be Kept?**

This is a common question, and fortunately, the answer is not "forever." A taxpayer has three years from the date the return was due or filed to amend it, and the IRS has three years from the same date to assess any additional tax. The possibility of either event means records should be kept a minimum of three years from the date the related return was filed. For example, the return for calendar year 2004 is due April 15, 2005. Therefore, records for year 2004 should be kept at least until April 15, 2008. Some tax practitioners recom-

mend that a copy of the tax return be kept for some longer period. If a fraudulent return or no return was filed, the IRS has a longer period of time to assess any tax.

All records related to the cost, depreciation, purchase/sale dates, and sales price of land and depreciable assets should be kept for three years *after* their sale or other disposition. Because land is often owned for many years, records for land purchases will often need to be kept for many years.

THE TAX SYSTEM AND TAX RATES[1]

The income tax system in the United States is based on a number of marginal tax rates (the additional tax on an additional dollar of income within a certain range). While the actual tax rates and the range of taxable income for each rate change from time to time, the rates have always been marginal rates. The table below shows the marginal tax rates for a married couple filing a joint tax return. It includes the rates for 2003 as well as the changes scheduled to take place in 2006.

	Marginal tax rates	
Taxable income*	**Year 2003**	**Year 2006**
$0–$12,000	10%	10%
$12,000–$47,450	15%	15%
$47,450–$114,650	27%	25%
$114,650–$174,700	30%	28%
$174,700–$311,950	35%	33%
over $311,950	38.6%	35%

*The dividing points between each marginal tax rate are adjusted upward each year by an amount equal to the inflation rate.

[1]As this book was going to print, a new tax bill was signed and became law. This bill changes, often temporarily, many rules and rates including marginal tax rates, tax brackets, tax rates on capital gain income, Section 179 expensing, and other items discussed in this chapter. The reader should always check current IRS regulations for the latest tax information.

Taxable income includes farm income, as well as income earned from other sources *minus* personal exemptions and deductions.

Most states also impose an income tax on individuals. Some states have progressive rates, that is, higher incomes are taxed at higher rates, while other states impose a flat rate on all levels of income. Due to the wide diversity of state income tax rules, the discussion in the rest of this chapter will apply only to Federal income taxes. In addition to income taxes, self-employed individuals such as farmers and ranchers are subject to self-employment tax. Only income from a farm or ranch or other business activity is subject to self-employment tax. Unlike income tax, all eligible income is subject to this tax. It is applied to income before personal exemptions or deductions are subtracted.

Self-employment tax includes both Social Security and Medicare taxes. For 2003, these combined tax rates were:

Income subject to self-employment tax*	**Rate**
$0–$87,000	15.3%
Over $87,000	2.9%

*These values are set by legislation and have generally increased every year.

As discussed above and because capital gains are not included, the income subject to self-employment tax will be different from what is subject to income taxes. However, the combination of the two taxes creates a number of income brackets and a wide range of total combined tax rates.

SOME POSSIBLE TAX MANAGEMENT STRATEGIES

A number of tax management strategies exist because of the marginal nature of the income tax rates and other features of the tax system. Tax management strategies are basically one of two types: those that reduce taxes, and those that may only postpone the payment of taxes. However, any tax payment that can be delayed until next year makes that sum of money available for business use during the coming year. That amount is available for investment or to reduce the amount of borrowing needed.

Form of Business Organization

As discussed in Chapter 14, the form of business organization will have an effect on the taxes paid over time. The different forms of business organization should be analyzed both when starting a farm or ranch business and whenever there is a major change in the business. While taxes may not be the only reason, nor always the most important reason, for changing the type of business organization, it is always a factor to consider in the decision.

Income Leveling

There are two reasons for trying to maintain a steady income level. First, years of higher than normal income may put the taxpayer in a higher marginal tax bracket. More tax will be paid over time than with a level income taxed at a lower rate. Second, in years of low income, some personal exemptions and deductions may be lost. These cannot be carried over to future tax years.

For the 2003 tax year, each taxpayer gets a personal exemption of $3,050, and the same amount for each dependent. There is also a standard deduction of $7,950 for a married couple filing a joint return.[2] Taxpayers who "itemize" their personal deductions may have an even larger deduction. Therefore, a married couple with two dependent children has exemptions and a minimum deduction of ($3,050 $\times$ 4) + $7,950 = $20,150. *Taxable income* is only the income above this amount. If the total family income is less than $20,150, no income taxes are due, but some (or all if income is $0 or less) of the allowable exemptions and deductions are lost. They cannot be carried over to use next year. Even though no income tax may be due, there may still be some self-employment tax.

Income Averaging

If it is impossible to maintain a fairly level taxable income from year to year, income averaging is allowed. It is particularly useful in a year when taxable income is well above average and the past three years were average or below, as income can be moved from high marginal tax rates to lower rates.

To use income averaging, farmers and ranchers select what amount of the current year's taxable income they want to use in the averaging process. That amount is deducted from the current year's taxable income and one-third of it is added to each of the past three years' taxable income. Income taxes are then recalculated for the past three years as well as for the current year. There are some limitations on what can and cannot be averaged and, because of other provisions of the tax code, the tax savings may be less than expected.

Deferring or Postponing Taxes

Taxes can be deferred or postponed by delaying taxable income until a later year. This saves

---

[2]Both the personal exemption and standard deduction are adjusted upward each year to account for the rate of inflation.

taxes in the current year, and the money saved can be invested or used to reduce borrowing during the next year. Taxable income can be deferred by increasing expenses in the current year, deferring sales until the next year, or both. Tax regulations limit the amount of prepaid expenses that can be deducted but still allow considerable flexibility for many farmers and ranchers.

To level income or defer taxes requires flexibility in the timing of income and expenses. The taxpayer must be able to make the expense deductible and the income taxable in the year of choice. Flexibility is greatest with the cash method of computing taxable income. With the accrual method, many of the timing decisions on purchases and sales are offset by a change in inventory values.

Taxpayers must be cautious and consider other than tax reasons when adjusting the timing of purchases and sales. It is very easy to make a poor economic or marketing decision in an effort to save or postpone a few dollars in taxes. Expected market trends are one factor to consider. For example, if commodity prices are expected to move lower into next year, a farmer may be ahead by selling now and paying the tax in the current year. Remember, the objective of tax management is to maximize long-run after-tax profit, not to minimize income taxes paid in any given year.

Net Operating Loss (NOL)

The wide fluctuations in farm prices and yields can cause a net operating loss some years, despite the best efforts to level out annual taxable income. Special provisions relating to net operating losses allow them to be used to reduce past and/or future taxable income. Any net operating loss from a farm business can first be used to offset taxable income from other sources. If the loss is greater than nonfarm income, any remaining qualifying farming loss can be carried back 5 years and then forward up to 20 years. Nonfarm losses can only be carried back 2

years. However, a taxpayer with a farming loss can elect to use the 2-year rather than the 5-year carry back.

When carried back, the net operating loss is used to reduce taxable income for these prior years, and a tax refund is requested. When carried forward, it is used to reduce taxable income, and therefore income taxes, in future years. A taxpayer can elect to forego the 2- or 5-year carry-back provision and apply the net operating loss only against future taxable income for up to 20 years. However, this election must be made at the time the tax return is filed, and it is irrevocable. In making this election, it is important to receive maximum advantage from the loss by applying it to years with above-average taxable income.

A net operating loss is basically an excess of allowable tax-deductible expenses over gross income. However, there are certain adjustments and special rules that apply when calculating a net operating loss. Expert tax advice may be necessary to properly calculate and use a net operating loss to its best advantage.

Tax-Free Exchanges

The sale of an asset such as land will generally incur some taxes whenever the sale price is above the asset's cost or tax basis (its value for tax purposes). If the land being sold is to be replaced by the purchase of another farm, a tax-free exchange can eliminate, or at least reduce, the tax caused by the sale. There are some specific regulations regarding tax-free exchanges, so competent tax advice is necessary to be sure the exchange qualifies as tax-free.

A tax-free exchange involves trading property. The farm you wish to own is purchased by the person who wants to purchase your farm. You then exchange or trade farms, with each party ending up with the farm they wish to own. Completion of a tax-free exchange simply transfers the basis on the original farm to the new farm, and no tax is due if the sale prices of both farms are equal. In practice, this rarely happens and some tax may be due. Again, good tax advice is necessary to compute any tax due on the exchange. Because of the strict requirements, tax-free exchanges do not fit every situation. However, they are something to consider whenever an eligible asset is to be replaced by a new or different one.

DEPRECIATION

Depreciation plays an important role in tax management for two reasons. First, it is a noncash but tax-deductible expense, which means it reduces taxable income without being a cash outflow. Second, some flexibility is permitted in calculating tax depreciation, making it another tool that can be used for income leveling and postponing taxes. A working knowledge of tax depreciation, depreciation methods, and the rules for their use is necessary for good tax management.

The difference between tax depreciation and other calculations of depreciation should be kept in mind throughout the following discussion. Tax depreciation is depreciation expense computed for tax purposes. Depreciation for accounting and management purposes would normally use one of the depreciation methods discussed in Chapter 4. As will be shown, the usual tax depreciation method is different and may use different useful lives and salvage values than would be used for management accounting. Therefore, there may be a large difference in the annual depreciation for tax versus management uses.

Tax Basis

Every business asset has a tax basis, which is simply the asset's value for tax purposes at any point in time. At the time of purchase, it is called a *beginning tax basis,* and when this value changes, it becomes an *adjusted tax basis.* Any asset purchased directly, either new or used, has a beginning tax basis equal to its purchase price. When a new asset is purchased by trading in a used asset plus cash, the basis is determined in a different manner. The old asset continues to be depreciated as if it was still owned until all eligible depreciation has been taken. Beginning basis on the new asset is equal to the cash paid to complete the trade.

For example, assume an old tractor with an adjusted tax basis of $10,000 plus $80,000 cash is traded for a new tractor which has a list price of $120,000. The $80,000 is the boot or the difference between the cost of the new tractor and the trade-in value of the old one. Even though it is no longer owned, the old tractor remains on the depreciation schedule being depreciated until its adjusted tax basis is zero. The beginning tax basis on the new tractor is $80,000, the cash boot paid. This may seem low for a tractor with a list price of $120,000 but it is the result of one or both possibilities. First, the old tractor had a trade-in value higher than its adjusted tax basis, which is likely given the "fast" depreciation for tax purposes. Second, the dealer was willing to sell the tractor for something less than full list price.

An asset's basis is adjusted downward each year by the amount of depreciation taken. The result is its adjusted tax basis. In the example above, if $8,500 of tax depreciation is taken the first year, the tractor would have an adjusted tax basis of $80,000 − $8,500 = $71,500 at the end of the year. The basis would be adjusted downward each following year by the amount of tax depreciation taken that year. Note that the adjusted basis is not the same as market value and may not be anywhere close to market value.

The basis on a nondepreciable asset such as land generally remains at its original cost for as long as it is owned. An exception would be some capital improvement to the land itself, when the cost of the improvement is not tax deductible in the current year nor depreciable. Terraces or earthen dams are possible examples. In this case, the cost of the improvement is *added* to the original basis to find the new adjusted basis.

An accurate record of the basis of each asset is important when it is sold or traded. The current adjusted basis is used to determine the gain or loss on the sale. Cost and basis information on assets such as land and buildings, which have a long life and are sold or traded infrequently, may need to be kept for many years. This is another reason for a complete, accurate, and permanent record-keeping system.

MACRS Depreciation

The current tax depreciation system is the Modified Accelerated Cost Recovery System, or MACRS. There are several alternatives under MACRS, but the discussion here will concentrate on the regular MACRS, which is used more frequently than the others. Alternative MACRS methods will be discussed briefly in a later section.

MACRS contains a number of property classes, which determine an asset's useful life. Each depreciable asset is assigned to a particular class, depending on the type of asset. Farm and ranch assets generally fall into the 3-, 5-, 7-, 10-, 15-, or 20-year classes. Some examples of assets in each class are:

3-year: breeding hogs

5-year: cars, pickups, breeding cattle and sheep, dairy cattle, computers, trucks

7-year: most machinery and equipment, fences, grain bins, silos, furniture

10-year: single-purpose agricultural and horticultural structures, trees bearing fruit or nuts

15-year: paved lots and drainage tile

20-year: general-purpose buildings such as machine sheds and hay barns.

These are only some examples. The tax regulations should be checked for the proper classes for other specific depreciable assets.

Another characteristic of MACRS is an assumed salvage value of zero for all assets. Therefore, taxpayers do not have to select a useful life and salvage value when using MACRS. This method also includes an automatic one-half year of depreciation in the year of purchase,

regardless of the purchase date. This is called the half-year convention, or half-year rule. It benefits a taxpayer who purchases an asset late in the year but is a disadvantage whenever an asset is purchased early in the year. This half-year rule applies unless more than 40 percent of the assets purchased in any year are placed in service in the last quarter of the year. In this case, the midquarter rule applies. It requires the depreciation on each newly purchased asset to begin in the middle of the quarter in which it was actually purchased.

MACRS depreciation is computed using fixed percentage recovery rates, which are shown in Table 16-1, for the 3-, 5-, and 7-year classes. For agricultural assets, these percentages are based on the use of a 150% declining balance for each successive year, until the percentage using straight-line depreciation results in a higher percentage. The rate for year 1 for the 5-year class was determined in the following manner. Remember that the straight-line percentage rate for a 5-year useful life is 20 percent:

$$20\% \times 1.5 \times 1/2 \text{ year} = 15\%$$

TABLE 16-1	**Regular MACRS Recovery Rates for Farm and Ranch Assets in the 3-, 5-, and 7-Year Classes (Half-Year Convention)**

	Recovery percentages		
Recovery year	3-year class	5-year class	7-year class
1	25.00	15.00	10.71
2	37.50	25.50	19.13
3	25.00	17.85	15.03
4	12.50	16.66	12.25
5		16.66	12.25
6		8.33	12.25
7			12.25
8			6.13

The same procedure will result in 25 percent recovery for year 1 of the 3-year class.

Regular MACRS depreciation for each year of an asset's assigned life is found by multiplying the asset's beginning tax basis by the appropriate percentage. For example, if a new pickup is purchased (5-year class) and has a beginning tax basis of $25,000, the depreciation would be computed as follows:

Year 1:	$25,000 × 15.00% = $3,750
Year 2:	$25,000 × 25.50% = $6,375
Year 3:	$25,000 × 17.85% = $4,463
Year 4:	$25,000 × 16.66% = $4,165
Year 5:	$25,000 × 16.66% = $4,165
Year 6:	$25,000 × 8.33% = $2,082

Depreciation for assets in other classes would be computed in a similar manner using the appropriate percentages for the class.

Alternative Depreciation Methods

Some flexibility is allowed in computing tax depreciation, because there are alternatives to the regular MACRS method discussed above. These alternatives all result in slower depreciation than regular MACRS, which is one reason they are not widely used. However, taxpayers who wish to delay some depreciation until later years can choose from one of the alternative methods:

- Regular MACRS class life or recovery period, with straight-line depreciation
- Alternative MACRS recovery periods (usually longer than regular MACRS) and 150% declining balance method
- Alternative MACRS recovery periods with straight-line depreciation

The latter would be the "slowest" of the four possible methods.

If an alternative method is used for one asset, all assets in that class purchased in the same year must be depreciated using the alternative method. Also, once an alternative method has been selected, it cannot be changed in a later year.

Expensing

Section 179 of the tax regulations provides for a special deduction called "expensing." While not specifically called depreciation, it has the same effect on basis and taxable income. Expensing is an *optional* deduction, which can be elected only in the year an asset is purchased. The maximum Section 179 expensing has been increased several times in the past and is currently $25,000 per year.

Property eligible for expensing is defined by the tax regulations somewhat differently than property eligible for depreciation. However, the general result is that most 3-, 5-, 7-, and 10-year class property is eligible. The amount eligible for expensing is equal to the cost or beginning basis on an outright purchase, but only the cash difference or "boot" is eligible when there is a trade.

Since expensing is treated as depreciation, it reduces an asset's tax basis. The beginning basis must be reduced by the amount of expensing taken before computing any remaining regular depreciation. From this procedure, it can be seen that an election to take expensing does not increase the total amount of depreciation taken over the asset's life. It only puts more depreciation in the first year (expensing plus regular depreciation for year 1), which leaves less annual depreciation for later years.

Assume a truck has been purchased for $38,000, which would be its beginning basis. If maximum expensing was taken (remember it is optional), the basis in the truck would be reduced to $38,000 − $25,000 = $13,000. This is all the depreciation remaining and annual MACRS depreciation for the first two years would be computed as follows:

Year 1:	$13,000 × 15.00% = $1,950
Year 2:	$13,000 × 25.50% = $3,315

Depreciation would continue for four more years using percentages for a 5-year class life asset. The combination of expensing and regular depreciation results in a tax expense of $25,000 + $1,950 = $26,950 in the first year. The choice to take expensing does not increase total depreciation over the life of the asset. It only moves more of the lifetime depreciation into the first year which leaves less depreciation in later years than if no expensing had been taken. Assets with a purchase price of $25,000 or less can be completely expensed in the year of purchase leaving no regular depreciation. Any remaining amount of the $25,000 annual limit can be applied to another eligible asset.

There is a limit on the amount of eligible assets a business can purchase in any one year and still take the maximum expensing. Purchases over $200,000 reduce the maximum expensing dollar for dollar, which means no expensing can be taken in any year where more than $225,000 of eligible assets are purchased.

Depreciation, Income Leveling, and Tax Deferral

The choices taxpayers are allowed in computing tax depreciation permit depreciation to be used as a means to level out taxable income or defer taxes. If eligible assets are purchased in a year when taxable income is high, expensing can be elected and the regular MACRS method used for depreciation to get maximum current tax deductions. In years when taxable income is below average, the optional expensing would not be taken and one of the alternative depreciation methods can be used. However, a taxpayer should always remember that due to the time value of money, a dollar of current tax savings is worth more than a dollar of tax savings in the future.

Depreciation Recapture

The combination of expensing, fast depreciation under the regular MACRS method, and the assumed zero salvage value means an asset's adjusted tax basis will often be less than its market value. Whenever a depreciable asset's selling price is more than its adjusted basis, the difference is called *depreciation recapture*. This amount represents excess depreciation taken, because the asset did not lose market value as fast as it was being depreciated for tax purposes. Depreciation recapture is included as taxable ordinary income for the year of sale, but it is not subject to social security taxes.

Depreciation recapture may be required on a direct sale of an asset but not if it is traded for a like asset. The asset traded in will continue to be depreciated as if it was still owned and the new asset will have a beginning tax basis equal to the cash boot paid.

CAPITAL GAINS

The tax regulations recognize two types of income: ordinary income and capital gain. Ordinary income includes wages and salaries, interest, dividends, cash rents, and revenue from the sale of crops and feeder livestock. Capital gain can result from the sale or exchange of certain types of qualified assets. In simplified terms, capital gain is the gain or profit made by selling an asset at a price that is higher than the original purchase price. Technically, it is the difference between the selling price and the higher of the asset's cost or adjusted basis. For depreciable assets, that means depreciation recapture applies first, and capital gain is only the amount by which the sales price exceeds the original cost or beginning basis. It is also possible to have a capital loss if the selling price is less than the adjusted basis.

Two types of assets can qualify for capital gain treatment. Capital assets include primarily nonbusiness investments such as stocks and bonds. Of more importance to most farmers and ranchers are Section 1231 assets, which are defined as property used in a trade or business. In a farm or ranch business, Section 1231 assets would include land, buildings, fences, machinery, equipment, breeding livestock, and other depreciable assets used in the business.

For income to qualify as long-term capital gain, the asset must have been owned for a minimum period of time, or else the gain will be short-term capital gain. The required holding period has changed from time to time but is currently 12 months for most assets. One exception is certain types of livestock, which will be discussed later.

As one example of long-term capital gain income, assume that farmland was purchased for $100,000 and sold 3 years later for $150,000. A $50,000 long-term capital gain would result from this transaction. In a similar manner, there can be a long-term capital gain (or loss) from the sale or exchange of other qualified assets, provided they have been owned long enough to qualify.

Taxation of Long-Term Capital Gains

The distinction between ordinary income and long-term capital gain income is important because of the different way the two types of income are taxed. The current tax advantages for long-term capital gain income are:

1. Capital gain income (both short and long term) is not subject to self-employment or social security tax.
2. With some exceptions not generally applicable to farms and ranches, long-term capital gain income is taxed at a maximum rate of 20 percent compared to a current maximum of 38.6 percent for ordinary income. Taxpayers in the 15 percent marginal income tax bracket pay 10 percent tax on long-term capital gain income and 8 percent on qualified property owned for more than 5 years.

Except for not being subject to social security tax, short-term capital gain income has not received any special tax treatment, as it is taxed the same as ordinary income. This makes knowledge of, and a record of, the required holding period an important factor. The tax advantages for long-term capital gain income will be lost by selling even a day short of the required holding period. Also, if there are short-term as well as long-term gains and losses in a given tax year, special rules apply for offsetting

the various gains with the losses. These rules should be studied carefully to obtain maximum advantage from the losses.

Capital Gain and Livestock

Farmers and ranchers generally have more frequent opportunities to receive capital gain income from livestock sales than from selling other assets. Certain livestock are classified as Section 1231 assets, and their sale or exchange may result in a long-term capital gain or loss, provided they were held for draft, dairy, breeding, or sporting purposes. In addition, cattle and horses must have been owned for at least 24 months, and hogs and sheep for 12 months. Eligibility depends on both requirements—the purpose and the holding period.

Raised Livestock

The above tax provision is of special importance to cash-basis farmers and ranchers who raise their own replacements for a breeding or dairy herd. Raised replacements do not have a tax basis established by a purchase price, nor do they have a basis established by an inventory value as would occur with accrual tax accounting. Therefore, cash-basis taxpayers have a zero basis on raised replacement animals. If they are used for draft, dairy, breeding, or sporting purposes and are held for the required length of time, the entire income from their sale is long-term capital gain (sale price minus the zero basis). This provision of the tax law makes cash-basis accounting attractive for taxpayers with beef, swine, sheep, or dairy breeding herds who raise their own replacements. An accrual-basis taxpayer loses most of this capital gain advantage on raised animals, because they have a tax basis derived from their last inventory value.

Purchased Livestock

It is also possible to receive capital gain income from the sale of purchased breeding livestock, but only if the selling price is above the original purchase price. For example, assume a beef cow was purchased for $500 and depreciated to an adjusted basis of $0 over the 5-year class life. If this cow is then sold for $600, the sale would result in $500 of depreciation recapture and $100 of long-term capital gain.

The capital gain that cash-basis taxpayers may receive from selling raised breeding and dairy animals does not necessarily mean that replacements should be raised rather than purchased. There are also some tax advantages from purchasing replacements. Purchased replacements can be depreciated, including the use of the optional expensing. A replacement method should be selected after carefully considering the costs and production factors, as well as the income tax effects. This is another example of the importance of income taxes and the various tax regulations in a manager's decision-making environment. To maximize long-run, after-tax profit, a manager is forced to consider income taxes along with the costs and technical production factors of the various alternatives being considered.

SUMMARY

A business can spend or invest only that portion of its profit that remains after income taxes are paid. Therefore, the usual goal of profit maximization should become maximization of long-run, after-tax profit. For a farm or ranch business, the tax management necessary to achieve this goal begins with selecting either the cash or accrual tax accounting method. It should continue on a year-round basis because of the many production and investment decisions that have tax consequences.

A number of tax management strategies are possible. They include ways to take full advantage of existing tax regulations and to postpone or defer taxes. Some examples include income leveling, tax deferral, income averaging, proper use of a net operating loss, and tax-free exchanges. Appropriate and full use of depreciation and the related expensing is another common tax management strategy.

Long-term capital gain income is not subject to self-employment taxes and is taxed at lower rates than ordinary income. This type of income may result from the sale or exchange of a capital asset or a Section 1231 asset. A common source of capital gain income for a cash-basis taxpayer is income from the sale of raised breeding or dairy animals. This and other sources of capital gain income may require careful planning to qualify the sale or exchange for the reduced tax rates applicable to long-term capital gain income.

QUESTIONS FOR REVIEW AND FURTHER THOUGHT

1. How does a farmer choose a tax accounting method? Can it be changed? How?
2. Which accounting method would you recommend for each of the following? Why?
   a. A crop farmer whose marketing policies cause wide variations in cash receipts and inventory values from year to year
   b. A rancher with a beef breeding herd who raises all the necessary replacement heifers
3. What advantage is there to deferring income taxes to next year? If they must be paid, why not pay them now?
4. Assume Fred Farmer purchases a new planter for $32,500 on March 1, 2004. Compute the maximum Section 179 expensing and regular depreciation for the year of purchase using the regular MACRS method and calculate the planter's adjusted tax basis at the end of the year.
5. Repeat the calculations in Question 4 without expensing, and compare annual depreciation.

6. In a year when farm prices and yields are very good, machinery and equipment dealers often experience large sales toward the end of the year. How would you explain this increase in sales?

7. Explain how a cash-basis farmer can raise or lower taxable income by purchase and selling decisions made in December. Can an accrual-basis farmer do the same thing? Why?

8. Assume that a rancher purchased a young herd bull for $3,000. Three years later, when the bull had an adjusted tax basis of $1,785, he was sold for $5,000. How much depreciation recapture and/or capital gain income resulted from this sale?

9. Suppose a farmer made a mistake and counted $10,000 of long-term capital gain income as ordinary income on the farm tax return. How much additional income and social security tax would this mistake cost if the farmer had taxable income of $45,000? Taxable income of $125,000? (Use the 2003 year tax rates in the "Tax System and Tax Rate" section.)

INVESTMENT ANALYSIS

CHAPTER OBJECTIVES

1. To explain the time value of money and its use in decision making and investment analysis

2. To illustrate the process of compounding or determining the future value of a sum of money or series of payments

3. To demonstrate the process of discounting or determining the present value of a future sum or series of payments

4. To discuss payback period, simple rate of return, net present value, and internal rate of return, and compare their usefulness for investment analysis

5. To show how to apply these concepts to different types of investment alternatives

6. To introduce how income taxes, inflation, and risk affect investment analysis

On a farm or ranch, capital can be used to finance not only annual operating inputs such as feed, seed, fertilizer, pesticides, and fuel, but capital assets such as land, machinery, buildings, and orchards, as well. Different methods should be used to analyze each type of investment because of differences in the timing of the expenses and their associated returns. Both expenses and income from investing in annual operating inputs occur within one production cycle, usually a year or less. In contrast, investing in a capital asset typically means a large initial expense, with the resulting returns spread over a number of future time periods. Partial budgeting can be used to analyze these investments by looking at changes in costs and returns for an average year. Enterprise and partial budgets recognize the time value of money by including opportunity costs on annual operating inputs, but the amounts are typically small due to the short investment period.

While time may be of minor importance when analyzing annual operating inputs, it becomes of major importance for capital assets. They usually require large sums of money, and the expenses and returns occur in different time periods spread over many years. The amounts may be very irregular, as well. A proper analysis of these capital investments requires careful consideration of the size and timing of the related cash flows.

There are other reasons for carefully analyzing potential capital investments before they are made. Decisions about operating inputs can be changed annually. However, capital investments are, by definition, long-lasting assets, and investment decisions are made less frequently. It is more difficult to change a capital investment decision once the asset is purchased or constructed. Therefore, more time and more accurate analytical techniques should be used when making these decisions.

TIME VALUE OF MONEY

We all would prefer to have a dollar today rather than a dollar at some time in the future. People instinctively recognize that a dollar today is worth more than a dollar at some future date. Why? Because a dollar received today can be invested to earn interest and will therefore increase to a dollar *plus* interest by the future date. In other words, the interest represents the opportunity cost of receiving the dollar in the future rather than now. This is the investment explanation of the time value of money. But there are other explanations of why interest is charged or paid on funds.

- Another explanation is from the viewpoint of *consumption.* If the dollar were to be spent for consumer goods such as a new TV, automobile, or furniture, we would prefer to have the dollar now so we could enjoy the new item immediately.
- *Risk* is another reason for preferring the dollar now rather than later. Some unforeseen circumstance could prevent us from collecting it in the future.
- Finally, *inflation* in the general cost of goods and services may diminish what a dollar can buy in the future compared to today.

This chapter will focus on the investment explanation for the time value of money. It is important in business investment decisions and for a manager who must make such decisions on a farm or ranch.

Terms, Definitions, and Abbreviations

A number of terms and abbreviations will be used in the discussion and equations throughout this section. They are as follows, with the abbreviations shown in parentheses:

Present Value (PV) The number of dollars available or invested at the current time, or the current value of some amount(s) to be received in the future.

Future Value (FV) The amount of money to be received at some future time, or the amount a present value will become at some future date when invested at a given interest rate.

Payment (PMT) The number of dollars to be paid or received at the end of each of a number of time periods.

Interest Rate (*i*) Also called the discount rate in some applications. It is the interest rate used to find present and future values, often equal to the opportunity cost of capital.

Time Periods (*n*) The number of time periods to be used for computing present and future values. Time periods are often a year in length but can be shorter, such as a month. The annual interest rate, *i,* must be adjusted to correspond to the length of the time periods; that is, a monthly interest rate must be used if the time periods are months.

Annuity A term used to describe a series of equal periodic payments (PMT). The payments may be either receipts or expenditures.

Future Values

The *future value* of money refers to the value of an investment at a specific date in the future. This concept assumes that the investment earns interest during each time period, which is then reinvested at the end of each period so it will also earn interest in later time periods. Therefore, the future value will include the original investment and the interest it has earned, plus interest on the accumulated interest. The procedure for determining future values when accumulated interest also earns interest is called *compounding.* It can be applied to a one-time lump sum investment (a PV) or to an investment that takes place through a series of payments (PMT) over time. Each case will be analyzed separately.

Future Value of a Present Value   Figure 17-1*a* graphically illustrates this situation. Starting with a given amount of money today, a PV, what will it be worth at some date in the future? The answer depends on three things: the PV, the interest rate it will earn, and the length of time it will be invested.

Assume you have just invested $1,000 in a savings account that earns 8 percent interest compounded annually. You would like to know the future value of this investment after 3 years. The data in Table 17-1 illustrate how the balance in the account changes from year to year.

Notice that all interest earned is allowed to accumulate in the account, so it also earns interest for the remaining years. This is the principle of compounding and shows the results of using

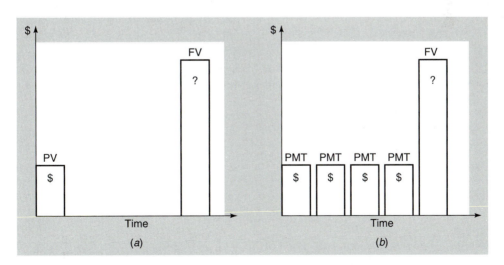

Figure 17-1   **Illustration of the concept of future value for a present value and for an annuity.**

TABLE 17-1	Future Value of $1,000			
Year	Value at beginning of year	Interest rate (%)	Interest earned ($)	Value at end of year ($)
1	1000.00	8	80.00	1,080.00
2	1080.00	8	86.40	1,166.40
3	1166.40	8	93.30	1,259.70

compound interest. In this example, a present value of $1,000 has a future value of $1,259.70 when invested at 8 percent interest for 3 years. Interest is compounded when the accumulated interest also earns interest in the following time periods.

If the interest had been withdrawn, only $1,000 would have earned interest each year. A total of $240.00 in interest would have been withdrawn, compared to the $259.70 of interest earned in this example.

This procedure for finding a future value can become very tedious if the investment is for a long period of time. Fortunately, the calculations can be simplified by using the mathematical equation

$$FV = PV (1 + i)^n$$

where the abbreviations are as defined earlier. Applied to the example, the calculations would be

$$FV = \$1,000 \times (1 + 0.08)^3$$
$$= \$1,000 \times (1.2597)$$
$$= \$1,259.70$$

giving the same future value as before.

This equation can be applied quickly and easily using a calculator that will raise numbers to a power, a financial calculator with built-in functions, or a computer spreadsheet. Without such tools, the equation can be difficult to use, particularly when n is large. To simplify the

calculation of future values, tables have been constructed giving values of $(1 + i)^n$ for different combinations of i and n. Appendix Table 2 is a table for finding the future value of a present value. Any future value can be found by multiplying the present value by the factor from the table that corresponds to the appropriate interest rate and length of the investment. The value for 8 percent and 3 years is 1.2597, which when multiplied by the $1,000 in the example, gives the same future value of $1,259.70 as before.

An interesting and useful rule of thumb is the "Rule of 72." The *approximate* time it takes for an investment to double can be found by dividing 72 by the interest rate. For example, an investment at 8 percent interest would double in approximately 9 years. (Notice that the table value is 1.999.) At 12 percent interest, a present value would double in approximately 6 years (table value of 1.9738).

The concept of future value can be useful in a number of ways. For example, what is the future value of $1,000 deposited in a savings account at 8 percent interest for 10 years? The Table 2 value is 2.1589, which gives a future value of $2,158.90. Another use is to estimate future asset values. For example, if land values are expected to increase at an annual compound rate of 6 percent, what will land with a present value of $1,500 per acre be worth 5 years from now? The table value is 1.3382, so the estimated future value is $1,500 × 1.3382, or $2,007.30 per acre.

Future Value of an Annuity   Figure 17-1*b* illustrates the concept of a future value of a regular series of payments. What is the future value of a number of payments (PMT) made at the end of each year for a given number of years? Each payment will earn interest from the time it is invested until the last payment is made. Suppose $1,000 is deposited at the end of each year in a savings account that pays 8 percent annual interest. What is the value of this investment at the end of 3 years? It can be calculated in the following manner:

First payment	$1,000 = 1,000(1 + 0.08)^2 = 1,166.40
Second payment	$1,000 = 1,000(1 + 0.08)^1 = 1,080.00
Third payment	$1,000 = 1,000(1 + 0.08)^0 = 1,000.00
Future value	$3,246.40

Because the money is deposited at the end of each year, the first $1,000 earns interest for only 2 years, the second $1,000 earns interest for 1 year, and the third $1,000 earns no interest. A total of $3,000.00 is invested, and the total interest earned is $246.40.

The future value of an annuity can be found using the above procedure, but an annuity with many payments requires many computations. An easier method is to use the equation

$$FV = PMT \times \frac{(1 + i)^n - 1}{i}$$

where PMT is the amount invested at the end of each time period. It is even easier to use table values such as those in Appendix Table 3, which cover a range of interest rates and time periods. Continuing with the same example, the table value for 8 percent interest and 3 years is

3.2464. Multiplying this factor by the annual payment, or annuity, of $1,000 confirms the previous future value of $3,246.40.

Present Values

The concept of *present value* refers to the value today of a sum of money to be received or paid in the future. Present values are found using a process called *discounting*. The future value is discounted back to the present to find its current or present value. Discounting is done because a sum to be received in the future is worth less than the same amount available today. A present value can be interpreted as the sum of money that would have to be invested now at the certain interest rate to equal a given future value on the same date. When used to find present values, the interest rate is often referred to as the *discount rate*.

Compounding and discounting are opposite, or inverse, procedures, as shown in Figure 17-2. A present value is compounded to find its future value, and a future value is discounted to find its present value. These inverse relations will become more apparent in the following discussion.

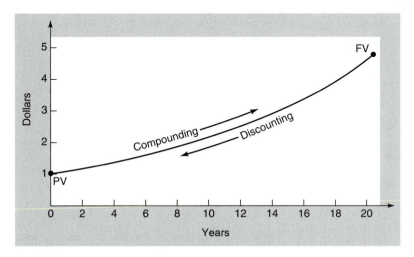

Figure 17-2 **Relation between compounding and discounting ($1 at 8 percent interest).**

Present Value of a Future Value   Figure 17-3*a* illustrates the concept of a present value. The future value is known, and the problem is to find the present value of that amount. Notice that Figure 17-3*a* is exactly like Figure 17-1*a,* except for the placement of the dollar sign and the question mark. This also illustrates the inverse relation between compounding and discounting.

The present value of a future value depends on the interest rate and the length of time before the payment will be received. Higher interest rates and longer time periods will reduce the present value, and vice versa. The equation to find the present value of a single payment to be received in the future is

$$PV = \frac{FV}{(1 + i)^n} \text{ or } FV \times \frac{1}{(1 + i)^n}$$

where the abbreviations are as defined earlier.

This equation can be used to find the present value of $1,000 to be received in 5 years using an interest rate of 8 percent. The calculations would be

$$PV = \$1,000 \times \frac{1}{(1 + 0.08)^5}$$
$$= \$1,000 \times (0.68058)$$
$$= \$680.58$$

A payment of $1,000 to be received in 5 years has a present value of $680.58 at 8 percent compound interest. Stated differently, $680.58 invested for 5 years at 8 percent compound interest would have a future value of $1,000. This again shows the inverse relation between compounding and discounting. A more practical explanation is that an investor should not pay more than $680.58 for an investment that will return $1,000 in 5 years, if there are other alternatives that will pay 8 percent interest or more. Investment analysis makes heavy use of present values, as will be shown later in this chapter.

Tables are also available to assist in calculating present values, such as Appendix Table 4. The factor for the appropriate interest rate and number of years is multiplied by the future value to obtain the present value. For example, the present value of $1,000 to be received in 5 years at 8 percent interest is equal to $1,000 multiplied by the table value of 0.68058, or $680.58.

Present Value of an Annuity   Figure 17-3*b* illustrates the common problem of determining the present value of an annuity or a number of payments to be received over time. Suppose a

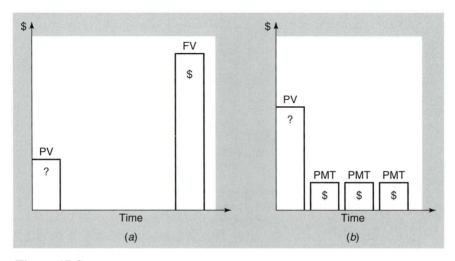

Figure 17-3   **Illustration of the concept of present value for a future value and for an annuity.**

payment of $1,000 will be received at the end of each year for 3 years, and the interest rate is 8 percent. The present value of this income stream or annuity can be found as shown in Table 17-2 using values from Appendix Table 4.

A present value of $2,577.10 represents the maximum an investor should pay for an investment that will return $1,000 at the end of each year for 3 years, if an 8 percent return is desired. Higher interest rates will reduce the present value, and vice versa.

The present value of an annuity can be found directly from the equation

$$PV = PMT \times \frac{1 - (1 + i)^{-n}}{i}$$

As before, it is much easier to use table values such as those in Appendix Table 5. The value corresponding to the proper interest rate and number of years is simply multiplied by the annual payment to find the present value of the annuity. The table value for 8 percent and 3 years is 2.5771, which is multiplied by $1,000 to find the present value of $2,577.10.

Present values are more useful than future values when analyzing investments. Not all investments have the same useful lives nor pattern of net cash flows. Future values that occur in different years and in different amounts are not directly comparable until they are discounted to a common point in time, such as the present.

INVESTMENT ANALYSIS

Investment, as the term will be used in this section, deals with other than short-term or annual expenditures. It refers to the addition of long-term or noncurrent assets to the business. Because these assets last a number of years, or indefinitely in the case of land, these investment decisions will have long-lasting consequences and often involve large sums of money. Such investments should be thoroughly analyzed before the investment decision is made.

TABLE 17-2	Value of an Annuity

Year	Amount ($)	Present value factor	Present value ($)
1	1,000	0.92593	925.93
2	1,000	0.85734	857.34
3	1,000	0.79383	793.83
		Total	2,577.10

Investment analysis, or *capital budgeting* as it is sometimes called, is the process of determining the profitability of an investment or comparing the profitability of two or more alternative investments. A thorough analysis of an investment requires four pieces of information: (1) the initial cost of the investment, (2) the annual net cash revenues realized, (3) the terminal or salvage value of the investment, and (4) the interest or discount rate to be used.

Initial Cost The *initial cost* of the investment should be the actual total expenditure for its purchase, not the list price nor just the down payment if it is being financed. Of the four types of information required, this will generally be the easiest to obtain.

Net Cash Revenues Net cash revenues or cash flows must be estimated for each time period in the life of the investment. Cash receipts minus cash expenses equals the additional *net cash revenues* generated by the proposed investment. Depreciation is not included, because it is a noncash expense and is already accounted for by the difference between the initial cost and the terminal value of the investment. Any interest and principal payments on a loan needed to finance the investment are also omitted from the calculation of net cash revenue. Investment analysis methods are used to determine the profitability of an investment without considering the method or amount of financing needed to purchase it. However, investment analysis techniques can be used to compare several different alternatives for financing an investment. An example is presented in Chapter 22.

Terminal Value   The *terminal value* will need to be estimated, and is usually the same as the salvage value for a depreciable asset. For a non-depreciable asset such as land, the terminal value should be the estimated market value of the asset at the time the investment is terminated. Or, if the land will be held indefinitely, its terminal value can be ignored, because the net cash revenues are assumed to go on forever (see Chapter 20).

Discount Rate   The *discount rate* is often one of the more difficult values to estimate. It is the opportunity cost of capital, representing the minimum rate of return required to justify the investment. If the proposed investment will not earn this minimum, the capital should be invested elsewhere. If funds will be borrowed to finance the investment, the discount rate can be set equal to the cost of borrowed capital. If a combination of borrowed and equity capital will be used, a weighted average of the interest rate on the loan and the equity opportunity cost should be used. Because risk must also be considered, the discount rate should be equal to the rate of return expected from an alternative investment of equal risk. Adjustment of the discount rate for income taxes, risk, and inflation will be discussed later in this chapter.

An example of the information needed for investment analysis is contained in Table 17-3. Investment A yields a fixed return of $3,000 per year for 5 years. Investment B also lasts 5 years, but net cash revenues are lower at first. They increase each year, though, and actually exceed those of investment A over the entire 5 years.

For simplicity, the terminal values are assumed to be zero. Whenever terminal values exist, they should be added to the net cash revenue for the last year, because they represent an additional cash receipt. The information for the two potential investments in Table 17-3 will be applied to illustrate four methods that can be used

TABLE 17-3	Net Cash Revenues for Two $10,000 Investments (No Terminal Value)

	Net cash revenues ($)	
Year	Investment A	Investment B
1	3,000	1,000
2	3,000	2,000
3	3,000	3,000
4	3,000	4,000
5	3,000	6,000
Total cash revenues	15,000	16,000
Less initial investment	−10,000	−10,000
Net cash revenues	5,000	6,000
Average net revenue/yr.	1,000	1,200

to analyze and compare investments. They are: (1) the payback period, (2) the simple rate of return, (3) the net present value, and (4) the internal rate of return.

Payback Period

The payback period is the number of years it would take an investment to return its original cost through the annual net cash revenues it generates. If net cash revenues are constant each year, the payback period can be calculated from the equation

$$P = \frac{C}{E}$$

where P is the payback period in years, C is the initial cost of the investment, and E is the expected annual cash revenue. For example, investment A in Table 17-3 would have a payback period of $10,000 divided by the net cash revenue of $3,000 per year, or 3.33 years.

When the annual net cash revenues are not equal, they should be summed year by year to find the year in which the total is equal to the amount of the investment. For investment B in Table 17-3, the payback period would be 4 years, because the accumulated net cash revenues reach

$10,000 in the fourth year. Investment A is preferred over investment B, because it has the shorter payback period.

The payback period method can be used to rank investments, as was done above. Limited capital can be invested first in the highest ranked investment and then on down the list until the investment capital is exhausted. Another application is to establish a maximum payback period and reject all investments with a longer payback. For example, a manager may select a 4-year payback as a standard and invest only in alternatives with a payback of 4 years or less.

The payback method is easy to use and quickly identifies the investments with the most immediate cash returns. However, it also has several serious disadvantages. This method ignores any cash flows that occur after the end of the payback period, as well as the timing of cash flows during the payback period. Selecting investment A using this method ignores the higher returns from investment B in years 4 and 5, as well as its greater total return. The payback method does not really measure profitability but is more a measure of how quickly the investment will contribute to the liquidity of the business. For these reasons, it should be used only to compare investments with similar lives and relatively constant net cash revenues.

Simple Rate of Return

The simple rate of return expresses the average annual net revenue as a percentage of the original investment. Net revenue is found by subtracting the average annual depreciation from the average annual net cash revenue. In Table 17-3, the net revenue for investment A is shown to be $1,000, and for investment B, $1,200. The simple rate of return is calculated from the equation

$$\text{Rate of return} = \frac{\text{average annual net revenue}}{\text{initial cost}}$$

Applying the equation to the example in Table 17-3 gives the following results:

$$\text{Investment A} = \frac{\$1,000}{\$10,000} = 10\%$$

$$\text{Investment B} = \frac{\$1,200}{\$10,000} = 12\%$$

This method would rank investment B higher than A, which is a different result than that obtained from the payback method. The simple rate of return method is a better indicator of profitability than the payback method, because it considers an investment's earnings over its entire life. Its biggest drawback is its use of average annual earnings, which fails to consider the size and timing of the earnings and can therefore cause errors in selecting investments. This is particularly true when there are increasing or decreasing net revenues. For example, investment A would have the same 10 percent rate of return if it had no net cash revenue the first 4 years and $15,000 in year 5, because the average return is still $3,000 per year. However, a consideration of the time value of money would show the present value of these investments to be greatly different. Because of this shortcoming, the simple rate of return method is not generally recommended for analyzing investments with variable annual net revenues.

Net Present Value

The net present value method is a preferred method of evaluation, because it does consider the time value of money as well as the size of the stream of cash flows over the entire life of the investment. It is also called the *discounted cash flow* method.

An investment's net present value is the sum of the present values for each year's net cash flow (or net cash revenue) minus the initial cost of the investment. The equation for finding the net present value of an investment is

$$\text{NPV} = \frac{P_1}{(1+i)^1} + \frac{P_2}{(1+i)^2} + \cdots + \frac{P_n}{(1+i)^n} - C$$

where *NPV* is net present value, P_n is the net cash flow in year *n, i* is the discount rate, and *C* is the initial cost of the investment. An example of calculating net present values using a 10 percent discount rate is shown in Table 17-4. The present value factors are table values from Appendix Table 4. Summing the NPVs for each investment and subtracting the initial cost of $10,000 results in a total NPV of $1,370 for investment A and $1,272 for investment B. In this example investment A would be preferred.

Investments with a positive net present value would be accepted using this procedure. Those with a negative net present value would be rejected, and a zero value would make the investor indifferent. The rationale behind accepting investments with a positive net present value can be explained two ways. First, it means that the actual rate of return on the investment is greater than the discount rate used in the calculations. In other words, the percent return is greater than the cost of capital. A second explanation is that the investor could afford to pay more for the investment and still achieve a rate of return equal to the discount rate used in calculating the net present value. In Table 17-4, an investor could afford an initial cost of up to $11,370 for investment A and $11,272 for investment B and still receive at least a 10 percent

return on invested capital. This method assumes that the annual net cash flows can be reinvested each period to earn a rate of return equal to the discount rate being used.

Both investments in Table 17-4 show a positive net present value using a 10 percent discount rate. In any determination of a present value, the selection of the discount rate has a major influence on the results. The net present values in Table 17-4 would have been lower if a higher discount rate had been used, and vice versa. At some higher discount rate, the net present values would fall to zero, and at an even higher rate, they would become negative. Therefore, care must be taken to select the appropriate discount rate.

Annual Equivalent and Capital Recovery

The net present value of an investment can be converted to an annual equivalent using the amortization factors found in Appendix Table 1. The annual equivalent is an annuity that has the same present value as the investment being analyzed.

For example, investment A in Table 17-4 had a net present value of $1,370, using a discount rate of 10 percent over 5 years. The amortization factor for 10 percent and 5 years is

	TABLE 17-4	**Net Present Value and Internal Rate of Return for Two Investments of $10,000 (10% Discount Rate and No Terminal Values)**

	Investment A			Investment B		
Year	**Net cash flow ($)**	**Present value factor**	**Present value ($)**	**Net cash flow ($)**	**Present value factor**	**Present value ($)**
1	3,000	0.909	2,727	1,000	0.909	909
2	3,000	0.826	2,478	2,000	0.826	1,652
3	3,000	0.751	2,253	3,000	0.751	2,253
4	3,000	0.683	2,049	4,000	0.683	2,732
5	3,000	0.621	1,863	6,000	0.621	3,726
		Total	11,370		Total	11,272
		Less initial cost	10,000		Less initial cost	10,000
		Net present value	1,370		Net present value	1,272
		Internal rate of return	15.2%		Internal rate of return	13.8%

0.2638 (Appendix Table 1). The annual equivalent is

$$\$1{,}370 \times 0.2638 = \$361.41$$

The annual equivalent for investment B is

$$\$1{,}272 \times 0.2638 = \$335.55$$

In other words, the net returns from investment A would be equivalent to receiving $361.41 per year for 5 years and reinvesting it at a 10 percent return. Investment B would be equivalent to receiving $335.55 per year, and would still be the less desirable alternative.

Computing the annual equivalent value makes it possible to compare investments with different lives. Of course, a different amortization factor would have to be used to convert each net present value. Implicitly, it is assumed that each investment could be repeated with the same results over time.

Annual equivalent values also can be used to convert the initial investment cost of a capital asset to an annual cost. This is an alternative to calculating depreciation and interest using the methods explained in Chapter 9, and is called *capital recovery*. The capital recovery value is the annual amount that recovers the initial cost of the asset over its life, including interest on the unrecovered balance (salvage value). An example is shown in Chapter 22.

Internal Rate of Return

The time value of money is also used in another method of analyzing investments, the *internal rate of return,* or IRR. It provides some information not available directly from the net present value method. Both investments A and B have a positive net present value using the 10 percent discount rate. But what is the actual rate of return on these investments? It is the discount rate that makes the net present value just equal to zero. The actual rate of return on an investment, with proper accounting for the time value of money, is the internal rate of return.

The equation for finding IRR is the same as for net present value, with two differences. First, *NPV* is set equal to zero, and second, the equation is solved for *i*. Because *NPV* is zero, the initial cost of the investment, C, is often moved to the left of the equality sign. In this arrangement, the solution to the equation is the interest rate, *i,* that sets the net present value of the investment's net cash flows equal to its initial cost. The IRR is usually calculated by using a computer program or a financial calculator.

The two investments shown in Table 17-4 both have a positive NPV when discounted at a rate of 10 percent. That means their IRR must be greater than 10 percent. In fact, investment A has an IRR of 15.2 percent, making it preferred again over B, which has an IRR of only 13.8 percent.

The internal rate of return method can be used several ways in investment analysis. Any investment with an IRR greater than the discount rate would be a profitable investment, that is, it has a positive net present value. Some investors select an arbitrary cutoff value for IRR and invest only in those projects with a higher value.

The IRR method has several limitations. It implicitly assumes that the annual net returns or cash flows from the investment can be reinvested each period to earn a return equal to the IRR. If the IRR is fairly high, this may not be possible, causing the IRR method to overestimate the actual rate of return. Another drawback is that IRR says nothing about the size of the investment. A small investment may have a high IRR but only a small NPV in absolute dollars.

FINANCIAL FEASIBILITY

The investment analysis methods discussed so far are methods to analyze *economic profitability*. They are meant to answer the question, "Is the investment profitable?" The question of how the investment was financed was ignored, except for calculating the discount rate. However, when the method and amount of financing used to make the investment are included in the analysis, investments identified as profitable may have years of negative cash flows. Thus, besides the profitability of an investment, an equally important question may be: "Is the investment *financially feasible?*" In other words, will the investment generate sufficient cash flows at the right time to meet the required cash outflows, including loan payments? This potential problem was discussed in Chapter 13 but is worthy of further discussion here. Determination of financial feasibility should be the final step in any investment analysis.

A potential problem is illustrated in Table 17-5 for the example investments A and B used throughout this chapter. Each investment is totally financed with a $10,000 loan at 8 percent interest, to be repaid over 5 years with equal annual principal payments, plus interest. Both interest and principal are included in the debt payment column, and are subtracted from the net cash revenue to find the net cash flow.

Investment A shows a positive net cash flow for each year, because the net cash revenue is greater than the debt payment. However, investment B has lower net cash revenues the first 2 years, which cause negative cash flows in those years. Both investments had positive net present values using a 10 percent discount rate. However, it is not unusual to find profitable investments that have negative cash flows in the early years if the net cash revenues start slowly and the investment requires a large amount of borrowed capital. The problem can be further compounded if the loan must be paid off in a relatively short period of time.

If a project such as investment B is undertaken, something must be done to make up for the negative cash flows. There are several possibilities, which can be used individually or in combination. First, some equity capital can be used for part or all of the initial cost of the investment, in order to reduce the size of the loan and the annual debt payments. Second, the payment schedule for the loan could be lengthened to make the debt payments more nearly equal to the net cash revenues. Arranging for smaller payments with a balloon payment at the end, or interest-only payments for the first few years, would be other possibilities.

If none of these alternatives is feasible, the cash deficits will have to be made up by using excess cash from other parts of the business or from savings. This will require a cash flow analysis of the entire business to see whether cash will be available in sufficient amounts and at the proper times to meet the temporary deficits from the investment.

Should cash flows related to financing ever be included in the calculation of the NPV of an investment? Generally not, because the decision

| TABLE 17-5 | Cash Flow Analysis of Investments A and B | | | | | |

	Investment A ($)			Investment B ($)		
Year	Net cash revenue	Debt payment*	Net cash flow	Net cash revenue	Debt payment*	Net cash flow
1	3,000	2,800	200	1,000	2,800	−1,800
2	3,000	2,640	360	2,000	2,640	−640
3	3,000	2,480	520	3,000	2,480	520
4	3,000	2,320	680	4,000	2,320	1,680
5	3,000	2,160	840	6,000	2,160	3,840

*Assumes a $10,000 loan at 8% interest with equal principal payments over 5 years.

to make an investment and the question of how to finance it should be considered separately. Occasionally, though, an investment and the method to finance it may be closely linked. An example would be comparing the purchase of a farm that can be obtained only with a seller installment contract, versus another farm that can be purchased only with a conventional loan. In this case, the net cash flows used to calculate the NPVs should include the down payment, if any, as well as the principal and interest payments, because the investment and the financing are tied together.

In some cases, it may be possible to finance the same investment several different ways. Once the investment is accepted, the financing alternatives can be compared by discounting the payment streams for each one and selecting the one with the *lowest* NPV. An example of this procedure to compare a lease-or-buy choice for acquiring a tractor is shown in Chapter 22.

INCOME TAXES, INFLATION, AND RISK

So far, only the basic procedures and methods used in investment analysis have been discussed. Several additional factors must be included in a thorough analysis of any investment or when comparing alternative investments. These include income taxes, inflation, and risk.

Income Taxes

The examples used to illustrate the methods of investment analysis did not consider the effects of income taxes on the net cash revenues. Taxes were omitted to simplify the introduction and discussion of investment analysis. However, many investments generate income that is taxable, as well as expenses that are tax deductible. Income taxes can substantially change the net cash revenues, depending on the investor's marginal tax bracket and the type of investment. Because different investments may affect income

taxes differently, they should be compared on an *after-tax basis.*

Income taxes will reduce the net cash revenues when taxable income generated by the investment exceeds deductions, and vice versa. The actual value of the change depends on the marginal tax bracket of the taxpayer. For example, when taxable net cash flows are positive, an investor in the 28 percent marginal tax bracket will have 28 percent less net cash revenues on an after-tax basis.

If all cash inflows are taxable and all cash outflows are tax deductible, then the NPV is simply reduced by the marginal tax rate. However, the net taxable income is often not the same as the net cash revenue. For example, depreciation is not included in calculating net cash revenue, because it is a noncash expense. However, depreciation is a tax-deductible expense that reduces taxable income and therefore income taxes. The tax savings resulting from any depreciation associated with a new investment should be added to the net cash revenues. Certain investments will further reduce income taxes if they are eligible for any special tax credits or deductions that may be in effect at the time of purchase.

Some cash inflows may be taxed differently than others. For example, capital gains income may be subject to a lower tax rate than ordinary income (see Chapter 16). Sales of cull breeding livestock or the terminal sale value of the original investment often give rise to capital gains, and should be adjusted by the appropriate marginal tax rate.

Whenever after-tax net cash revenues are used, it is important that an *after-tax discount rate* also be used. This is because interest on borrowed capital is usually tax deductible, and earnings on alternative investments are usually taxable. The after-tax discount rate can be found from the equation

$$r = i\,(1 - t)$$

where r is the after-tax discount rate, i is the before-tax discount rate, and t is the marginal tax rate. The marginal tax rate was discussed in Chapter 16. It consists of the amount of added federal income tax, state and local income tax, and self-employment tax due on each added dollar of taxable income.

For example, what is the after-tax discount rate if the before-tax discount rate is 12 percent and the taxpayer is in the 28 percent marginal tax bracket? The answer is

$$12\% \times (100\% - 28\%) = 8.64\%$$

Notice that the after-tax cost of borrowed capital can be found from the same equation, with r and i equal to after- and before-tax interest rates. Interest is a tax-deductible business expense, and every dollar of interest expense would reduce taxes by 28 cents for a taxpayer in the 28 percent marginal tax bracket. An investor paying 12 percent interest on borrowed capital would have an 8.64 percent after-tax cost of capital, as computed from the equation above.

Inflation

Inflation is a general increase in price levels over time, and it affects three factors in investment analysis: net cash revenues, terminal value, and the discount rate. Net cash revenues will change over time due to changes in the input and output prices, even if they increase at different rates. Inflation will also cause terminal values to be higher than would be expected otherwise.

When the net cash revenues and terminal value are adjusted for inflation, the discount rate should also include the expected rate of inflation. An interest or discount rate can be thought of as consisting of at least two parts: (1) a real interest rate, or the interest rate that would exist in the absence of inflation, either actual or expected; and (2) an adjustment for inflation, or an inflation premium. During inflationary periods, a dollar received in the future will purchase fewer goods and services than it will at the present time. The inflation premium compensates

for this reduced purchasing power through a higher interest or discount rate. An inflation-adjusted discount rate can be estimated in the following manner:

Real discount rate	5%
Expected inflation rate	+ 3%
Adjusted discount rate	= 8%

The inflation-adjusted discount rate is also called the *nominal* discount rate. If the cost of capital is used for the discount rate, that cost generally already includes an expectation of inflation, and the adjustment has been made implicitly.

In summary, there are two basic ways to treat inflation in capital budgeting.

1. Estimate net cash revenues for each year, and the terminal value, using current prices, and then discount them using a *real* discount rate (the nominal rate minus the expected inflation rate).[1] This eliminates inflation from all elements of the equation.

2. Increase the net cash revenues for each year and the terminal value to reflect the expected inflation rate, and then use an inflation-adjusted (nominal) discount rate. This method incorporates inflation into all parts of the equation. If the same inflation rate is used throughout, the net present value will be the same under either method. However, if some revenues or expenses are assumed to inflate at different rates than others, the inflation-adjusted NPV will be different from the real NPV, and is the more accurate measure. For example, energy-related costs may be projected to increase faster than the general rate of inflation in the economy, while tax savings from depreciation do not inflate at all.

---

[1] A more precise method for converting a nominal rate to a real rate is $[(1 + i)/(1 + f)] - 1$, where i is the nominal rate and f is the inflation rate. However, the value $(i - f)$ gives a close approximation to the real rate.

Risk

Risk exists in investments because the estimated net cash revenues and the terminal value depend on future production, prices, and costs, which can be highly variable and difficult to predict. Unexpected changes in these values can quickly make a potentially profitable investment unprofitable. The longer the life of the investment, the more difficult it is to estimate future costs and revenues accurately.

One method of incorporating risk into the analysis is to add a risk premium to the discount rate. Investments with greater risk have a higher risk premium. This is based on the concept that the greater the risk associated with a particular investment, the higher the expected return must be before an investor would be willing to accept that risk. In other words, there is a positive correlation between the degree of risk involved and the return an investor would demand before accepting that risk.

The risk-free rate of return can be defined as the rate of return that would exist for an investment with a guaranteed net return. Insured savings accounts and U.S. government securities are typically considered to be nearly risk-free investments. Continuing with the example from the last section, risk can be added to the discount rate as follows:

Risk-free real rate of return	4%
Expected inflation rate	+ 3%
Risk premium	+ 1%
Adjusted discount rate	= 8%

Investments with a greater amount of risk would be assigned a higher risk premium, and vice versa. This procedure is consistent with an earlier statement that the discount rate should be the rate of return expected from an alternative investment with equal risk. However, the risk premium is a subjective estimate, and different individuals may use different estimates for the same investment. Interest rates on borrowed funds typically have some risk premium already built in to reflect the possibility that a borrower will not repay the loan.

Adding a risk premium to the discount rate does not eliminate risk. It is simply a way to build in a margin for error. The higher the risk premium used, the higher the net cash revenues will have to be to yield a positive NPV.

Sensitivity Analysis

Given the uncertainty that may exist about the future prices and costs used to estimate net cash revenues and the terminal value, it is often useful to look at what would happen to net present value or IRR if the prices and costs were different. *Sensitivity analysis* is a process of asking several "What if?" questions. What if the net cash revenues were higher or lower? What if the timing of net cash revenues were different? What if the discount rate were higher or lower? Sensitivity analysis involves changing one or more values and recalculating the NPV or IRR. Recalculating for a number of different values gives the investor an idea of how "sensitive" the results, and therefore the investment decision, are to changes in the values being used. A sensitivity analysis will often give the investor better insight into the probability that the investment will be profitable, the effects of changes in prices or the discount rate, and therefore the amount of risk involved.

Recalculating the NPV, annual equivalent, and IRR can be tedious and time-consuming, particularly for investments with a long life and variable net cash revenues. However, with specialized financial software or spreadsheet programs, the results for many different combinations of net cash revenues, terminal values, and discount rates can be compared quickly and easily.

The appendix to this chapter presents a complete example of analyzing an investment using the methods discussed. Variable cash flows, tax effects and inflation are all incorporated.

SUMMARY

The future value of a sum of money is greater than its present value because of the interest it can earn over time. Future values are found through a process called compounding. The present value of a sum of money is smaller than its future value, because money invested today at compound interest will grow into the larger future value. Discounting is the process of finding present values for amounts to be received in the future, and is the inverse of compounding.

Investments can be analyzed by any of four methods: payback period, simple rate of return, net present value, and internal rate of return. The first two are easy to use but have the disadvantage of not accurately incorporating the time value of money into the analysis. This may cause errors in selecting or ranking alternative investments. The net present value method is widely used, because it properly accounts for the time value of money. An investment with a positive net present value is profitable, because the present value of the cash inflows exceeds the present value of the cash outflows. The internal rate of return (IRR) method also considers the time value of money. It represents the discount rate at which the NPV of an investment is exactly zero. An IRR that is higher than the normal discount rate indicates a profitable investment.

All four methods of investment analysis require estimation of net cash revenues over the life of the investment, as well as terminal values. The net present value method also requires selecting a discount rate. Both the net cash revenues and the discount rate should be on an after-tax basis in a practical application of these methods. The cash flows and discount rate can also be adjusted for risk and inflation. A final step in analyzing any investment should be a financial feasibility analysis, particularly when a large amount of borrowed capital is used to finance the investment. Alternative financing methods for the same investment can also be compared by using net present value calculations.

QUESTIONS FOR REVIEW AND FURTHER THOUGHT

1. Put the concepts of future value and present value into your own words. How would you explain these concepts to someone hearing about them for the first time?

2. Explain the difference between compounding and discounting.

3. Assume someone wishes to have $80,000 ten years from now as a college education fund for a child.
   a. How much money would have to be invested today at 6 percent compound interest? At 8 percent?
   b. How much would have to be invested annually at 6 percent compound interest? At 8 percent?

4. If farmland is currently worth $1,750 per acre and is expected to increase in value at a rate of 5 percent annually, what will it be worth in 5 years? In 10 years? In 20 years?

5. If you require a 7 percent rate of return, how much could you afford to pay for an acre of land that has expected annual net cash revenues of $60 per acre for 10 years, and an expected selling price of $2,500 per acre at the end of 10 years?

6. Assume you have only $20,000 to invest and must choose between the two investments below. Analyze each using all four methods discussed in this chapter and an 8 percent opportunity cost for capital (discount rate). Which investment would you select? Why?

	Investment A ($)	Investment B ($)
Initial cost	20,000	20,000
Net cash revenues:		
Year 1	6,000	5,000
Year 2	6,000	5,000
Year 3	6,000	5,000
Year 4	6,000	5,000
Year 5	6,000	5,000
Terminal value	0	8,000

7. Discuss economic profitability and financial feasibility. How are they different? Why should both be considered when analyzing a potential investment?

8. What two approaches can be used to account for the effects of income taxes in investment analysis?

9. What two approaches can be used to account for the effects of inflation in investment analysis?

10. Why would capital budgeting be useful in analyzing an investment of establishing an orchard where the trees would not become productive until 6 years after planting?

11. What advantages would present value techniques have over partial budgeting for analyzing the orchard investment in Question 10?

| APPENDIX. AN EXAMPLE OF INVESTMENT ANALYSIS |

J oe and Sheila Mason have 5 acres of sloping land that is not suitable for crop production. It is cleared and well fenced, however. They are considering raising Christmas trees on it. Their teenage children would be able to help out, and the income would be useful when they are ready for college. However, Joe and Sheila aren't sure if the Christmas tree project would really be better than simply putting money into a savings account each year. With help from a forestry specialist and their farm financial advisor, they put together an investment analysis using net present value techniques.

INITIAL COST

Their first task is to estimate the *initial cost* for 5 acres of Christmas trees. Their budget for establishing the trees is as follows:

	Quantity per Acre	$/acre	Total Cost
Machinery expense			
Sprayer	3 times over	6.00	$ 30
Mower	3 times over	19.20	96
Tree planter	1 time over	3.00	15
			$ 141
Labor @ $7.00 per hour			
Spraying	2 hours	14.00	$ 70
Mowing	4 hours	28.00	140
Planting	3 hours	21.00	105
			$ 315
Supplies			
Burn-down herbicide		15.00	$ 75
Pre-emergent herbicide		18.00	90
Pesticides		15.00	75
Fall herbicide		15.00	75
Tree seedlings @ $.20	767	153.40	767
			$1,082
Rental of tree planter @ $6.00	3 hours	18.00	90
Total Initial Cost			$1,628

ESTIMATING CASH EXPENSES AND REVENUES

Joe and Sheila Mason now want to project their cash expenses and revenue for the next 8 years for their Christmas tree investment. Each year they will mow around the trees and spray weeds. The first two years they will have to replant some trees by hand. Starting in the third year they will have to shear (prune) the trees each year. They plan to pay a wage of $7 per hour to family members as well as hired workers. No land cost is included because they feel that land costs would not be affected by the investment.

In years 6, 7, and 8 they expect to begin selling trees. By the end of the eighth year they expect to have sold the last of their trees. Following is a year-by-year summary of their expected production expenses.

Year	Machinery Operating	Seedlings	Pesticides, Herbicides	Labor	Retailing	Total
1	$115	$192	$240	$ 332	$ 0	$ 879
2	115	10	240	675	0	1,040
3	115	0	240	1,088	0	1,443
4	115	0	240	1,311	0	1,666
5	115	0	240	1,535	0	1,890
6	115	0	240	2,060	535	2,950
7	115	0	240	3,329	2,000	5,684
8	115	0	240	1,844	850	3,049

They estimate that they can receive an average selling price of $18 per tree, with the largest portion of the sales occurring in year 7. Following is a year-by-year estimate of their cash income.

Year	Trees Sold	Income
6	535	$ 9,630
7	2,000	36,000
8	850	15,300

THE DISCOUNT RATE

Now that the Masons have estimated their net cash flows from their Christmas tree investment, they must choose a discount rate for calculating the present values. They plan to finance about 60 percent of the costs out of their own savings and borrow the rest. They estimate that their savings account will earn interest at an average rate of 5 percent annually. Their lender anticipates that the average interest rate on their borrowed funds will be about 10 percent. Thus, their *weighted cost of capital* is

$$(5\% \times .60) + (10\% \times .40) = 7.0\%$$

Joe and Sheila did not incorporate any effects from inflation on wages, the prices of inputs, or the selling price of Christmas trees when they estimated their cash inflows and outflows. Thus, their estimates are *real* values. To adjust their discount rate to a real value also, they subtract 2 percent, their anticipated annual rate of inflation in prices over the next eight years, from their weighted cost of capital to get a *real discount rate:*

$$7.0\% - 2.0\% = 5.0\%$$

The Masons are generally in the 28 percent marginal tax bracket for their federal income taxes, and the 5 percent bracket for state income taxes. In addition, they will have to pay self-employment tax at the rate of about 15 percent on their profits. Their total *marginal tax rate* is

$$28\% + 5\% + 15\% = 48\%$$

Of course, until they begin selling trees, they will show a negative taxable income, so their taxes will be decreased by this rate. Their *after-tax real discount rate* becomes

$$5.0\% \times (1.00 - .48) = 2.6\%$$

Finally, the Masons realize that there are several sources of risk associated with this investment, such as death, loss of trees, and fluctuating selling prices. They decide that they would like to earn at least a 3 percent higher return after taxes on this investment compared to their savings account, to compensate for the added risk. Thus, their final *risk-adjusted real discount rate* is

$$2.6\% + 3.0\% = 5.6\%$$

Net Cash Revenues

Combining the expected cash income for each of the 8 years with the expected cash expenses provides the net cash revenue estimates shown below. For tax purposes, the Masons can deduct the initial cost of establishing the trees, $1,628, as a depletion allowance when they begin selling trees. Since they expect to sell 3,385 trees, they can deduct ($1,628 ÷ 3,385) = $0.48 for each tree sold in years 6, 7, and 8. Their total tax deductible expenses (cash expenses plus depletion) are shown in the last column.

Year	Cash Income	Cash Expense	Net Cash Revenue	Income Tax Depletion	Total Tax Deduction
1	$ 0	$ 879	$− 879	$ 0	$ 879
2	0	1,040	− 1,040	0	1,040
3	0	1,443	− 1,443	0	1,443
4	0	1,666	− 1,666	0	1,666
5	0	1,890	− 1,890	0	1,890
6	9,630	2,950	6,680	257	− 6,423
7	36,000	5,684	30,316	962	− 29,354
8	15,300	3,049	12,251	409	− 11,842

NET PRESENT VALUE

The Masons can deduct the cash operating expenses plus depletion from their business income tax return each year. Given their marginal tax rate of 48 percent, income tax savings are equal to 48 percent of their total tax deductions each year. Once they begin selling trees, their cash income will exceed their deductible costs. They will have to pay additional taxes (negative savings) equal to 48 percent of the difference between income and deductible expenses. The Masons' after-tax, net cash revenue is shown in the table below, along with its present value.

Year	Net Cash Revenue	Income Tax Savings	After-tax Net Cash Revenue	Discount Factor[2]	Present Value
1	$ − 879	$ 422	$− 457	.947	$ − 433
2	− 1,040	499	− 541	.897	− 485
3	− 1,443	693	− 750	.849	− 637
4	− 1,666	800	− 866	.804	− 697
5	− 1,890	907	− 983	.761	− 748
6	6,680	− 3,083	3,597	.721	2,594
7	30,316	− 14,090	16,226	.683	11,081
8	12,251	− 5,684	6,567	.647	4,247
				Present Value	$14,922
				Internal rate of return	19.3%

[2]The discount factor is equal to $(1 + i)^{-n}$ where i is the discount rate and n is the year.

The present value of each year's after-tax net cash revenue is found by multiplying by the discount factor for that year, which is based on the Masons' estimated annual discount rate of 5.6 percent. When the present values for all 8 years are summed, they equal $14,922. Subtracting the initial cost of $1,628 from this amount shows a net present value from the Christmas tree enterprise of $13,294. This means that the net value of the revenue generated over the next 8 years is equal to receiving $13,294 today, tax-free. When they calculate the projected internal rate of return on their investment it turns out to be slightly over 19 percent, substantially higher than their cost of capital.

The Masons decide that this is more than enough to compensate them for their risk and management and decide to go ahead with the project.

FARM BUSINESS ANALYSIS

CHAPTER OBJECTIVES

1. To show how farm business analysis contributes to the control function of management

2. To suggest standards of comparison to use in analyzing the farm business

3. To outline a procedure for locating economic or financial problem areas in the farm or ranch business

4. To review measures that can be used to analyze the solvency, liquidity, and profitability of the business

5. To identify measures of business size

6. To illustrate the concept of economic efficiency and show how it is affected by physical efficiency, product prices, and input prices

7. To demonstrate how enterprise analysis can be used to identify strong and weak areas of production in a farm business

Farm managers and economists have always been interested in the reasons why some farms have higher net incomes than others. The observation and study of these differences and their causes began in the early 1900s and marked the beginning of farm management and farm business analysis. Differences in management are a common explanation for the different net farm incomes, but this explanation is not very complete. Unless the specific differences in management can be identified, there can be no precise recommendations for improving net income on the farms with "poor management."

The differences in profitability of similar farms can be illustrated in several ways. Table 18-1 shows some observed differences in the returns to management per farm on a group of Iowa grain and livestock farms. Similar differences can be found in other states and for other types of farm and ranches. Notice that having high gross sales does not ensure profitability. The average return to management per farm was −$80,473 for the low-profit farms in the highest sales group, while the high-profit farms in the lowest sales group had an average return of $829. Larger farm size simply means a greater disparity in profits between the top and bottom farms.

Several questions beg for an answer. What accounts for the differences in profits? What can be done to improve returns on the less profitable farms? Differences in the operator's goals or the quality of resources available may be partial answers. However, more detailed answers can be found only by performing a complete farm business analysis. This is part of the control function of management.

Control is critical for the entire farm business as well as for individual enterprises. Three steps in the control function are: (1) establishing standards for comparing results, (2) measuring the actual performance of the business, and (3) taking corrective action to improve the performance once problem areas have been identified. Improvement in performance is the payoff for the time spent controlling the business.

TYPES OF ANALYSIS

A farm business analysis can be divided into five areas of investigation:

1. *Profitability*   Profitability is analyzed by comparing income and expenses. High net farm income is usually an important goal of the farm manager, though not necessarily the only one.
2. *Farm Size*   Not having adequate resources is often a cause for low profits. Growing too rapidly or exceeding the size that the operator can manage effectively can also reduce profits.
3. *Efficiency*   Low profitability can often be traced to inefficient use of resources in one or more areas of the business. Both economic and physical efficiency measures should be examined.

TABLE 18-1	**Returns to Management per Farm by Level of Gross Sales**		
	Returns to management per farm ($)		
Gross sales ($)	**Low third**	**Middle third**	**High third**
40,000 to 100,000	− $56,883	− $20,868	$ 829
100,000 to 250,000	− 69,091	− 22,425	13,305
Over 250,000	− 80,473	− 5,331	83,542

Source: Iowa Farm Business Association, 2001.

Box 18-1 **Farm Business Analysis—An Ancient Art**

The following conclusions about farm business analysis are reprinted from *Iowa Farm Management Surveys,* a study of 965 farms in north central Iowa summarized by H. B. Munger in 1913:

CONCLUSIONS

"From a business standpoint a farm cannot be called successful that does not pay operating expenses, a current mortgage rate of interest on capital and a fair return as pay for the farmer's labor and management. By analyzing the business of large numbers of farms in one area, the reasons why some farms are more profitable than others stand out clearly.

"The first consideration is to have a type of farming adapted to the region. Farmers doing the wrong thing cannot expect to make labor incomes. Change in the demands of the market, increase in value of farm land, and other factors make necessary a readjustment of the crops and livestock raised. A farmer in this region for the period of this investigation has better chances of success who had more corn and hogs than the average, and less pasture and beef cattle. Most dairy farmers were doing well.

"For efficient farming in Iowa, where the usual crops and livestock are raised, at least 160 acres is necessary in order to use labor, horses, and machinery to good advantage. Many farmers who own less than 160 acres rent additional land in order to utilize efficiently labor, teams, and tools.

"Good crop yields are fundamental to successful farming. Regardless of other factors, most farms with poor crops were not highly profitable. One should aim to get at least one-fifth better yields than the average of other farms on similar soil.

"In a region where most of the farm receipts come from the sale of livestock, the returns per animal unit are of highest importance. No matter how good the crops may be, if they are fed to unproductive livestock the chances for success are small.

"The highest profits come from a well-balanced business. One should have a farm large enough to provide for the economical use of labor, horses, and up-to-date machinery. He should raise better crops than the average and feed most of them to livestock that can use them profitably. Each of these factors is important, and the combination of all on an individual farm brings the highest success."

4. *Financial* Financial analysis concentrates on the capital position of the business, including solvency, liquidity, and changes in owner's equity.
5. *Enterprises* The efficiency and profitability of each individual enterprise should be analyzed. Low profits in some enterprises may offset high profits in others.

STANDARDS OF COMPARISON

The remainder of this chapter will examine some measures and ratios that can be used to conduct a complete farm analysis. Once a measure has been calculated, the problem becomes one of evaluating the result. Is the value good, bad, or average? Compared with what? Can it be improved? These questions emphasize the need to have some standards against which the measures and ratios can be compared. There are three basic standards to use in analyzing farm business results: budgets, comparative farms, and historical trends.

Budgets

In budget analysis, the measures or ratios are compared against budgeted goals or objectives identified during the planning process. When results in some area consistently fall short of the budgeted objectives, that area needs additional managerial effort or the budgets need to be more realistic.

Comparative Farms

Unfortunately, poor weather or low prices may prevent a farm from reaching its goals, even in years when no serious management problems exist. Therefore, a second source of useful standards for comparison is actual results from other farms of similar size and type, for the same year. These may not represent ideal standards but will indicate if the farm being analyzed was above or below average given the weather and market conditions that existed.

There are organized record-keeping services available through lending agencies, accounting firms, farm cooperatives, and university extension services. One of the advantages of participating in such a service is the annual record summary provided for all farms using the service. These summaries contain averages and ranges of values for different analysis measures for all farms, as well as for groups of farms sorted by size and type. Such results provide an excellent set of comparative standards as long as the farms being compared are similar and in the same general geographic area.

Historical Trends

When using trend analysis, the manager records the various measures and ratios for the same farm for a number of years and observes any trends in the results. The manager looks for improvement or deterioration in each measure over time. Those areas showing a decline deserve further analysis to determine the causes. This method does not compare results against any particular standards, only against previous performance. The objective becomes one of showing improvement over the results of the past. However, trend analysis must take into account year-to-year changes in weather, prices, and other random variables.

Any of the analytical measures discussed in the remainder of this chapter can be used with budget, comparative, and trend analysis. These measures are shown in tables that contain results from an actual farm as well as average values for a group of similar farms for the same year and state.[1]

DIAGNOSING A FARM BUSINESS PROBLEM

A complete whole-farm business analysis can be carried out by using a systematic procedure to identify the source of a problem. Figure 18-1 illustrates a procedure that can make the process more efficient.

Profitability is generally the first area of concern. Low net farm income or returns to management can have many causes. The farm or ranch may not be large enough to generate the level of production needed for an adequate income. Measures of farm size should be compared to those of other farms supporting the same number of operators. If the level of production is too low, it may be because not enough resources are employed. The operator should look for ways to farm more acres, increase the labor supply, expand livestock, or obtain more capital. If farm size cannot be increased, then fixed costs such as machinery and building depreciation, interest, and general farm overhead costs should be evaluated carefully. Steps should be taken to reduce the costs that have the least effect on the level of production. Off-farm employment or other types of self-employment may also have to be considered as a way to improve family income.

If adequate resources are available but production levels are low, then resources are not being used efficiently. Computing several *economic efficiency* measures can identify areas of inefficient resource use, such as capital, labor, or land. Poor economic efficiency may be due to

---

[1] The analysis measures discussed in the remainder of this chapter include those recommended by the Farm Financial Standards Council, as well as a number of others. Those recommended by the Council are identified by the abbreviation (FFSC) shown after their name.

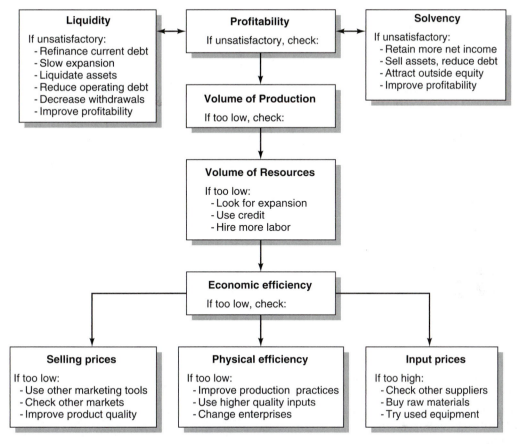

Figure 18-1   Procedure for diagnosing a farm business problem.

low physical efficiency, low selling prices, and/or high input costs. If efficiency measures for the current enterprises cannot be improved, it may mean the wrong enterprises are in the current farm plan. Enterprise analysis and enterprise budgeting can be used to identify more profitable enterprises and develop a new whole-farm plan.

This procedure should help managers isolate and identify the causes of a profitability problem quickly and systematically. However, even profitable farms need to be concerned about *liquidity* and *solvency.* If cash flow always seems to be tight, even when net farm income is satisfactory, the operation may need to refinance some current debt, slow down expansion, sell off some assets, reduce nonfarm withdrawals, or

pay down operating debt. If solvency is not as strong as the operator would like it to be, more of the net farm income may need to be retained in the business each year, or some assets can be sold to reduce debt levels. The three areas of farm business management—profitability, solvency, and liquidity—are all closely related. Outstanding farm managers do an above-average job in all of them.

MEASURES OF PROFITABILITY

Profitability analysis starts with the income statement, as was explained in Chapter 6 where several measures of profitability were discussed. These measures are shown in Table 18-2, along

TABLE 18-2	Measures of Farm Profitability		

Item	Average of comparison group	Our farm	Comments
1. Net farm income	$ 35,123	$ 10,203	Much lower
2. Return to labor and management	$ −1,477	$ −6,133	Lower
3. Return to management	$−21,477	$−26,533	Lower
4. Rate of return on farm assets (ROA)	3.1%	2.9%	About the same
5. Rate of return on farm equity (ROE)	0.4%	−4.7%	Lower, probably due to high financial leverage
6. Operating profit margin ratio	9.2%	12.3%	Higher

Source: Illinois Farm Business Farm Management Association (dairy farms).

with values from an actual dairy farm and average values from a group of Illinois dairy farms in the same year.

Net Farm Income (FFSC) Net farm income is a measure of return to the equity capital, unpaid labor, and management contributed by the owner/operator to the farm business. Net farm income also includes gains or losses from selling capital assets. One method of establishing a goal for net farm income is to estimate the income that the owner's labor, management, and capital could earn in nonfarm uses. In other words, the total opportunity cost for these factors of production becomes the minimum goal or standard for net farm income. In any year for which the income statement shows a lower net farm income, one or more of these factors did not earn its opportunity cost. Another goal would be to reach a specific income level at some time in the future, such as at the end of 5 years. Progress toward this goal can be measured over time.

Although net farm income is influenced heavily by the profitability of the farm, it also depends on what proportion of the total resources utilized are contributed by the operator.

Replacing borrowed capital with equity capital, rented land with owned land, or hired labor with operator labor can all improve net farm income. Figure 18-2 shows how the gross farm revenue "pie" is divided among the parties who supply resources to the farm or ranch business. The size of the farmer's slice of the pie depends on how many of the resources he or she contributes, as well as the size of the pie. Other resource contributors, such as lenders, landlords, employees, and input suppliers must be paid first. What is left over is net farm income. Farms with high net incomes are often those that have accumulated a large equity over time, or that depend heavily on operator and family labor.

Return to Labor and Management Some farm businesses have more assets or borrow more money than others. As shown in Chapter 6, a fair comparison requires that the interest expense incurred by the business be added back to net farm income from operations to get adjusted net farm income. Next, an opportunity cost for the investment in total farm assets is subtracted. What remains is the return to the unpaid labor and management utilized in the business. One goal or standard for return to labor

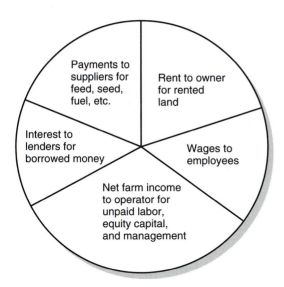

Figure 18-2   **How is the total revenue pie divided?**

and management is the opportunity cost of the operator's labor and management in another use.

Return to Management   The return to management is the portion of adjusted net farm income that remains after the opportunity costs of both unpaid labor and equity capital have been subtracted. It represents the residual return to the owner for the management input, and it can be highly variable from year to year. Negative returns to management in years of low prices or production are not unusual, but the goal should be to earn a positive return over the long run.

Rate of Return on Farm Assets (FFSC)
The rate of return on farm assets (ROA) was discussed in Chapter 6. This value can be compared to rates of return on other long-term investments, although differences in risk and potential gains in asset values must also be considered. It also allows comparison among farms of different sizes and types. Because ROA includes a return to both equity capital and debt capital (interest on loans), it is independent of how highly leveraged the business is. It includes

only net income from operations, though, and not capital gain or losses.

Rate of Return on Farm Equity (FFSC)
The return on equity (ROE) is perhaps more indicative of financial progress than ROA, as it measures the percent return to the owner's business equity. If the farm business has no debts, the rate of return on farm equity will be equal to the rate of return on farm assets, because they are one and the same. When debt capital is used, though, interest must be paid. The dollars available as a return on equity then are only that part of net farm income remaining after interest is paid. Opportunity costs for labor and management are also subtracted before computing the rate of return to equity. See Chapter 6 for details on how to compute both ROA and ROE.

Operating Profit Margin Ratio (FFSC)   As explained in Chapter 6, the operating profit margin ratio (OPMR) measures the proportion of gross revenue that is left after paying all expenses. It is found by dividing the dollar return to farm assets (adjusted net farm income minus the opportunity cost of labor and management) by the gross revenue of the farm.

Farms with large investments in fixed assets such as land and buildings, and few operating expenses such as cash rent, will generally show higher operating profit margin ratios. Conversely, farms with more rented assets will often have a higher rate of return on farm assets (ROA) but a lower OPMR. Rented and owned assets are substitutes for each other. As more of one is used, the rate of return to the other will generally increase. Thus, these two measures of profitability must be considered together. An operation may be high in one ratio and low in the other simply because of the mix of owned and rented assets used. If both measures are below the average or the goal, then profitability problems are evident.

General Comments   There are two cautions regarding the procedures used to calculate returns to labor, management, assets, and equity.

The first is the somewhat arbitrary nature of estimating the opportunity costs used in the calculations. There are no rigid rules to follow, and individuals may have different opinions about the appropriate values to use. Obviously, the calculated return to any factor will change if different opportunity costs are used for the others. The second caution is that these values are *average* returns to the factors, not the marginal returns. For example, the rate of return to assets is the average rate of return on all capital invested in the business, not the marginal return on the last dollar invested. The marginal rates of return would be much more useful for planning purposes, especially for new investments.

Nevertheless, if the opportunity costs are estimated carefully and used consistently from year to year, and the average nature of the result is kept in mind, these measures can provide a satisfactory means for making historical and between-farm comparisons of profitability. They are, however, less useful for evaluating marginal changes in the organization and operation of the farm business.

The measures of profitability included in Table 18-2 can be used to identify the existence of a problem. However, these measures do not identify the exact source of the problem. Further analysis is needed to clarify the cause of the problem and suggest the corrective action

needed. The next three sections proceed with a more in-depth analysis of the farm business to pinpoint more specific problem areas.

MEASURES OF SIZE

An income or profitability problem can exist in any single year due to low selling prices or low crop yields. If the problem persists even in years with above-average prices and yields, it may be caused by insufficient farm size. Historically, net farm income has been highly correlated with farm size. Small farms sometimes make inefficient use of fixed capital investments and labor. Small farms that do make efficient use of resources may still fall short of generating enough net income to support the operator and family, though. Several common measures of farm size are shown in Table 18-3. Some capture the quantity or value of production generated, while others measure the volume of resources used.

Quantity of Sales   For specialized farms, the number of units of output sold annually is a convenient measure of size. Examples are 5,000 head of farrow-to-finish hogs or 25,000 boxes of apples. On a diversified farm, it may be necessary to record the units sold from several different enterprises to more accurately reflect farm size. Where a crop share or livestock share lease

	TABLE 18-3	**Measures of Farm Size**	

Item	Average of comparison group	Our farm	Comments
1. Gross revenue	$288,504	$213,538	Lower
2. Value of farm production	$216,055	$148,197	Much lower
3. Total farm assets	$823,832	$738,631	Smaller
4. Total acres farmed	414 acres	295 acres	Smaller
5. Number of cows milked	89 cows	52 cows	Smaller
6. Total labor	1.8 persons	1.0 person	Smaller

Source: Illinois Farm Business Farm Management Association (dairy farms).

is being used, both the owner's and tenant's shares of production should be included when making comparisons with other farms.

Total or Gross Revenue The volume of production from a farm producing several different products can be measured by total revenue. Table 6-2 shows the calculations where *total revenue* is defined as total sales and other cash income, adjusted for changes in inventories, accounts receivable, and any gains/losses on the sale of breeding livestock.

Value of Farm Production Some livestock operations buy feeder livestock and feed, while others produce all their own feed and animals. To make a fair comparison of size of business, the value of feed and livestock purchases is subtracted from gross revenue to arrive at the *value of farm production.* In this measure, only the value added to purchased feed and livestock is credited to the value of production. It is calculated as follows:

- Total (gross) revenue
- *minus* livestock purchases
- *minus* feed purchases
- *equals* value of farm production

Value of farm production for a given farm may vary substantially from year to year, depending on weather, disease, pests, and product prices. However, it is a convenient way to compare the size of different types of farms.

Total Farm Assets Another measure of size is the total capital invested in land, buildings, machinery, livestock, crops, and other assets. The market value of total farm assets from the balance sheet is the easiest place to obtain this value. With all values in dollar terms, this measure allows an easy comparison of farm size across different farm types. However, it does not take into account the value of rented or leased assets.

Total Acres Farmed The number of acres farmed includes both owned and rented land. It is useful for comparing similar types of farms but is not particularly good for comparing farms with different types of land. A California farm with 200 acres of irrigated vegetables is obviously not similar in size to an operation with 200 acres of dryland wheat in western Kansas, or to a 200-acre cattle feedlot in Texas. Number of acres is best used for comparing the size of crop farms of the same general type in an area with similar soil resources and climate.

Livestock Numbers It is common to speak of a 500-cow ranch or a 250-cow dairy farm or a 500-sow hog farm. These measures of size are useful for comparing size among farms with the same class of livestock.

Total Labor Used Labor is a resource common to all farms. Terms such as a one- person or two-person farm are frequently used to describe farm size. A full-time worker equivalent can be calculated by summing the months of labor provided by operator, family, and hired labor and dividing the result by 12. For example, if the operator works 12 months, family members provide 4 months of labor, and 5 months of labor is hired, the full-time worker equivalent is 21 months divided by 12, or 1.75 years of labor. This measure of size is affected by the amount of labor-saving technology used and should therefore be interpreted carefully when comparing farm sizes.

EFFICIENCY MEASURES

Some of the measures of farm size presented here are measures of *production,* such as sales, gross revenue, and value of farm production. Others such as land area, value of total assets, labor employed, and number of breeding livestock owned measure the quantity of *resources* used. Will the farms or ranches with the most resources also have the most production? Generally speaking, yes, but there will always be exceptions. Some managers are able to generate more production or use fewer resources than their neighbors because they use their resources more efficiently. A general definition for efficiency is the quantity or value of production achieved per unit of resource employed.

$$\text{Efficiency} = \frac{\text{production}}{\text{resources used}}$$

If a comparison with other farms or with a budget goal shows that an operation has an adequate volume of resources but is not reaching its production goals, then some resources are not being used efficiently. Because a farm or ranch business may use many types of resources, there are many ways to measure both economic and physical efficiency. Some of the more useful and common ones will be discussed here.

Economic Efficiency

Economic efficiency measures are computed either as dollar values per unit of resource or as some rate or percentage relating to capital use. Measures of economic efficiency must be interpreted carefully. They measure average values, not the marginal values or the effect that small changes would have on overall profit. Some of the measures to be discussed are shown in Table 18-4.

Asset Turnover Ratio (FFSC) This ratio measures how efficiently capital invested in farm assets is being used. It is found by dividing the gross revenue generated by the market value of total farm assets. For example, an asset turnover ratio equal to 0.25, or 25 percent, indicates that gross revenue for one year was equal to 25 percent of the total capital invested in the business. At this rate, it would take 4 years to produce agricultural products with a value equal to the total assets.

The asset turnover ratio will vary by farm type. Dairy, hog, and poultry farms will generally have higher rates, beef cow farms tend to be

TABLE 18-4	Measures of Economic Efficiency		

Item	Average of comparison group	Our farm	Comments
Capital efficiency:			
1. Asset turnover ratio	.30	.20	Lower
2. Operating expense ratio	.62	.59	About the same
3. Depreciation ratio	.13	.15	Slightly higher
4. Interest expense ratio	.10	.19	Much higher
5. Net farm income from operations ratio	.15	.07	Lower
Livestock efficiency:			
4. Production per $100 of feed fed	$183	$141	Much lower
5. Feed cost per 100 lb of milk	$7.39	$9.33	Higher
Land efficiency:			
6. Crop value per acre	$294	$324	Higher
Labor efficiency:			
7. Gross revenue per person	$120,030	$117,616	About the same
Machinery efficiency:			
8. Machinery cost per crop acre	$68	$55	Lower

Source: Illinois Farm Business Farm Management Association (dairy farms).

lower, and crop farms usually have intermediate values. Farms or ranches that own most of their resources will generally have lower asset turnover ratios than those that rent land and other assets. Therefore, the asset turnover ratio should be compared only among farms of the same general type.

Operating Expense Ratio (FFSC) Four operational ratios are recommended to show what percent of gross revenue went for operating expenses, depreciation, interest, and net income. The operating expense ratio is computed by dividing total operating expenses (excluding depreciation) by gross revenue. Farms with a high proportion of rented land and machinery or hired labor will tend to have higher operating expense ratios.

Depreciation Expense Ratio (FFSC) This ratio is computed by dividing total depreciation expense by gross revenue. Farms with a relatively large investment in newer machinery, equipment, and buildings will have higher depreciation expense ratios. A higher than average ratio may indicate underutilized capital assets.

Interest Expense Ratio (FFSC) Total farm interest expense (adjusted for accrued interest payable at the beginning and end of the year) is divided by gross revenue to find this ratio. Ratios higher than average, or higher than desired, may indicate too much dependence on borrowed capital or high interest rates on existing debt.

Net Farm Income from Operations Ratio (FFSC) Dividing net farm income from operations by gross revenue measures what percent of gross revenue is left after paying all expenses (but before subtracting any opportunity costs).

Note that these four operational ratios will sum to 1.00, or 100 percent. They can also be calculated using the value of farm production as a base instead of gross revenue. In that case, the cost of purchased feed and livestock should not be included in operating expenses. The asset turnover ratio can also be calculated using the

value of farm production. This is more accurate when comparing operations that purchase large quantities of feed and livestock against those that do not.

Livestock Production per $100 of Feed Fed
This is a common measure of economic efficiency in livestock production. It is calculated by dividing the value of livestock production during a period by the total value of all the feed fed in the same period, and multiplying the result by 100. The value of livestock production is equal to:

- Total cash income from livestock
- *plus* or *minus* livestock inventory changes
- *minus* value of livestock purchased
- *equals* value of livestock production

Livestock production per $100 of feed fed is computed from the following:

$$\frac{\text{value of livestock production}}{\text{value of feed fed}} \times 100$$

A ratio equal to $100 indicates that the livestock just paid for the feed consumed but no other expenses. The return per $100 of feed fed also depends on the type of livestock, as shown in the following table:

Livestock enterprise	Production per $100 of feed fed (10-year average, 1992–2001)
Beef cow	$153
Dairy	197
Feeding cattle	138
Hogs (farrow to finish)	160

Source: Iowa Farm Costs and Returns, Iowa State University Extension Publication FM-1789, 1992–2001.

Higher values for some livestock enterprises do not necessarily mean that these enterprises are more profitable. Enterprises with higher building and labor costs, or other nonfeed costs such as dairy and feeder pig production, need higher feed returns in order to be as profitable as enterprises with low nonfeed costs, such as finishing feeder animals. Because of these differences, the production per $100 of feed fed should be calculated for each individual livestock enterprise and compared only with values for the same enterprise.

Feed Cost per 100 Pounds of Gain   The feed cost per 100 pounds of gain (or per 100 pounds of milk for a dairy enterprise) is found by dividing the total feed cost for each enterprise by the total pounds of production, and multiplying the result by 100. The total pounds of production for the year should be equal to the pounds sold and consumed minus the pounds purchased, and plus or minus any inventory changes. This measure is affected by feed prices and feed conversion efficiency, and values should be compared only among farms with the same livestock enterprise.

Crop Value per Acre   This value measures the intensity of crop production and whether or not high-value crops are included in the crop plan. It is calculated by dividing the total value of all crops produced during the year by the number of crop acres. The result does not take into account production costs and therefore does not measure profit.

Gross Revenue per Person   The efficiency with which labor is being utilized can be measured by dividing gross revenue for the year by the number of full-time equivalents of labor used, including hired workers. Since capital and labor can be substituted for each other, labor and capital efficiency measures should be evaluated jointly.

Machinery Cost per Crop Acre   To calculate this measure, the total of all annual costs related to machinery is divided by the number of crop acres. Total annual machinery costs include all ownership and operating costs, plus the cost of custom work hired and machinery lease and rental payments. Cost for any machinery used mainly for livestock should be excluded.

This is another measure that can be either too high or too low. A high value indicates that machinery investment may be excessive, while very low values may indicate that the machinery is too old and unreliable, or too small for the acres being farmed. There is an important interaction between machinery investment, labor requirements, timeliness of field operations, and the amount of custom work hired. These factors are examined more thoroughly in Chapter 22.

Physical Efficiency

Poor economic efficiency can result from several types of problems. *Physical efficiency,* such as the rate at which seed, fertilizer, and water are converted into crops, or the rate at which feed is turned into livestock products, may be too low.

Physical efficiency measures include bushels harvested per acre, pigs weaned per sow, and pounds of milk sold per cow. An example of physical efficiency measures is shown in Table 18-5. Just like the measures of economic efficiency, they are average values and must be interpreted

TABLE 18-5	Measures of Physical Efficiency		
Item	Average of comparison group	Our farm	Comments
1. Corn yield per acre	135 bu.	156 bu.	Higher
2. Milk production per cow per year	18,191 lb.	18,660 lb.	About the same

Source: Illinois Farm Business Farm Management Association (dairy farms).

as such. An average value should not be used in place of a marginal value when determining the profit-maximizing input or output level.

Selling Prices

The average price received for a product may be lower than desired, for several reasons. In some years, supply and demand conditions cause prices in general to be low. However, if the average price received is lower than for other similar farms in the same year, the operator may want to look for different marketing outlets or spend some time improving marketing skills. Poor product quality also can cause selling prices to be below average.

Purchase Prices

Sometimes physical efficiency is good and selling prices are acceptable, but profits are low because resources were acquired at high prices. Examples are buying land at a price above what it will support, paying high cash rents, paying high prices for seed and chemical products, buying overly expensive machinery, purchasing feeder livestock at high prices, borrowing money at high interest rates, and so forth. Input costs can be reduced by comparing prices among different suppliers, buying used or smaller equipment, buying large quantities when possible, and substituting

raw materials for more refined or more convenient products.

Physical efficiency, selling price, and purchase price all combine to determine economic efficiency, as shown in the following equations:

$$\frac{\text{Economic}}{\text{efficiency}} = \frac{\text{value of product produced}}{\text{cost of resources used}}$$

$$\text{or} = \frac{\text{quantity of product} \times \text{selling price}}{\text{quantity of resource} \times \text{purchase price}}$$

FINANCIAL MEASURES

Even profitable farms may suffer from tight cash flows or high debt loads. The financial portion of a complete farm analysis is designed to measure the solvency and liquidity of the business and to identify weaknesses in the structure or mix of the various types of assets and liabilities. The balance sheet and the income statement are the primary sources of data for calculating the measures related to the financial position of the business. These financial measures and ratios were presented in Chapters 5 and 6 but will be reviewed briefly to show how they fit into a complete farm analysis.

It is convenient to compare the actual financial performance to values from a comparison group on a form such as that shown in Table 18-6. Actual values should be compared

TABLE 18-6	Analyzing the Financial Condition of the Business		

Item	Average of comparison group	Our farm	Comments
1. Net worth	$601,688	$299,853	About half
2. Debt/asset ratio	.28	.59	Much higher
3. Current ratio	1.46	1.54	Slightly higher
4. Working capital	$157,311	$21,722	Much lower
5. Term debt and capital lease coverage ratio	1.54	1.16	Much lower

Source: Illinois Farm Business Farm Management Association (dairy farms).

to historical values, as well, to identify any trends that may be developing.

Solvency

Solvency refers to the value of assets owned by the business compared to the amount of liabilities, or the relation between debt and equity capital.

Debt/Asset Ratio (FFSC)   The debt/asset ratio is a common measure of business solvency and is calculated by dividing total farm liabilities by total farm assets, using current market values for assets:

$$\text{Debt/asset ratio} = \frac{\text{total farm liabilities}}{\text{total farm assets}}$$

Smaller values are usually preferred to larger ones, as they indicate a better chance of maintaining the solvency of the business should it ever be faced with a period of adverse economic conditions. However, low debt/asset ratios may also indicate that a manager is reluctant to use debt capital to take advantage of profitable investment opportunities. In this case, the low ratio would be at the expense of developing a potentially larger business and generating a higher net farm income. A more complete discussion of the use of debt and equity capital is found in Chapter 19.

Two other common measures of solvency are the equity/asset ratio (FFSC) and the debt/equity ratio (FFSC). These were discussed in Chapter 5. They are simply mathematical transformations of the debt/asset ratio and provide the same information.

Change in Equity   An important indicator of financial progress for a business is an owner's equity that is increasing over time. This is an indication that the business is earning more net income than is being withdrawn. Reinvesting profits allows for acquiring more assets or reducing debt. As discussed in Chapter 5, equity should be measured both with and without the benefit of inflation and its effect on the value of assets such as land. This requires a balance sheet prepared on both a cost and a market basis, so that any increase in equity can be separated into the result of profits generated by the production activities during the accounting period, and the effect of changes in asset values.

Liquidity

The ability of a business to meet its cash flow obligations as they come due is called *liquidity*. Maintaining liquidity is important to keep the financial transactions of the business running smoothly.

Current Ratio (FFSC)   This ratio is a quick indicator of a farm's liquidity. The current ratio is calculated by dividing current farm assets by current farm liabilities:

$$\text{Current ratio} = \frac{\text{current farm assets}}{\text{current farm liabilities}}$$

The current (or short-term) assets will be sold or turned into marketable products in the near future. They will generate cash to pay the current liabilities; that is, the debt obligations that come due in the next 12 months.

Working Capital (FFSC)   Another measure of liquidity, called working capital, is the difference between current assets and current liabilities. It represents the excess dollars that would be available from current assets after current liabilities have been paid.

$$\begin{array}{c}\text{Working}\\\text{capital}\end{array} = \left(\begin{array}{c}\text{current farm}\\\text{assets}\end{array}\right) - \left(\begin{array}{c}\text{current farm}\\\text{liabilities}\end{array}\right)$$

These liquidity measures are easy to calculate but fail to take into account upcoming operating costs, overhead expenses, capital replacement needs, and nonbusiness expenditures. They also do not include sales from products yet to be produced and do not consider the timing of cash receipts and expenditures throughout the year. An operation that concentrates its production during one or two periods of the year, such as a cash grain farm, needs to have a high current ratio at the beginning of the year, because no new production will be available for sale until near the end of the year. On the other hand, dairy producers or other livestock operations with continuous sales throughout the year can operate safely with lower current ratios and working capital margins.

Chapter 13 discussed how to construct a cash flow budget and how to use it to manage liquidity. Although developing a cash flow budget takes more time than computing a current ratio, it takes into account all sources and uses of cash, as well as the timing of their occurrence throughout the year. Every farm or ranch that has any doubts about being able to meet cash flow needs should routinely construct cash flow budgets and monitor them against actual cash flows throughout the year. This allows early detection of liquidity problems.

Tests for Liquidity Problems   When overall cash flow becomes tight, there are several simple tests that can be carried out to try to identify the source of the problem.

1. Analyze the debt structure by calculating the percent of total farm liabilities classified as current and noncurrent. Do the same for total farm assets. If the debt structure is "top heavy," that is, a larger proportion of the farm's liabilities are in the current category than is true for its assets, it may be advisable to convert some of the current debt to a longer-term liability. This will reduce the annual principal payments and improve liquidity.
2. Compare the changes in the current assets over time, particularly crop and livestock inventories. This can be done by looking at the annual year-end balance sheets. Building up inventories can cause temporary cash flow shortfalls. On the other hand, if cash flow has been met by reducing current inventories for several

successive years, then the liquidity problem has only been postponed.
3. Compare purchases of capital assets to sales and depreciation. Continually increasing the inventory of capital assets may not leave enough cash to meet other obligations. On the other hand, liquidating capital assets may help meet cash flow needs in the short run, but could affect profitability in the future.
4. Compare the amount of operating debt carried over from one year to the next for a period of several years. Borrowing more than is repaid each year will eventually reduce borrowing capacity, increase interest costs, and make future repayment even more difficult. This only hides the real problem of low profitability.
5. Compare the cash flow needed for land (principal and interest payments, property taxes, and cash rent) per acre to a typical cash rental rate for similar land. Taking on too much land debt, trying to repay it too quickly, or paying high cash rents can cause severe liquidity problems.
6. Compare withdrawals for family living and personal income taxes to the opportunity cost of unpaid labor. The business may be trying to support more people than are fully employed, or at a higher level than is possible from net farm income.

Any of these symptoms can cause poor liquidity, even for a profitable farm, and can eventually lead to severe financial problems.

Measures of Repayment Capacity

Two measures are recommended to evaluate the debt repayment capacity of the business. The first is the *term debt and capital lease coverage ratio* (FFSC), where term debt refers to liabilities that have scheduled, amortized payments. It is the ratio of the cash available for term debt payments for the past year, divided by the total of the term debt payments due in the next year,

---

Box 18-2 Farm Business Analysis Case Study

The example farm that is illustrated in Table 18-2 through Table 18-6 had a net farm income that was considerably lower than the average net farm income for dairy farms in its state that year. It was also below average in return to labor and management, and return to management. It did, however, have a return on assets (ROA) nearly equal to the group average. Its return on equity (ROE) was lower, though, probably because the farm was more highly leveraged than average.

The example farm had considerably less value of production than average, as well as fewer resources. It was milking only 52 cows versus the group average of 89. It also had fewer total assets, fewer acres, and less labor. Part of the profitability problem was likely due to not having enough productive assets.

The farm's asset turnover ratio was lower than average. Only 7 percent of its value of farm production went to net income, while 19 percent went to interest expense. Apparently some of the low

net income was due to high debt or high interest rates. The farm also generated below average returns per dollar of feed fed, and feed costs per hundredweight of milk produced were higher than average. However, the value of crop production per acre was above average, probably due to an above average corn yield.

The farm's financial situation is very tight, due to a high debt-to-asset ratio. It has only a small amount of working capital, although the current ratio is about average. Due to a high overall debt load, the farm's term debt and capital lease coverage ratio is much lower than average.

Several areas of the business deserve attention, in order to improve profitability: (1) reduce debt load as quickly as possible, although selling off assets is probably not advisable due to the small size of the business already; (2) reduce feed costs; (3) increase the size of the business, possibly by renting additional crop land or expanding the dairy herd.

---

including principal and interest, on amortized loans plus capital lease payments. The cash available is computed by:

- net farm income from operations
- *plus* total nonfarm income
- *plus* depreciation expense
- *plus* interest paid on term debt and capital leases
- *minus* withdrawals for family living and personal income taxes

A ratio greater than 1.0 means that there is sufficient cash flow to meet the term debt payments. A projected ratio for the coming year can also be computed, using information from a cash flow budget.

The second measure is called the *capital replacement and term debt repayment margin* (FFSC). It is simply the dollar difference between the cash available and the total term debt payments due. These two measures are similar

in concept to the current ratio and working capital measures but take into account all cash inflows and outflows.

ENTERPRISE ANALYSIS

A profitability analysis done for a whole farm may indicate a problem, but the source of the problem is often difficult to find if there are many enterprises. Several enterprises may be highly profitable, while others are unprofitable or only marginally profitable. Enterprise analysis can identify the less profitable enterprises, so that some type of corrective action can be taken.

An enterprise analysis consists of allocating all income and expenses among the individual enterprises being carried out. Manual farm accounting systems often use a separate column for each enterprise. Computer systems assign enterprise code numbers that are entered with each transaction. The end result is similar to an income statement for each enterprise, showing its gross revenue, expenses, and net income. Data gathered during an enterprise analysis are extremely useful in developing enterprise budgets for future years. They can also be used to calculate the total cost of production, which is helpful for making marketing decisions. Some analyses compute a cash flow break-even cost, which is useful for financial management, as well as an economic break-even cost.

The accounting period for an enterprise analysis may be different from the whole farm accounting year. It is often more logical to summarize costs and returns over the production cycle of the enterprise, such as from when the first inputs are purchased until the last bushel of grain or head of livestock is sold. This period may include parts of several accounting years, or be less than a full year. Annual ownership costs should be allocated according to the number of production cycles that occur each year.

Crop Enterprise Analysis

An example of an enterprise analysis for peanuts is contained in Table 18-7. Values are shown both for the whole farm and per acre. The farm planted 125 acres to peanuts, or 40 percent of its total crop acres. The first step is to calculate the total income, beginning with sales, during the accounting period. For crops that are also used for livestock feed, the value of any amounts fed must also be included in total income. Finally, any crop insurance payments or government program benefits received should also be added.

Costs are summarized next. The total costs of items such as seed, fertilizer, and pesticides are relatively easy to calculate from farm records or to estimate based on the quantities actually used. Costs such as fuel, machinery repairs, depreciation, interest, and labor are more difficult to allocate fairly among enterprises, unless records are kept on the hours of machinery and labor use by each enterprise. If not, they can simply be allocated equally over all planted acres. In the example, it was estimated that the

peanuts took 35 percent of the field time used, so 35 percent of the machinery and labor costs were charged to peanuts.

Overhead and miscellaneous costs such as property and liability insurance, consultant fees, office expenses, and general building upkeep are also difficult to assign to individual enterprises. They are often allocated in the same proportion as the contribution of each enterprise to gross revenue or to all other expenses. In the example, 35 percent of the farm overhead costs and 40 percent of the land charge were assigned to this crop.

The example in Table 18-7 indicates that the peanut enterprise had a total profit above all costs of $29,291, or $234.32 per acre. These values can be compared against similar values for other crop enterprises to determine which are contributing the most to overall farm profit. If any crop shows a consistent loss for several years, action should be taken either to improve its profitability or to shift resources to the production of some other more profitable crop.

One word of caution is needed, however. Enterprise analysis does not identify or value any complementary or detrimental interactions between enterprises. Where the presence of one enterprise significantly affects the performance of another, a farming systems approach must be used in which various whole-farm plans are compared.

Livestock Enterprise Analysis

An analysis of a livestock enterprise is conducted in the same manner as for a crop enterprise. However, several special problems should be noted. Inventory changes are likely to be more important in determining income for a livestock enterprise and should be carefully estimated to avoid biasing the results. Current market prices at the beginning and end of the analysis period can be used to value market livestock, but per-head values for breeding livestock should be held constant. Another problem is how to handle farm-raised feed fed to livestock. The amounts of grain, hay, and silage fed must be measured or estimated and then valued

TABLE 18-7	Example of an Enterprise Analysis for Peanuts (125 acres)

	Farm total	Percent to peanuts	Peanuts total	Per acre
Peanut production (125 acres)			387,650 lb.	3,101 lb.
Income:				
Sales of peanuts	$119,643	100%	$119,643	$957.14
USDA payments	$ 2,500	100%	2,500	20.00
Total income			$122,143	$977.14
Variable costs:				
Seed	$ 9,782	100%	$ 9,782	$ 78.26
Fertilizer	2,418	100%	2,418	19.34
Lime	4,800	40%	1,920	15.36
Pesticides	23,654	100%	23,654	189.23
Fuel, lubrication	9,680	35%	3,388	27.10
Machinery repairs	13,264	35%	4,642	37.14
Hauling and drying	8,170	100%	8,170	65.36
Insurance, miscellaneous	7,806	35%	2,732	21.86
Labor	20,000	35%	7,000	56.00
Interest on variable costs	5,471	35%	1,915	15.32
Fixed costs:				
Machinery ownership	$ 49,100	35%	$ 17,185	$137.48
Land rent	15,000	40%	6,000	48.00
Farm overhead	11,560	35%	4,046	32.37
Total costs			$ 92,852	$742.82
Profit			$ 29,291	$234.32
Average income per pound				$ 0.32
Average cost per pound				$ 0.24
Average profit per pound				$ 0.08

according to market prices during the feeding period minus transportation costs to market. Pasture can be valued using either the opportunity cost of the land itself or the income that could have been received from renting the pasture to someone else.

Internal Transactions

Table 18-8 shows an example of a net income statement using enterprise accounting, including several *internal transactions*. Net overhead (mis-

cellaneous income minus unallocated expenses) was assigned to enterprises in the same proportion as gross revenue. Machinery costs were recorded separately, then divided between crops and livestock according to estimated hours of use for each one. The value of farm-raised feed was entered as a cost to the livestock enterprise that consumed it and as income to the crop enterprise that produced it. Other internal transactions that can occur in enterprise accounting include transferring manure value from a livestock to a crop

	Whole farm	Crops	Cattle	Machinery	Overhead
TABLE 18-8	**Income Statement Example with Enterprise Accounting**				

Income

	Whole farm	Crops	Cattle	Machinery	Overhead
Cash crop sales	$ 42,644	$ 42,644			
Cash livestock sales	72,271		72,271		
Government payments	2,100	2,100			
Miscellaneous income	3,369				$3,369
Home consumption	427		427		
Livestock inventory change	(2,870)		(2,870)		
Crop inventory change	13,835	13,835			
Total revenue	$131,776	$58,579	$69,828	$ 0	$3,369
Expenses					
Crop inputs	16,971	16,971			
Machine hire	4,693			4,693	
Fuel, lubrication	4,356			4,356	
Machinery repairs	3,780			3,780	
Building repairs	3,224	1,156	2,068		
Purchased feed	6,031		6,031		
Insurance, property taxes	3,462				3,462
Utilities	2,056	456	1,600		
Interest paid	19,433	15,000	3,000		1,433
Livestock health, supplies	1,228		1,228		
Miscellaneous	4,021				4,021
Depreciation	19,058	1,688	3,351	12,944	1,075
Total expenses	$ 88,313	$35,271	$17,278	$25,773	$9,991
Net farm income	$ 43,463	$23,308	$52,550	($25,773)	($6,622)
Internal transactions					
Raised crops fed	0	39,500	(39,500)		
Manure credit	0	(4,500)	4,500		
Machine work (allocated by hours)	0	(23,773)	(2,000)	25,773	
Allocation of net overhead (proportional to gross revenue)	0	(2,357)	(4,265)	0	6,622
Net farm income by enterprise	$ 43,463	$32,178	$11,285	$ 0	$ 0

enterprise and transferring weaned livestock from a breeding to a feeding enterprise.

Verifying Inventories

To compute many physical efficiency measures for both crop and livestock enterprises, an accurate count of the quantity of production is needed. This is usually measured in bushels, tons, pounds, or head. A check on the accuracy of such inventory numbers can be made using the general rule that *sources must equal uses.*

Table 18-9 shows the relevant measures of sources and uses for a crop enterprise. Most of the quantities and values can be measured

TABLE 18-9	Verifying Crop Inventories (grain sorghum)		

Sources	Quantity (cwt.)	Value per cwt.	Value ($)
Beginning inventory	3,100	$2.85	$ 8,835
Purchased	None		
Produced	11,200		$ 34,813[*]
Total	14,300		$ 43,648

Uses	Quantity (cwt.)	Value per cwt.	Value ($)
Ending inventory	5,300	$3.10	$ 16,430
Sold	1,470	$3.25	$ 4,778
Used for seed	none		
Used for feed	7,480	$3.00	$ 22,440
Spoilage, other losses	50	$ .00	
Total	14,300		$ 43,648

[*]Equal to total value of crop increase ($43,648 − $8,835).

directly or derived from sales receipts or purchase records. However, if all the quantities or values but one are known, the unknown one can be found by calculating the difference between the sum of the sources and the sum of the uses. This is sometimes done to estimate the quantity or value of feed fed to livestock, or the amount of crop produced. However, greater accuracy will be achieved if all these quantities are measured directly and the equality relation is used to verify their accuracy; that is, that the sources total is equal to the uses total.

The total value of the crop produced can be found by subtracting the value of the beginning inventory and purchases from the total value of all "uses." The value of crop produced includes any gain or loss on inventories due to price changes.

Livestock inventories can be verified with a similar procedure by adding an additional column for weight, as shown in Table 18-10. For livestock, physical quantities are often measured in pounds or hundredweight as well as number of head. Calves, pigs, or lambs lost to death are assumed to have an ending weight and value of zero. Some animals may also enter or leave the inventory when they are reclassified, such as female offspring selected for breeding stock replacement, or calves transferred from the cow herd to the feedlot.

The blanks for "produced" weight and value can be calculated by subtracting the total of the other "sources" from the total for "uses." The differences are the total pounds of weight gain and the total value of livestock increase, respectively.

FARM BUSINESS ANALYSIS AND ACCOUNTING

Most farm accounting programs also calculate many of the measures and ratios used for farm business analysis. It is particularly convenient if most of the required values can be drawn directly from the accounting data that have already been entered and summarized. Many farm software vendors offer separate modules that keep detailed inventory records and calculate analysis ratios for specific enterprises. Some are stand-alone packages, while others interface with the whole-farm accounting routines. Several programs also offer trend analysis and graphics.

| TABLE 18-10 | Verifying Livestock Inventories (cattle) | | | |

Sources	Head	Weight (cwt.)	Value per cwt.	Value ($)
Beginning inventory	315	1,890	$80	$151,200
Purchased	265	1,908	$75	$143,100
Produced	175	2,843*		$194,398**
Reclassified in	0	0		0
Total	755	6,641		$488,698

Uses	Head	Weight (cwt.)	Value per cwt.	Value ($)
Ending inventory	296	1,702	$84	$142,968
Sold	415	4,648	$70	$325,360
Death loss	11	xxx		xxx
Reclassified out	30	255	$70	$17,850
Home used	3	36	$70	$2,520
Total	755	6,641		$488,698

*Equal to total hundredweight of gain produced (6,641 − 1,890 − 1,908).
**Equal to total value of livestock increase ($488,698 − $151,200 − $143,100).

Most whole-farm accounting programs also have the ability to perform basic enterprise analysis. Receipts and variable expenses are identified by enterprise as they are entered, usually by a code number. Some programs also have a procedure for automatically allocating overhead costs among enterprises. The computer can then quickly sort through all receipts and expenses, collect and organize those that belong to a particular enterprise, and present the results in total dollars, cost per acre, or cost per unit of output. To do business and enterprise analysis correctly, the software must track physical quantities of inputs and products, as well as monetary values, and allow for internal transactions among enterprises, such as for farm-raised feed. An additional code will allow enterprises to be summarized by production period as well as by accounting year.

SUMMARY

A whole-farm business analysis is much like a complete medical examination. It should be conducted periodically to see whether symptoms exist that indicate the business is not functioning as it should. Standards for comparison can be budgeted goals, results from other similar farms, or past results from the same farm. A systematic approach can be used to compare these standards to actual results.

Profitability can be measured by net farm income and returns to labor, management, total assets, and equity. If an income or profitability problem is found, the first area to look at is farm size. An analysis of size checks to see if there are enough resources to generate an adequate net income, and how many of the resources are being contributed by the operator. Next, various measures relating to efficient use of machinery, labor, capital, and other inputs can be computed. Economic efficiency depends on physical efficiency, the selling price of products, and the purchase price of inputs. Finally, enterprise analysis can be used to identify the less profitable production areas on the farm so they can be improved or eliminated.

Financial analysis uses balance-sheet values to evaluate business solvency and liquidity. The various measures calculated for a whole-farm analysis are part of the control function of management. As such, they should be used to identify and isolate a problem before it has a serious negative impact on the business.

QUESTIONS FOR REVIEW AND FURTHER THOUGHT

1. What are the steps in the control function of farm management?

2. What types of standards or values can be used to compare and evaluate efficiency and profitability measures for an individual farm? What are the advantages and disadvantages of each?

3. What are the differences between "gross revenue" and "value of farm production"? Between "net farm income" and "return to management"?

4. Use the following data to calculate the profitability and efficiency measures listed below:

Gross revenue		$185,000
Value of farm production		$167,000
Net farm income		$ 48,000
Interest expense		$ 18,000
Value of unpaid labor		$ 31,000
Opportunity cost on farm equity		$ 17,300
Total asset value:	beginning	$400,000
	ending	$430,000
Farm equity:	beginning	$340,000
	ending	$352,000

   a. Rate of return on assets   _____ %
   b. Rate of return on equity   _____ %
   c. Asset turnover ratio   _____
   d. Return to management $_____
   e. Net farm income from operations ratio   _____

5. What three general factors determine economic efficiency?

6. Tell whether each of the following business analysis measures relates to volume of business, profitability, economic efficiency, or physical efficiency.
   a. Pounds of cotton harvested per acre
   b. Return to management
   c. Livestock return per $100 of feed fed
   d. Gross revenue
   e. Total farm assets
   f. Asset turnover ratio

7. Would a crop farm that cash rents 1,200 crop acres and owns 240 acres be more likely to have a better than average *asset turnover ratio* or *operating expense ratio*? Why?

8. Explain the difference between "solvency" and "liquidity." List three tests that can be used to diagnose liquidity problems.

9. Select an enterprise with which you are familiar, and list as many appropriate physical efficiency measures as you can. Which measures do you think have the greatest impact on the profitability of that enterprise?

10. Given the inventory, purchase, and sales data following for a beef feedlot:

	Head	Weight (lb)	Value ($)
Beginning inventory	850	765,000	$ 612,000
Ending inventory	1,115	936,600	730,548
Purchases	1,642	1,018,040	865,334
Sales	1,340	1,586,000	1,064,630
Death loss	____	xxx	xxx
Production increase	0	_____	_____

   a. What was the apparent death loss, in head?
   b. How much beef was produced, in pounds?
   c. What was the value of this production increase?

11. What is the purpose of enterprise analysis?

12. What are "internal transactions"?

13. How can overhead expenses be allocated among enterprises?

ACQUIRING RESOURCES FOR MANAGEMENT

Although developing farm management skills is the central theme of this book, few people make their living in agriculture from their management skills alone. A large portion of the revenue generated from ranching and farming goes to the providers of the physical, financial, and human resources needed for agricultural production to take place. How much of these resources are available to the manager and how they are obtained may make the difference between operating a business at a profit or a loss.

Net farm income is the return to all resources contributed by the operators. One key to improving it is to increase the quantity or quality of resources owned by the operator over time. Some resources are contributed by the operator and family. Others are obtained through borrowing, renting, or hiring. Determining the proper mix of owned and nonowned resources to use is a key management decision. This requires a long-term, strategic plan.

Chapter 19 discusses *capital* as an agricultural production resource. Capital itself does not produce agricultural products, but it can be used to purchase or rent other resources that do. Sources of capital include the operator's equity, borrowing, rented assets, or contributions by partners or investors. The use of credit to acquire capital assets is common in agriculture, but it requires careful planning and control to use it profitably and judiciously.

In dollar terms, *land* is the most valuable resource used in agricultural production. Controlling and using farmland is the subject of Chapter 20. Buying and valuing land is discussed, along with the various types of leases and their advantages and disadvantages. Owning and using land for agricultural production requires attention to resource conservation and environmental sustainability as well as profits.

Human resources in agriculture have evolved from doing hard physical labor to carrying out highly skilled tasks using sophisticated equipment and technology. Although labor used in agricultural production has been declining, its productivity has increased greatly. Chapter 21 explains the concepts of planning and managing both hired labor and the operator's own time.

Mechanization has changed the occupation of farming more than any other innovation during the last century. It has caused a rapid increase in productivity per person, which has resulted in less labor in agriculture and larger farms. *Machinery* represents a large capital investment on many farms, and Chapter 22 explores alternatives for acquiring the use of machinery services. Methods for computing and controlling machinery costs and for improving machinery efficiency are also discussed.

CAPITAL AND THE USE OF CREDIT

CHAPTER OBJECTIVES

1. To point out the importance of capital in agriculture

2. To illustrate the optimal use and allocation of capital

3. To compare different sources of capital and credit in agriculture

4. To describe different types of loans used in ranching and farming

5. To show how to set up various repayment plans for loans

6. To explain how to establish and develop credit worthiness

7. To examine factors that affect the liquidity and solvency of a farm business

Many people think of *capital* as cash, balances in checking and savings accounts, and other types of liquid funds. This is a narrow definition of capital. Capital also includes money invested in livestock, machinery, buildings, land, and any other assets that are bought and sold.

Agriculture has one of the largest capital investments per worker of any major U.S. industry. This helps make farm operators very productive. Figure 19-1 shows the changes that have taken place in the amount of capital invested in agriculture in the United States. During the 1970s, the nominal value of total agricultural assets grew rapidly. Increases in land prices accounted for much of this increase. Likewise, as land prices and inflation declined in many parts of the United States in the early 1980s, total asset value declined also. Since then, it has grown again, but at a more modest rate.

The capital investment *per farm* has risen even more rapidly than total investment in agriculture, because the number of farms and ranches in the United States has been declining. Many full-time commercial farms and ranches have capital investments over $1,000,000. Land accounts for much of this investment, but buildings, livestock, and machinery are also important. This large capital investment per farm requires a sound understanding of financial management principles to compete in today's international economy.

Credit is important to capital acquisition and use. It is the ability to borrow money with a promise to return the money in the future and pay interest for its use. The use of credit allows farmers and ranchers to acquire productive assets and pay for them later with the income they generate.

ECONOMICS OF CAPITAL USE

Broadly defined, capital is the money invested in the physical inputs used in agricultural production. It is needed to purchase or rent productive assets, pay for labor and other inputs, and finance family living and other personal expenditures. Capital use can be analyzed using the economic principles discussed in Chapters 7 and 8. The basic questions to be answered are:

1. How much total capital should be used?
2. How should limited capital be allocated among its many potential uses?

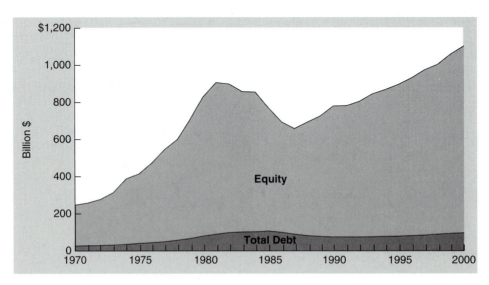

Figure 19-1   **Capital investment in U.S. agriculture, January 1.**
(*Source:* U.S. Department of Agriculture).

Total Capital Use

When unlimited capital is available, the problem is how much capital the business should use. In Chapter 7, the question of how much input to use was answered by finding the input level where the marginal value product (MVP) was just equal to the marginal input cost (MIC). The same principle can be applied to the use of capital.

Figure 19-2 is a graphical presentation of MVP and MIC, where MVP is declining, as occurs whenever diminishing marginal returns exist. The MVP is the additional net return, before interest payments, that results from an additional capital investment. This added return can be estimated using the partial budgeting techniques described in Chapter 11. The MVP is calculated in a manner similar to the return on assets (ROA), except a factor of 1.00 is added, and ROA measures average rather than marginal return.

Marginal input cost is equal to the additional dollar of capital invested plus the interest that must be paid to use it. Therefore, MIC is equal to $1 + i$, where i is the rate of interest on borrowed funds, or the opportunity cost of investing the farm's own capital. In this example, profit will be maximized by using the amount of capital represented by a, where MVP is equal to MIC; that is, the investment generates just enough revenue to repay the initial capital outlay plus the interest cost.

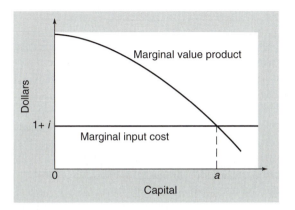

Figure 19-2   Using marginal principles to determine optimal capital use.

In some cases, the interest rate increases as more capital is used, such as when a lender classifies a borrower as being in a higher risk category. At the point where this happens, the MIC curve would rise to a higher level. The optimum amount of capital to use would then be less than when the interest rate and MIC are constant.

Allocation of Limited Capital

Many businesses do not have sufficient capital of their own or may not be able to borrow enough to reach the point where MVP is equal to MIC for the total capital being used. In other words, capital is limited to something less than the amount that will maximize total profit. The problem now becomes one of allocating limited capital among its alternative uses. This can be accomplished by using the equal marginal principle, which was discussed in Chapter 7.

Use of the equal marginal principle results in capital being allocated among alternative uses in such a way that the marginal value product of the last dollar is equal in all uses. Even if no additional capital is available for investment, there may be opportunities for shifting capital between uses to more nearly equate the marginal value products.

This principle is often difficult to apply in an actual farm situation. First, there may be insufficient information available to calculate the marginal value products accurately. Prices and costs are constantly changing. Second, some alternative uses may require large lump-sum investments of an all-or-nothing nature, such as livestock buildings, greenhouses, or irrigation systems. Third, capital invested in assets such as land or buildings cannot easily be shifted to other uses. This makes it difficult to equate the marginal value products for these alternatives with others where capital can be invested dollar by dollar. Nonetheless, difficulties encountered in applying the equal marginal principle should not discourage its use. Whenever limited capital can be reallocated to make the marginal value products more nearly equal, total profit will be increased.

Box 19-1   **The 1980's Farm Debt Crisis: What Went Wrong?**

The decade of the 1980s brought financial disaster to many U.S. farmers. Was it caused by irresponsible financial management, or by a combination of economic forces over which farmers had no control?

From the end of World War II until the early 1970s, U.S. agriculture increased production by rapidly adopting new technologies. The size of the farms increased as the number of farm units decreased by half. Yet, low profit margins and memories of the Great Depression caused farmers to be very conservative in their use of credit.

Suddenly, beginning in 1973, the U.S. farm economy boomed. World demand for agricultural commodities, combined with production shortfalls, caused prices of most farm products to double in less than two years. Farmers saw their net incomes soar, and looked to reinvest their windfalls in more land, machinery, and livestock. A plentiful supply of credit and single-digit interest rates fueled the boom (see Fig. 19-1). Lenders were only too willing to extend credit when asset values, especially land prices, were increasing at rates of 20 to 30 percent annually (see Fig. 20-1).

In the early 1980s, a series of events caused an abrupt turnaround in the farm economy. The Federal Reserve System made a decision to reduce the rate of inflation in the economy (the annual increase in the Consumer Price Index peaked at 13.5 percent in 1980). It restricted the growth of the national money supply, and interest rates suddenly rose to 15 to 20 percent and higher. The subsequent increase in the international value of the dollar caused world demand for U.S. products to decrease, and farm commodity prices plummeted. Many farmers who had borrowed large sums of money through variable interest rate loans now found themselves faced with lower incomes and higher payments.

As the demand for farmland dried up and cash-strapped farmers tried to sell land, real estate values dropped quickly. Lenders moved to foreclose on delinquent loans before collateral values became less than the loan balances. Two severe droughts in the Midwest in 1983 and 1988 further eroded many farmers' financial positions.

An estimated 200,000 to 300,000 farms failed financially during the 1980s. Many rural banks and input suppliers also closed their doors. Many of the farms that survived did so with help from family members, through negotiated write-off of debt by lenders, or because they had been very conservative with their use of credit.

Gradually, interest rates decreased, commodity prices rebounded, and farm asset markets stabilized. However, a whole generation of farmers, ranchers, and agricultural lenders had their attitudes toward capital and credit use profoundly influenced by the Farm Crisis of the 1980s.

SOURCES OF CAPITAL

Because capital consists of cash and assets purchased with cash, it is relatively easy to intermingle capital from different sources. An important part of farm financial management is the ability to obtain capital from several sources and combine it in the proper proportions in various uses.

Owner Equity

The farmer's own capital is called owner equity or net worth. It is calculated as the difference between the total assets and the total liabilities of the business, as shown on the balance sheet discussed in Chapter 5. There are several ways the operator can secure or accumulate equity. Most farmers begin with a contribution of original capital acquired through savings, gifts, or inheritances. As the farm or ranch generates profits in excess of what is withdrawn to pay personal expenses and taxes, retained earnings can be reinvested in the business. Some operators may have outside earnings, such as a non-farm job or other investment income, which they can invest in their farming operation.

Assets that are already owned may increase in value through inflation or changes in demand. This does not increase the amount or productivity of the physical assets, but additional cash can be obtained by either selling the assets or using them as collateral for a loan.

Outside Equity

Some investors may be willing to contribute capital to a farm or ranch without being the operator. Under some types of share lease agreements, the landowner contributes operating capital to buy seed and fertilizer, or even provides equipment and breeding livestock, as explained in Chapter 20. Larger agricultural operations may include limited or silent partners who contribute capital but do not participate in management. Farms that are incorporated may sell stock to outside investors. These arrangements increase the pool of capital available to the business but also obligate the business to share earnings with the investors.

Leasing

It is often cheaper to gain the use of capital assets by leasing or renting rather than owning them. Short-term leases make it easier for the operator to change the amount and kind of assets used from year to year. However, this also creates more uncertainty about the availability of assets such as land and discourages making long-term improvements. The advantages and disadvantages of leasing farm assets are discussed in more detail in Chapters 20 and 22.

Contracting

Farmers or ranchers who have very restricted access to capital or credit, or who wish to limit their financial risk, may contract their services to agricultural investors. Examples include custom feeding of cattle, finishing pigs on contract, contract broiler or egg production, and custom crop farming. Typically, the operator provides labor and management and some of the equipment or buildings, while the investor pays for the other inputs. The operator receives a fixed payment per unit of production. Any special skills the operator has can be leveraged over more units of production without increasing financial risk. However, potential returns per unit for contract operations may be lower than for a well-managed owner-operated business.

Credit

After owner's equity, capital obtained through credit is the second largest source of farm capital. Borrowed money can provide a means to:

- quickly increase business size,
- improve the efficiency of other resources,
- spread out the purchase of capital assets over time,
- withstand temporary periods of negative cash flow.

Figure 19-3 shows the use of credit by U.S. farmers and ranchers for real estate and non-real estate purposes since 1970. Farm debt increased rapidly in the late 1970s and early 1980s. However, in the mid-1980s, many farmers sold assets to reduce debt, or had loan payments forgiven when they were unable to pay. Since then, farm liabilities have again grown, but at a slow rate.

Real estate debt, or borrowing against land and buildings, accounts for approximately one-half of the total debt. Borrowing against livestock, machinery, and grain inventories accounts for the other half. Comparing total farm debt against total farm assets indicates that U.S. agriculture is in sound financial condition (see Figure 19-1). However, this does not mean that every individual farmer and rancher is in sound financial condition. There are always individuals and businesses who have more debt than they can easily service.

TYPES OF LOANS

Agricultural loans can be classified by their length of repayment, use of the funds, and type of security pledged. All of them include certain terms used in the credit industry. A prospective

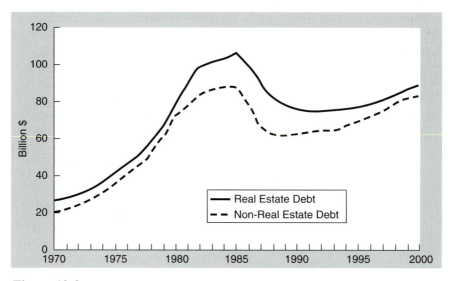

Figure 19-3 **Total U.S. farm debt by type, January 1.**
(*Source:* U.S. Department of Agriculture).

borrower needs to be familiar with these terms to communicate effectively with lenders.

Length of Repayment

Classifying loans by the length of the repayment period is widely used when preparing balance sheets, as was discussed in Chapter 5. Three time classifications are commonly used in agricultural lending.

Short-Term Loans Short-term loans are generally used to purchase inputs needed to operate through the current production cycle. Purchases of fertilizer, seed, feeder livestock, and feed are examples. Wages and rents are also financed with short-term credit. Repayment is due when the crops are harvested and sold or when the feeder livestock are sold. Short-term loans are also called *production* or *operating loans,* and are listed under current liabilities on the farm balance sheet.

Intermediate-Term Loans When a loan is repayable over more than one year but less than 10 years, it is classified as an intermediate loan.

One or more payments are usually due each year. Intermediate loans are often used for the purchase of machinery, breeding and dairy livestock, and some buildings. These assets will be used in production for several years and cannot be expected to pay for themselves in one year or less.

Long-Term Loans A loan with a term of 10 years or longer is classified as a long-term loan. Assets with a long or indefinite life, such as land and buildings, are often purchased with funds from long-term loans. Loans for the purchase of land may be made for a term of as long as 20 to 40 years, for example. Annual or semiannual payments normally are required throughout the term of the loan.

The Farm Financial Standards Council recommends that both intermediate and long-term loans be shown as *noncurrent liabilities* on the balance sheet.

Use

The use or purpose of the funds is another common criteria for classifying loans.

Real Estate Loans   This category includes loans for the purchase of real estate such as land and buildings, or where real estate assets serve as security for the loan. Real estate loans are typically long-term loans.

Non–Real Estate Loans   All business loans other than real estate loans are included in this category, and are usually short-term or intermediate-term loans. Crops, livestock, machinery or other non–real estate assets may be pledged as security.

Personal Loans   These are nonbusiness loans used to purchase personal assets such as homes, vehicles, and appliances.

Security

The security for a loan refers to the assets pledged to the lender to ensure loan repayment. If the borrower is unable to make the necessary principal and interest payments on the loan, the lender has the legal right to take possession of the mortgaged assets. These assets can be sold by the lender, and the proceeds used to pay off the loan. Assets that are pledged or mortgaged as security are called loan *collateral.*

Secured Loans   With secured loans, some asset is mortgaged to provide collateral for the loan. Lenders obviously favor secured loans, because they have greater assurance that the loan will be repaid. Intermediate and long-term loans are usually secured by a specific asset, such as a tractor or a parcel of land. Some loans are backed by a blanket security statement, which can even include assets acquired or produced after the loan is obtained, such as growing crops.

Unsecured Loans   A borrower with good credit and a history of prompt loan repayment may be able to borrow some money with only "a promise to repay," or without pledging any specific collateral. This would be an unsecured loan, also called a *signature loan,* as the borrower's signature is the only security provided the lender. Most lending practices and banking regulations discourage making unsecured loans.

Repayment Plans

There are many types and variations of repayment plans used for agricultural credit. Lenders try to fit repayment to the purpose of the loan, the type of collateral used to secure the loan, and the borrower's projected cash flow.

When a loan is negotiated, the borrower and the lender should be in agreement about when it is to be repaid. In each case, the total interest paid will increase if the money is borrowed for a longer period of time. The fundamental equation for calculating interest is

$$I = P \times i \times T$$

where I is the amount of interest to pay, P is the principal or amount of money borrowed or currently owed, i is the interest rate per time period, and T is the number of time periods over which interest accrues.

Single Payment   A single-payment loan has all the principal payable in one lump sum when the loan is due, plus interest. Short-term or operating loans are usually of this type. Single-payment loans require good cash flow planning to ensure that sufficient cash will be available when the loan is due.

The interest paid on a loan with a single payment is called *simple interest.* For example, if $40,000 is borrowed for exactly one year at 12 percent annual interest, the single payment would be $44,800, including $40,000 principal and $4,800 interest. If the loan is repaid in less than or more than one year, interest would be computed for only the actual time the money was borrowed.

$$\$40,000 \times 12\% \times 1 \text{ year} = \$4,800$$

Line of Credit   The use of single-payment loans often means having more money borrowed than the operator really needs at one time, or having to take out several individual loans. As an alternative, some lenders allow a borrower to negotiate a *line of credit*. Loan funds are transferred into the farm account as needed, up to an approved maximum amount. When farm income is received, the borrower pays the accumulated interest on the loan first, then applies the rest of the funds to the principal. There is no fixed repayment schedule or amount.

Table 19-1 contains an example of a line of credit. The amounts borrowed are $40,000 on February 1 and $20,000 on April 1, at 9 percent annual interest. On September 1, a payment of $32,850 is made to the lender. The interest due is calculated as follows:

$$\$40,000 \times 9\% \times 2/12 = \$\ \ 600$$
$$\$60,000 \times 9\% \times 5/12 = \underline{\$2,250}$$
$$\$2,850$$

Note that interest was charged on $40,000 for two months and on $60,000 for five months. The remaining $30,000 is used to reduce the principal balance from $60,000 to $30,000.

On October 1, the interest rate is reduced to 8 percent. Another payment is made on December 1, for $24,000. The interest calculation for this payment is as follows:

$$\$30,000 \times 9\% \times 1/12 = \$225$$
$$\$30,000 \times 8\% \times 2/12 = \underline{\$400}$$
$$\$625$$

Interest accrues at 9 percent for one month and 8 percent for two months. After the interest is paid, the remaining $23,375 goes toward reducing the outstanding principal balance.

A line of credit reduces the borrower's interest costs and results in less time spent in the loan approval process. However, the borrower must exercise more discipline in deciding how to use the loan funds and when and how much to borrow or repay.

Amortized An amortized loan is one that has periodic interest and principal payments. It may also be called an *installment loan*. As the principal is repaid, and the loan balance declines, the interest payments also decline. Assume that $10,000 is borrowed, and the repayment schedule is $5,000 in 6 months and the remaining $5,000 at the end of one year. The interest calculations would be as follows:

First payment:	$10,000 at 12% for ½ year = $600
Second payment:	$5,000 at 12% for ½ year = $300
Total interest:	$900

The total payments would be $5,600 and $5,300.

Notice that interest is paid only on the unpaid loan balance, and then only for the length of time that amount was still borrowed. The total interest is less than if the whole loan was repaid at the end of the year, because only $5,000 was outstanding for the last half of the year.

TABLE 19-1	Illustration of a Line of Credit

Date	Amount borrowed	Interest rate	Amount repaid	Interest paid	Principal paid	Outstanding balance
Feb. 1	$40,000	9%	$ 0	$ 0	$ 0	$40,000
April 1	20,000	9%	0	0	0	60,000
Sept. 1	0	9%	32,850	2,850	30,000	30,000
Oct. 1	0	8%	0	0	0	30,000
Dec. 1	0	8%	24,000	625	23,375	6,625

There are two types of amortization plans: the *equal principal payment* and the *equal total payment.* An amortized loan with equal principal payments has the same amount of principal due on each payment date, plus interest on the unpaid balance. For example, a 10-year, 8 percent loan of $100,000 would have annual principal payments of $10,000. Because the loan balance decreases with each principal payment, the interest payments also decrease, as shown in Figure 19-4.

Borrowers often find the first few loan payments the most difficult to make, because a new or expanded business may take some time to generate its maximum potential cash flow. For this reason, many long-term loans have an amortized repayment schedule with equal total payments, in which all payments are for the same amount. Figure 19-4 also shows the amount of principal and interest paid each year under this plan for a $100,000 loan. A large portion of the

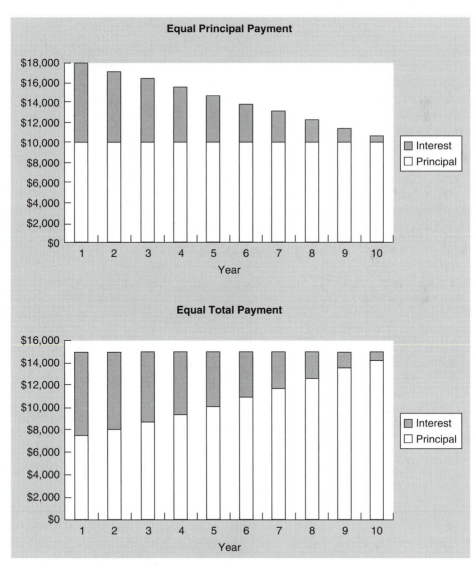

Figure 19-4   **Loan repayment under two types of amortization.**

total loan payment is interest in the early years, but the interest decreases and the principal increases with each payment, making the last payment mostly principal.

To calculate the total loan payment for an amortized loan with equal total payments, a table of amortization factors can be used. These factors are contained in Table 1 in the Appendix. The annual payment will depend on both the interest rate and the length of the loan. For example, the amortization factor for a 10-year loan at 8 percent interest is 0.14903. This factor is multiplied by the amount of the loan to find the total annual payment. A loan of $100,000 would have an annual payment of (0.14903 × $100,000), or $14,903. A financial calculator or computer program can also be used to compute the size of the payment.

Table 19-2 shows the actual principal and interest payments under each plan. The equal total payment method has a smaller total loan payment than the equal principal payment loan in the first 4 years. However, a total of $49,030 in interest is paid, compared to only $44,000 under the first plan. The advantage of the lower initial payments is partially offset by more total interest being paid over the life of the loan, because the principal is being reduced at a slower rate in the early years.

Balloon Payment Loan   Some amortization schedules are set up with lower periodic payments, such that not all of the principal is repaid by the end of the loan period. For example, half the principal may be paid with the periodic payments, with the other half due at the end of the loan period. In some cases, the periodic payments may be interest only, with all the principal due at the end of the loan period. These types of loans are called *balloon payment loans,* because the last payment "balloons" in size. They have the advantage of smaller periodic payments but the disadvantage of a large final payment, and more total interest cost. Balloon loans often require some form of refinancing to make the final payment.

The payments for a balloon loan are calculated by amortizing the amount of principal to be repaid with periodic payments using either the equal principal or equal total payment method explained above. Then, interest on the principal included in the balloon payment is added to each periodic payment. Table 19-3

| TABLE 19-2 | Amortization of a $100,000 Loan Over 10 Years at 8% Interest |

	Equal principal payments				Equal total payments			
Year	Principal paid	Interest paid	Total payment	Principal remaining	Total payment	Interest paid	Principal paid	Principal remaining
1	$ 10,000	$ 8,000	$ 18,000	$90,000	$ 14,903	$ 8,000	$ 6,903	$93,097
2	10,000	7,200	17,200	80,000	14,903	7,448	7,455	85,642
3	10,000	6,400	16,400	70,000	14,903	6,851	8,052	77,590
4	10,000	5,600	15,600	60,000	14,903	6,207	8,696	68,895
5	10,000	4,800	14,800	50,000	14,903	5,512	9,391	59,503
6	10,000	4,000	14,000	40,000	14,903	4,760	10,143	49,360
7	10,000	3,200	13,200	30,000	14,903	3,949	10,954	38,406
8	10,000	2,400	12,400	20,000	14,903	3,073	11,830	26,576
9	10,000	1,600	11,600	10,000	14,903	2,126	12,777	13,799
10	10,000	800	10,800	0	14,903	1,104	13,799	0
	$100,000	$44,000	$144,000		$149,030	$49,030	$100,000	

shows the annual payments for an equal total payment loan with a 25 percent balloon payment at the end of the tenth year, amortized with equal payments in years one through nine.

THE COST OF BORROWING

Lenders use several different methods of charging interest, which makes comparisons difficult. The true interest rate, or *annual percentage rate* (APR), should be stated in the loan agreement. Some lenders charge loan closing fees, sometimes called "points," appraisal fees, or other fees for making a loan. These fees, as well as interest rates, affect the total cost of credit.

One way to compare the cost of various credit plans is to calculate the dollar amount to be repaid in each time period (principal, interest, and other fees) and find the discounted present value of the series of payments, as described in Chapter 17. If several financing alternatives are being compared, the same discount rate should be used for all of them. Subtracting the original loan from the present value of the payments leaves the net present value, or true cost, of the

loan. The true cost can also be expressed in percentage terms, by calculating the *internal rate of return* (IRR) to the lender. These same methods can be used to calculate the true cost of a financial lease under which an operator leases equipment for several years and then purchases it. An example appears in Chapter 22.

Fixed-rate loans have an interest rate that remains the same for the entire length of the loan. However, some lenders do not like to make long-term loans at a fixed interest rate, because the rate they must pay to obtain loan funds may change. Borrowers do not like to borrow long-term at a fixed rate if they think interest rates may decrease. Predicting future interest rates is difficult for both borrowers and lenders.

For this reason, variable-rate loans have been developed, which allow for adjustment of the interest rate periodically, often annually. There may be limits on how often the rate can be changed, the maximum change on a single adjustment, and maximum and minimum rates. For example, the Farm Credit System offers a variable rate tied to the average interest rate on the bonds they issue to raise their loan capital. They also make loans with a fixed rate for a period of several years, with the understanding that this rate will be adjusted for the next time period.

Loans with a fixed interest rate typically carry a higher initial interest rate than variable-rate loans, to protect the lender against future increases in rates. The borrower must weigh this higher but known rate against the possibility that the variable rate could eventually go even higher than the fixed rate. Of course, the variable rate could also decrease. Variable-rate loans are one way to ensure that the interest rate is always near the current market rate.

SOURCES OF LOAN FUNDS

Farmers and ranchers borrow money from many different sources. Some lending institutions specialize in certain types of loans, and some provide other financial services in addition to lending money. The more important sources of

TABLE 19-3	Amortization of a $100,000 Loan Over 10 Years at 8% Interest, with a Balloon Payment

Year	Principal paid	Interest paid	Total payment	Principal remaining
1	$ 5,177	$ 8,000	$ 13,177	$94,823
2	5,591	7,586	13,177	89,232
3	6,039	7,138	13,177	83,193
4	6,522	6,655	13,177	76,671
5	7,044	6,134	13,177	69,627
6	7,607	5,570	13,177	62,020
7	8,215	4,962	13,177	53,805
8	8,873	4,304	13,177	44,932
9	9,583	3,595	13,177	35,349
10	35,349	2,828	38,177	0
	$100,000	$56,772	$156,772	

funds for both real estate and non-real estate agricultural loans are shown in Figure 19-5.

Commercial Banks

Commercial banks are the single largest source of non–real estate loans for agriculture, and they also provide some real estate loans. Banks limit their long-term loans in order to maintain the liquidity needed to meet customers' cash requirements and unexpected withdrawals of deposits. However, the use of variable interest rates, balloon payments, and secondary mortgages has allowed banks to increase their share of the farm real estate loan market.

The large share of agricultural loans held by banks is due partially to the large number of local banks in rural communities. This proximity to their customers allows bank personnel to become well acquainted with customers and their individual needs. Banks also provide other financial services such as checking and savings accounts, which make it convenient for their farm and ranch customers to take care of all their financial business at one location.

Sometimes smaller banks cannot legally extend enough credit to finance a large-scale ranch or farm operator. They may arrange for credit to be supplied through a correspondent bank. Banks may arrange for secondary mortgages on real estate loans that are too large or have too long a term for the bank itself.

Farm Credit System

The Farm Credit System was established by the U.S. Congress in 1916 to provide an additional source of funds for agricultural loans. Government funds were used initially to organize and operate the system, but it is now a private cooperative owned by its member/borrowers. The system is supervised, audited, and regulated by the Farm Credit Administration, which is an independent Federal government agency.

The Farm Credit System obtains loan funds by selling bonds and notes in the national money markets. Proceeds from these sales are made available to the six regional Farm Credit Banks located across the country. These regional banks then provide funds to local associations, which in turn, initiate and supervise loans to farmers and ranchers. Loans from the Farm Credit System may be used to purchase livestock, machinery, buildings, rural homes, and land. Short-term operating credit is also available.

Life Insurance Companies

Life insurance companies acquire funds from the premiums paid on life insurance policies and from other earnings and reserve funds. They place these funds in various investments, including long-term agricultural real estate loans. The amount of money these companies make available for agricultural loans may vary from year to year, depending on the rate of

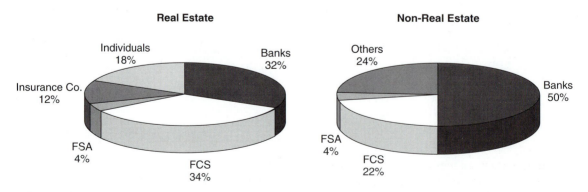

Figure 19-5 **Market share of U.S. farm debt, January 1, 2002.**
(*Source:* U.S. Department of Agriculture).

return from agricultural loans compared to alternative investments. Life insurance companies generally prefer large farm real estate loans, often over $500,000.

Farm Service Agency

The Farm Service Agency (FSA) is part of the U.S. Department of Agriculture, and has offices in most agricultural counties. A farm loan program division of FSA makes both farm ownership and farm operating loans. Most direct loans made by FSA are now granted to beginning farmers. The FSA also has authority to make emergency loans to qualified farmers and ranchers in officially declared disaster areas. These are temporary loans used to restore normal operations after a natural disaster such as flood or drought. Over the years, FSA has moved away from direct loans made from congressional appropriations and toward more *guaranteed* loans. With the latter, a private lender provides the loan funds, and FSA guarantees up to 95 percent repayment in case of default by the borrower.

To be eligible for an FSA loan or guarantee, the borrower must operate a family-size or smaller farm or ranch, receive a substantial portion of total family income from farming or ranching, and be unable to obtain conventional financing from other lending institutions. The last requirement does not mean FSA borrowers are always poor credit risks. Many are beginning farmers who simply do not have enough equity to borrow capital from other sources. As soon as borrowers improve their financial condition to the point where funds can be obtained from a commercial source, they must switch to another lender.

Individuals and Suppliers

Individuals, farm supply stores, dealers, and others are important sources of both real estate and non–real estate agricultural loans, as shown in Figure 19-5. Non–real estate loans can come from friends or relatives, accounts payable at farm supply stores, input manufacturers, or special agricultural credit companies. Many suppliers allow customers 30, 60, or 90 days to pay their accounts before any interest is charged, and may finance a purchase for a longer period with interest. This policy is essentially a loan, and the total balances in these accounts can be quite large at certain times of the year. Farm equipment and automobile dealers also provide loans by financing purchases themselves or through an affiliated finance company.

The relatively large portion of real estate debt owed to individuals and others comes mostly from seller-financed land sales. Many land sales utilize a land purchase installment contract, in which the seller transfers possession of the land and the buyer makes periodic loan payments directly to the seller. This differs from a cash sale, where the buyer borrows from a commercial lender, pays the seller cash for the full purchase price, and then makes periodic payments to the lender. A land purchase contract can have some income tax benefits for the seller, and the buyer may be able to negotiate a lower down payment, lower interest rate, and more flexible repayment terms.

Other Sources

The Commodity Credit Corporation (CCC) is a Federal government entity that provides some short-term loans using stored grain or cotton as collateral. These loans are made at a fixed rate per bushel or pound, and usually carry a below-market interest rate.

The Small Business Administration (SBA) also makes some agricultural loans and has an emergency loan program for farmers in designated disaster areas.

ESTABLISHING AND DEVELOPING CREDIT

When trying to establish or develop credit, it is useful to look at it from the lender's viewpoint. What does a lender consider when making a decision on a loan application? Why can one business borrow more money than another or receive different interest rates and repayment terms? A borrower should be aware of the need to demonstrate and communicate credit worthiness to lenders. Some of the more important factors that go into making loan decisions are the following:

- Personal character
- Management ability
- Financial position and progress over time
- Repayment capacity
- Purpose of the loan
- Collateral

When using these factors as a guide for establishing and developing credit, a prospective borrower should remember that lenders want to make loans. That is their business. However, they are looking for profitable and safe loans that will be repaid.

Personal Character   Honesty, integrity, judgment, reputation, and other personal characteristics of the loan applicant are always considered by lenders. Credit can be quickly lost by being untruthful in business dealings and slow to meet financial obligations. If the lender is not acquainted with the borrower, character references usually will be requested and checked. To maintain a good credit record, borrowers should promptly inform lenders of any changes in their financial condition or farming operation that might affect loan repayment. An honest and open relationship with lenders is necessary to maintain credit worthiness.

Management Ability   A lender must try to evaluate a borrower's management ability. Established farmers and ranchers will be evaluated on their past records, but beginners can be judged only on their background, education, and training. These factors affect profitability and therefore the ability to repay a loan. Lenders often rate poor management ability as the number one reason for borrowers getting into financial difficulty.

Financial Position   Accurate, well-prepared balance sheets and income statements are needed to document the current financial position of the business, and its profitability. Lenders can learn much about a business from these records. A track record of good financial progress over time can be just as important as the current financial position.

Repayment Capacity   Having a profitable business does not guarantee repayment capacity. There must be enough cash income to meet family living expenses and income tax payments, as well as the interest and principal payments on loans. Repayment capacity is best measured by the cash flow generated by the business. A cash flow budget projected for one or more years should be completed before borrowing large amounts and establishing rigid repayment schedules. Too often, money is borrowed for a profitable business only to find that cash flow in the early years is not sufficient to make interest and principal payments. A longer-loan term or a more flexible repayment schedule may solve the problem if it is identified in time.

Purpose of the Loan   Loans that are *self-liquidating* may be easier to obtain. Self-liquidating loans are for items such as fertilizer, seed, and feeder livestock, where the loan can be repaid from the sale of the crops or livestock. On the other hand, capital asset loans are those used to purchase long-term tangible assets such as land or machinery, which must generate additional revenue without being sold themselves. Capital asset loans may require extra collateral.

Collateral   Land, buildings, livestock, machinery, stored grain, and growing crops can all be used as loan collateral. The amount and type of collateral available are important factors in a loan request. Loans should not be made or requested unless repayment can be projected from farm income. However, lenders still ask for collateral to support a loan request. If the unexpected happens and the loan is in default, it may be the lender's only means of recovering the loan funds.

LIQUIDITY

The ability of a business to meet cash flow obligations as they come due is called liquidity. Maintaining liquidity in order to meet loan payments as they become due is an important part of establishing and improving credit. Several measures of liquidity that can be calculated from the balance sheet were discussed in Chapter 5.

Factors Affecting Liquidity

A farm or ranch that is profitable will usually have adequate liquidity in the long run. However, even profitable businesses experience cash flow problems at times, due to several factors.

Business Growth   Holding back ever-increasing inventories of young breeding livestock or feed reduces the volume of production sold in the short run. Construction of new buildings or purchases of land or machinery require large cash outlays up front, but may not generate additional cash income for months or years. Moreover, when new technology is involved, several production cycles may pass before an efficient level of operation is achieved. All these factors produce temporary cash shortfalls and need to be considered when planning financing needs and repayment plans.

Nonbusiness Income and Expenditures   These are especially important for family farms and ranches. Basic family living expenses must be met even when agricultural profits are low. At certain stages of their lives, farm families may have high educational or medical expenses. Generally, nonbusiness expenditures should be postponed until surplus cash is available. However, reinvesting every dollar earned into the farm business may eventually cause family stress, and keep personal goals from being reached.

Over the years, farm and ranch families have received more and more of their total income from off-farm employment and investments. A regular and dependable source of outside income may not only stabilize resources for family living expenses but also help support the farm during periods of negative cash flow.

Debt Characteristics   The rates and terms of credit may affect cash flow as much as the amount of debt incurred. Seeking out the lowest available interest rate obviously reduces debt payments. Using longer-term debt or balloon payment loans also reduces short-term financial obligations.

Debt Structure   Structure refers to the distribution of debt among current, intermediate, and long-term liabilities. Generally, loan repayment terms should correspond to the class of assets that they were used to acquire. Financing intermediate or long-term assets with short-term debt often leads to repayment problems and negative cash flows.

Financial Contingency Plan

No matter how well the manager budgets or how efficiently the business is managed, there will be periods in which cash flow is negative. Figure 19-3 illustrated how the amount of debt used by farmers and ranchers expanded dramatically from 1975 to 1983. However, interest rates climbed sharply in the early 1980s and reached levels that were not anticipated when many loans were made. This, combined with generally lower agricultural prices, produced what was known as the "farm financial crisis" of the 1980s.

The lessons learned during this period have made many farmers and lenders more conservative about the use of credit. Every operation should have a *financial contingency plan* to provide for unexpected cash flow shortfalls. In some cases, actions that are not profitable in the long run may have to be taken to cover cash flow obligations in the short run. The following steps are possible financial contingency actions:

1. Maintain savings or stored crops and livestock in a form that can be easily turned into cash and that carries a low risk of loss.

2. Maintain a credit reserve or some unused borrowing capacity for both current and noncurrent borrowing. A long-term credit reserve can be used to refinance excess current liabilities if the need arises.

3. Prepay debt in years when cash income is above average.

4. Reduce nonfarm expenditures or increase nonfarm earnings when farm cash flow is tight. Acquiring additional education or work experience may make obtaining off-farm employment easier.

5. Carry adequate insurance coverage against crop losses, casualties, medical problems, and civil liabilities.

6. Sell off less productive assets to raise cash. The marginal cost, marginal revenue principle should be employed to identify assets that will have the least negative effect on total farm profits. In some cases, cash flow can be improved without reducing efficiency by selling assets and then leasing them back, thereby maintaining the size of the operation.

7. Rely on relatives or other personal contacts for emergency financing or for the use of machinery or buildings at little or no cost.

8. Declare bankruptcy, and work out a plan to gradually repay creditors while continuing to farm, or conduct an orderly disposal of assets and cancellation of debts.

These actions are not substitutes for operating a profitable business. Some of them may even reduce the farm's profitability. But any of them can be applied, depending on the severity of the farm's financial condition, as a means to continue operating until profits increase.

SOLVENCY

While liquidity management focuses on cash flow, solvency refers to the amount of debt capital (loans) used relative to equity capital, and the security available to back it up. The debt/asset ratio and other measures of solvency were discussed in Chapter 5.

Leverage and the Use of Credit

Using a combination of equity capital and borrowed capital permits owning a larger business than would be otherwise possible. The degree to which borrowed capital is used to supplement and extend equity capital is called *leverage*. Leverage increases with increases in the debt/asset ratio. A debt/asset ratio of 0.50 indicates that one-half of the total capital used in the business is borrowed, and one-half is equity capital.

When the return on borrowed capital is greater than the interest rate, profits will increase and equity will grow. Higher leverage will increase profits and equity even faster. For example, assume a firm has $100,000 in equity capital, as shown in the upper part of Table 19-4. If $100,000 more is borrowed at 10 percent interest, the debt/asset ratio is 0.50. If the business earns a return on assets (ROA) of 15 percent, or $30,000, and the interest cost of $10,000 is paid, the remaining $20,000 is the return on equity (ROE). The rate of return on equity is 20 percent. Increasing the leverage, or debt/asset ratio, to 0.67 results in an even higher return to equity, as shown in the right-hand column of Table 19-4.

However, there is another side to the coin. If the return on assets is less than the interest rate on borrowed capital, return on equity is decreased by using increased leverage, and can even be negative. This is illustrated in the lower part of Table 19-4, where only a 5 percent ROA is assumed, and ROE decreases to a negative 5 percent under high leverage. Thus, high leverage can substantially increase the financial risk of the farm or ranch when interest rates are high or return on farm assets is low.

Maximum Debt/Asset Ratio

Most lenders use the debt/asset ratio or some variation of it to measure solvency. However, they do not always agree on what constitutes a

TABLE 19-4	Illustration of the Principle of Increasing Leverage

	Debt/asset ratio			
	0.00	**0.33**	**0.50**	**0.67**
Equity capital ($)	100,000	100,000	100,000	100,000
Borrowed capital ($)	0	50,000	100,000	200,000
Total assets ($)	100,000	150,000	200,000	300,000
Good Year				
Return on assets (15%)	15,000	22,500	30,000	45,000
Interest paid (10%)	0	5,000	10,000	20,000
Return on equity ($)	15,000	17,500	20,000	25,000
Return on equity (%)	15.0	17.5	20.0	25.0
Poor Year				
Return on assets (5%)	5,000	7,500	10,000	15,000
Interest paid (10%)	0	5,000	10,000	20,000
Return on equity ($)	5,000	2,500	0	−5,000
Return on equity (%)	5.0	2.5	0.0	−5.0

"safe" ratio. Profitability of the business and the cost of borrowed funds also enter into the decision. A simple relation among these factors can be expressed as follows:

$$\text{Maximum debt/asset ratio} = \frac{\text{ROA}}{\text{IR}}$$

where Maximum debt/asset ratio = the highest debt/asset ratio the business can sustain on its own, ROA = percent return on total assets after deducting the value of unpaid labor, and IR = average interest rate on the farm's debt.

For example, a farm that earns a 6 percent average return on assets and pays an average interest rate of 12 percent on its debt could service a maximum debt/asset ratio of 0.50 without reducing equity. In other words, the return on each $1 of assets pays the interest on $.50 of debt, but no principal. If the ROA dropped to 3 percent and the average interest rate rose to 15 percent, only a 0.20 debt/asset ratio could be sustained.

The "maximum" debt/asset ratio is the level at which return on equity is just equal to zero. Debt is serviced without reducing equity or utilizing outside income. Obviously, farms that earn a higher return on assets or that can borrow money at a lower interest rate can safely carry a higher debt load. The wise manager will maintain a margin of safety, however.

Inflation and Capital Gains

In the United States, the rate of inflation has generally been maintained at a low level, although it did surpass 10 percent for several years in the 1970s. Some countries have experienced inflation rates of over 100 percent per year. With high inflation, managers often prefer to hold tangible assets such as land that will increase in value, rather than financial assets such as cash, savings accounts, or bonds. However, tangible assets may also lose value when economic conditions change, as shown in Figure 19-1.

When farmers invest in intermediate or long-term assets, both their degree of solvency and growth in market-value equity become closely tied to changes in the values of these capital assets. However, changes in asset values alone have little effect on liquidity or cash flow, unless assets are sold. Thus, during the land boom of the 1970s, many farmland owners found their equity rising rapidly without a corresponding increase in cash income. A few sold land and turned the asset appreciation into capital gains, although some cash was lost to income tax payments. Others used their new equity as collateral to borrow money, but had difficulty generating enough cash flow to repay the loans.

In general, the longer the life of a farm asset, the lower its rate of cash return. Nevertheless, many operators have a goal to own land and buildings, with the expectation that they will increase in value over time. The prudent financial manager must carefully balance the need for short-term liquidity against the security and growth potential of long-term investments.

SUMMARY

Capital includes money invested in machinery, livestock, buildings, and other assets, as well as cash and bank account balances. Sources of capital that are available to farmers include the operator's own equity, equity from outside investors, leased or contracted assets, and borrowed funds. Today's farm managers must be skilled in acquiring, organizing, and using capital, whether it is equity capital or borrowed capital. The economic principles discussed in Chapters 7 and 8 can be used to determine how much total capital can be used profitably and how to allocate limited capital among alternative uses.

Agricultural loans are available for the purchase of both real estate and non–real estate assets and for paying operating costs. They can be repaid over periods ranging from less than a year to as long as 40 years or more, in a single payment or with several types of amortized payments. Interest rates, loan terms, and repayment schedules vary from lender to lender and by loan type. Borrowers should compare the annual percentage rate of interest, loan fees, variable rate provisions, and other loan terms when shopping for credit.

Agricultural loans are available from commercial banks, the Farm Credit System, the Farm Service Agency, life insurance companies, individuals, and other sources. Some lenders specialize in certain types of loans, but all are interested in the credit worthiness of a prospective borrower. Borrowers should work at improving their credit by maintaining good personal character, improving management skills, demonstrating adequate financial progress and repayment capacity, and providing sufficient collateral.

Liquidity or cash flow management is affected by business growth, nonbusiness income and expenses, and the characteristics and structure of the debt held. A financial contingency plan should be formulated to provide for unexpected cash flow shortages. Solvency refers to the degree to which farm liabilities are secured by assets. Increased leverage can increase the rate at which equity grows but also increases the risk of losing equity. The amount of debt that a farm can support depends on the rate of return earned on assets and the interest rate paid on debt capital. Inflation increases the market value of assets but does not contribute to cash flow unless the assets are sold.

QUESTIONS FOR REVIEW AND FURTHER THOUGHT

1. What economic principles are used to determine: (a) how much capital to use and (b) how to allocate a limited amount of capital?

2. What is the single largest source of capital used in U.S. agriculture? What other sources are used?

3. Define the following terms:
   a. Secured loan
   b. Amortized loan
   c. Real estate loan
   d. Collateral
   e. Line of credit
   f. Balloon payment

4. How are loans classified as short-term, intermediate-term, and long-term loans? List the types of assets that might serve as collateral for each.

5. Assume that a $200,000 loan will be repaid in 30 annual payments at 9 percent annual interest on the outstanding balance. How much principal and interest will be due in the first payment if the loan is amortized with equal principal payments? If it is amortized with equal total payments? How would these figures change for the second payment in each case? (Use Appendix Table 1 to find the amortization factor for the equal total payment case.)

6. What are the advantages and disadvantages of a 10-year balloon payment loan versus a completely amortized loan of the same amount and interest rate?

7. Identify the different sources of agricultural loans in your home town or home county. Which types of loans does each lender specialize in? You might interview several lending institutions to learn more about their lending policies and procedures.

8. Select one agricultural lender and find out the rates and terms currently available for an intermediate or long-term loan. Are both fixed- and variable-interest-rate loans available? What loan closing fees or other charges must be paid?

9. Assume you are a beginning farmer or rancher and need capital to purchase breeding livestock. What information and material would you need to provide a lender to improve your chances of getting a loan? What lenders might be willing to finance you?

10. List several reasons why a request for a loan by one farm operator might be denied while a similar request from another operator is approved.

11. Explain the difference between liquidity and profitability. Give three reasons why a profitable farm could experience liquidity problems.

LAND—CONTROL AND USE

CHAPTER OBJECTIVES

1. To explore the unique characteristics of land and its use in agriculture

2. To compare the advantages and disadvantages of owning and renting land

3. To explain important factors in land purchase decisions, methods of land valuation, and the legal aspects of a land purchase

4. To compare the characteristics of cash, crop share, livestock share, and other leasing arrangements

5. To demonstrate how an equitable share lease arrangement can be developed that will encourage efficient input use

6. To discuss profitable land management systems that conserve resources and sustain the environment

Agriculture uses large areas of land, a fact that distinguishes it from most other industries. Land is the basic resource that supports the production of all agricultural commodities, including livestock, since they depend on land to produce the forage and grain they consume. Land is the single most valuable asset in the balance sheet of U.S. agriculture, accounting for nearly three-fourths of the value of total assets. The index of land values shown in Figure 20-1a shows a

steady upward trend during the 1960s, followed by a very sharp rise in the 1970s, fueled by high grain prices and rapid inflation. In the 1980s, high interest rates, drought, and low commodity prices brought financial difficulty for many farmers and ranchers. Those who had high debt loads often could not meet their payments. The result was a sudden decline in land values as farmers tried to liquidate farmland, and buyers left the market. Many farmers and ranchers were

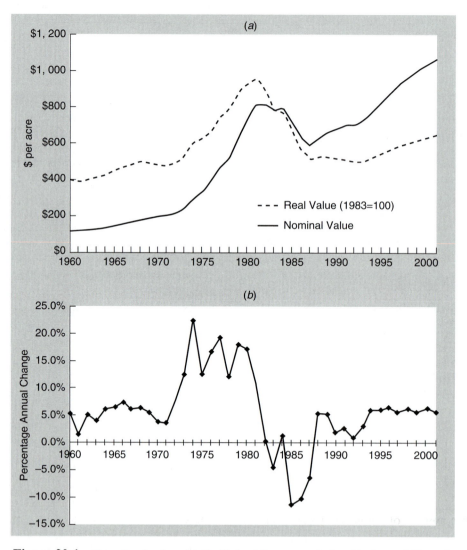

Figure 20-1   Farmland values in the United States; excludes Alaska and Hawaii.
(*Source:* U.S. Department of Agriculture).

forced to sell land at a loss or saw their equity decline as the land on their balance sheets decreased in market value. In the late 1980s, land values began to recover, and values have risen slowly but steadily since. The dashed line shows that even *real* land values (corrected for inflation) rose sharply in the 1970s, then fell. Figure 20-1*b* indicates that land values have increased in most of the years since 1960, except for a decline in the mid-1980s.

A number of factors determine the market value of farmland. The potential profit per acre from crop production is the most fundamental force behind land values. Increased crop yields made possible by the many technological advances introduced in the past century have contributed to this. In most years, increases in farmland values have kept pace with or exceeded the rate of inflation in the United States, making ownership of land a good hedge against inflation. This factor attracts nonfarm investors, as well as farmers, into the land market. Interest rates and the rates of return available from other investments also affect the demand for and the price of farmland. Likewise, competing uses for agricultural land such as recreation and housing may push the market price of land higher than its use in agriculture would justify.

While nonfarm investors have been important purchasers of land in some areas, the large majority of land buyers are still farmers and ranchers who desire to expand the size of their operations to increase their incomes and take advantage of larger-scale technologies. They recognize the economies of size they can achieve with a larger operation, the opportunity for greater profits, and the effect of increasing land values on their equity.

THE ECONOMICS OF LAND USE AND MANAGEMENT

Land has a number of unique characteristics not found in other agricultural or nonagricultural resources. These characteristics greatly influence the economics of land use and management.

Characteristics of Land

Land is a permanent resource that does not depreciate or wear out, provided soil fertility is maintained and appropriate conservation measures are used. Proper management not only will maintain the inherent productivity of land but can even improve it. Land is productive in its native state, producing stands of timber and native grasses, but the management efforts of farmers and ranchers have improved the agricultural productivity of many types of land. This has been accomplished through land clearing, drainage, good conservation practices, irrigation, the introduction of new and improved plant species, and the use of limestone and fertilizer. Land use often changes as a result of these improvements.

Each tract of land has a legal description, which identifies its particular location, size, and shape. Land is immobile and cannot be moved to be combined with other resources. Machinery, seed, fertilizer, water, and other inputs must be transported to the land and combined with it to produce crops and livestock.

Not only is land a unique resource in general, but each farm or specific parcel of land is unique. Any piece of land larger than several acres often contains two or more distinct soil types, each with its own set of characteristics. Topography, drainage, organic material, and the existence of natural hazards such as flooding, wind and water erosion, and rock outcrops are other factors that combine to make land resources different from farm to farm.

The supply of land suitable for agricultural production is essentially fixed, although small amounts may be brought into production by clearing and draining, or may be lost to nonfarm uses. This makes the price of land very sensitive to changes in the demand for its products. Unlike other agricultural inputs, additional land cannot be manufactured when the demand increases. Therefore, changes in the profitability of agricultural production are eventually factored into land prices and rents, and the land owner receives the economic benefits or losses.

Planning Land Use

The difference in land resources among farms explains why one of the first steps in whole-farm planning is to make a complete inventory of the land, including soil types, drainage, slope, and fertility. Without this information, the most profitable farm plan cannot be developed. The potential livestock and crop enterprises, yields, fertility requirements, and necessary conservation practices are related directly to the nature of the land resources available. Whole-farm planning often involves maximizing the return to the most limited resource. The fixed nature of land in the short run makes it the beginning of most farm planning efforts.

Land use is affected by regional differences in land productivity. Figure 20-2 shows the wide range in farmland values across regions of the United States. However, the most profitable use for land also depends on relative commodity prices and production technology. Both can change over time and bring about changes in

land use. Cotton production moved westward out of the southeastern United States and was often replaced by pasture and livestock production. Soybeans were once a minor crop but have become the second most important crop in the Midwest and Mississippi Delta region. The development of irrigation has transformed former livestock grazing areas into important crop production regions. An increase in grain production in the Southern Plains region of Texas and Oklahoma encouraged the development of large-scale cattle feeding operations there. All these changes can be traced to changes in relative prices of products and inputs, new technology, and competing uses for land.

CONTROLLING LAND—OWN OR LEASE?

How much land to control and how to acquire it are two of the most important decisions to be made by any farmer or rancher. Errors made at

Figure 20-2   Average value of farmland per acre, by region.
Source: National Agricultural Statistics Service, USDA (2002).

this point may plague the business for many years. Too little land may mean the business is too small to utilize other resources fully. At the other extreme, too much land may require borrowing a large amount of money, cause serious cash flow problems, and overextend the operator's management and machinery capacity. Either situation can result in financial stress and eventual failure of the business.

Land acquisition should be thought of in terms of control and not just ownership. Control can be achieved by either ownership or leasing. Over 46 percent of U.S. farmland is leased by the operator. Many farmers find a combination of ownership and leasing to be desirable, particularly when capital is limited. They often own some land and buildings to have a permanent "home base" and then lease additional land to attain the desired farm size. These part owners accounted for 61 percent of the U.S. farms with sales of over $50,000 in 1999. Another 26 percent were full owners, and 13 percent were full tenants who owned no land.

Ownership

Owning land is an important goal for many ranchers and farmers, regardless of the economics involved. A certain amount of pride, satisfaction, and prestige is derived from owning land. It also provides a tangible estate to pass on to one's heirs.

Owning land has the following advantages:

1. **Security of Tenure** Landownership eliminates the uncertainty of losing a lease and having the size of the business reduced unexpectedly. It also ensures that the operator will receive the benefits of any long-term improvements made to the land.
2. **Loan Collateral** Accumulated equity in land provides an excellent source of collateral when borrowing money. Increasing land values over time have provided substantial equity for landowners, although some of this equity can be lost with a decline in land values such as occurred during the mid-1980s.

3. **Management Independence and Freedom** Landowners are free to make their own decisions about enterprise combinations, conservation measures, fertilizer levels, and other choices, without consulting with a landlord or professional farm manager.
4. **Hedge Against Inflation** Over the long run, land provides an excellent hedge against inflation, because increases in land values have tended to equal or exceed the rate of inflation. However, land values do not necessarily increase every year.
5. **Pride of Ownership** Owning and improving one's own property is a source of pride. It assures a future benefit from years of labor and investment.

Ownership of land controlled by the business can also have some disadvantages. These disadvantages are primarily related to the capital position of the business. Possible disadvantages are:

1. **Cash Flow** A large debt load associated with purchasing farmland can create serious cash flow problems. The cash earnings from the land may not be sufficient to meet the required principal and interest payments, as well as the other cash obligations of the business.
2. **Lower Return on Capital** Where capital is limited, there may be alternative uses for it with a higher return than investing in land. Machinery, livestock, and annual operating inputs such as fertilizer, seed, and feed, often produce a higher rate of return on investment than land.
3. **Less Working Capital** A large investment or heavy debt load on land may restrict the amount of working capital that is available, severely limiting the volume of production, choice of enterprises, input levels, and profits.
4. **Size Limits** A combination of limited capital and a desire to own all the land to be operated will limit the size of the business. A small size may prevent the use

of certain technologies and result in higher average costs.

The disadvantages of landownership are more likely to affect the beginning farmer with limited capital. With the accumulation of capital and borrowing capacity over time, they may become less and less important. Older farmers tend to own more land and have less debt than younger farmers.

Leasing Land

Beginning farmers or ranchers are often advised to lease land. With limited capital, leasing is a means of controlling more acres. Other advantages of leasing land are:

1. **More Working Capital** When capital is not tied up in land purchases, more is available to purchase machinery, livestock, and annual operating inputs.
2. **Additional Management** A beginning farmer may be short of management skills. Management assistance can be provided by a knowledgeable landlord or professional farm manager employed by the landlord.
3. **More Flexible Size** Lease contracts are often for only one year or, at most, several years. Year-to-year changes in business size or location can be easily accomplished by giving up old leases or leasing additional land.
4. **More Flexible Financial Obligations** Lease payments are more flexible than mortgage payments, which may be fixed for a long period. The value of share rent will automatically vary with crop yields and prices. Cash rents are less flexible but can be negotiated each time the lease is renewed, taking into account current and projected economic conditions.

Leasing land also has disadvantages, which take on particular significance when all the land operated is leased. These disadvantages are:

1. **Uncertainty** Given the short term of many leases, the danger always exists that all or part of the land being farmed can be lost on short notice. This possibility discourages long-term investments and contributes to a general feeling of uncertainty about the future of the business.
2. **Poor Facilities** Some landlords are reluctant to invest money in buildings and other improvements. Tenants cannot justify investing in improvements that are attached to someone else's property. Thus, family housing, livestock facilities, grain storage, fences, and machinery housing may be obsolete, in poor condition, or nonexistent.
3. **Slow Equity Accumulation** Without landownership, equity can be accumulated only in machinery, livestock, or cash savings. In periods of rising land values, tenants may have to pay higher rents without accumulating any equity.

There is no clear advantage for either owning or leasing. Control is still the important factor, as income can be obtained from either owned or leased land. In the final analysis, the proper combination of owned and leased land is the one providing enough land to fully use the available labor, machinery, management, and working capital without creating excessive financial risk.

BUYING LAND

The purchase of a farm or ranch is an important decision that often involves large sums of money. A land purchase will have long-run effects on both the liquidity and solvency of the business.

The first step in a land purchase decision is to determine the value of the parcel under consideration. While income potential is the most important determinant of land value, many other factors also contribute:

1. **Soil, Topography, and Climate** These factors combine to affect the crop and livestock production potential and therefore the expected income stream.

2. **Buildings and Improvements** The number, size, condition, and usefulness of buildings, fences, storage structures, and other improvements will affect the value of a parcel of land. A neat, attractive farmstead with a modern house can add many dollars to a farm's value, while rundown, obsolete buildings may detract from it. Farm buildings and improvements are eligible for depreciation expense, creating a potential income tax savings.

3. **Size** Small and medium-sized farms may sell for a higher price per acre than large farms. A smaller total purchase price puts a farm within the financial reach of a larger number of buyers. Several neighboring farmers may consider the purchase of a small farm a good way to increase the size of their business, and bid up the price.

4. **Markets** Proximity to a number of markets will reduce transportation costs, increase competition for the farm's products, and possibly raise net selling prices.

5. **Community** A farm located in an area of well-managed, well-kept farms or in a community where land seldom changes hands will have a higher selling price.

6. **Location** Location with respect to schools, churches, towns, recreational facilities, paved roads, and farm input suppliers will also affect value. A prospective purchaser who plans to live on the farm will pay particular attention to the community and services provided in the general area.

7. **Competing Uses** Land that is close to urban or recreational areas, or has mineral deposits, may have a higher value than other agricultural land due to its other potential uses. Potential capital gains may outweigh income from agricultural production.

8. **Agricultural Program Characteristics** Some parcels of land have historical cropping patterns and yields that affect the level of benefits available for certain USDA commodity programs. The existence of special conservation contracts and easements also affects earning potential and, ultimately, market values.

A property under consideration for purchase should be inspected thoroughly to identify any problem areas that might reduce its value. Buildings should be inspected for structural soundness, and soil types properly identified. Drainage and erosion problems, sand and gravel areas, rock outcroppings, and other soil hazards should be noted and yield expectations adjusted accordingly. Environmental hazards will also detract from the value of a property.

Land Appraisal

An appraisal is a systematic process for estimating the current market value of a given piece of real estate. There are business firms that provide appraisal services for a fee. They provide their clients with a detailed analysis of the factors that determine the value of the property and conclude the report with an estimate of its current value. Two basic methods that are used to appraise the value of income-producing property such as farmland are the income capitalization method and the market data method.

Income Capitalization   This method uses investment analysis tools to estimate the present value of the future income stream from the land. It requires an estimate of the expected annual net income, the selection of a discount rate, and computation of the present value of an annuity, as outlined in Chapter 17. It is usually assumed that net income from land has an infinite life. Therefore, the present value equation for a perpetual or infinite stream of income is

$$V = \frac{R}{d}$$

where R is the average annual net return, d is the discount rate, or capitalization rate as it is called in appraisal work, and V is the estimated land value.

The first step is to estimate the net income stream, *R,* that can be obtained from the property. This requires determining potential long-term yields, selling prices, and production costs for the most profitable enterprise combination. *R* is the net return to investment in land and is equal to the farm's estimated gross income minus the costs of all resources used to generate it. These costs include land ownership costs such as property taxes, depreciation, and maintenance. The opportunity cost of the capital invested in the land, and any interest cost that might be incurred on a loan used to purchase it, are not subtracted, because the goal is to estimate the returns to land investment.

The results of such a procedure are shown in the top portion of Table 20-1 for an example 160-acre tract of farmland. It has 150 tillable acres after deducting for roads, ditches, and waterways. The long-run crop plan is assumed to be 100 acres of corn and 50 acres of soybeans, generating an annual gross income of $47,150. Yield and price estimates are important. They strongly influence the final estimate of net income. Both items must be estimated accurately to arrive at an accurate estimate of value. County yield averages and yields based on soil types are useful starting places for yield estimates. Prices should reflect the appraiser's best estimate of long-run prices after a careful review of past price levels. Annual expenses for the crop plan total $33,250, which makes the expected net return equal to $13,900 per year.

The second step in the income capitalization method is to select the discount or capitalization rate. The effect of the capitalization rate on value is readily apparent, so the appropriate rate must be selected carefully. The estimate of the net income does not include any expectation of inflation. Therefore, as explained in Chapter 17, the discount rate should be based on the real interest rate; that is, the nominal or actual rate of interest minus the anticipated rate of inflation.

Actual appraisal practice is to first estimate the average rate of return to land investment in the area for similar farms at recent sale prices. Using this rate in the capitalization procedure will give an appraised value that can be compared to the recent selling prices for other farms. This procedure often results in a capitalization rate of 3 to 6 percent, which is within the range of historical rates of return to farmland based on current market values. Anticipation of long-term increases in land values helps explain why landowners have historically been willing to accept a current cash rate of return on land that is lower than the rates for other investments that do not have the potential for appreciation in value.

The third step is to divide the expected net return by the capitalization rate. This process is shown in the bottom portion of Table 20-1 for three different capitalization rates. The estimated value of the land in the example, using a 6 percent capitalization rate, is $13,900 divided by 0.06 = $231,667, or about $1,448 per acre.

Market Data   The second appraisal approach compares the characteristics of land parcels that have been sold recently with the tract of land being appraised. Prices of the comparable sales are adjusted for differences in factors such as soil type, productivity, buildings and improvements, size, proximity to markets, the community, location, and competing uses, as discussed earlier. Three additional factors need to be considered when comparable sales values are being adjusted to reflect the "fair market value" of the farm being appraised.

1. **Financing** The method and terms of the financing arrangements for the purchase will affect selling price. Land may be sold with an installment purchase contract where the seller provides the financing. The terms of a contract may include a smaller down payment and/or a lower interest rate than conventional mortgage financing, which will increase the price a buyer is willing to pay for a given piece of land.
2. **Relationships** If the buyer and seller are closely related, such as parents selling to their children, the selling price is often below the price that would be agreed on by unrelated parties.

TABLE 20-1	**Estimated Annual Income and Expenses for Appraisal Purposes** *(Hypothetical 160-Acre Tract with 150 Tillable Acres)*

Income	Acres	Yield (bu)	Price	Total
Corn	100	135	$2.50	$33,750
Soybeans	50	40	6.70	13,400
Total Income				47,150

Expenses				
Fertilizer				6,500
Seed				3,350
Pesticides				4,500
Trucking				500
Drying				1,300
Labor and management				4,500
Machinery:				
Ownership costs				5,500
Operating costs				3,000
Property taxes and insurance				2,800
Repairs and maintenance				500
Depreciation of buildings, fences				800
Total expenses				$33,250
Annual net income				$13,900

Capitalization of income:

Capitalization rate		Total value		Per acre
8% ($13,900 / 0.08) =		$173,750		$1,086
6% ($13,900 / 0.06) =		$231,667		$1,448
4% ($13,900 / 0.04) =		$347,500		$2,172

3. **Time of sale** The more time that has passed since the comparable sale took place, the more likely it is that the price must be adjusted to reflect current market conditions. It may be adjusted up or down, depending on whether land values have been rising or falling in recent years.

Financial Feasibility Analysis

A financial feasibility, or cash flow, analysis should be performed on any prospective land purchase that involves repayment of a loan or installment contract. It is not a method for deter-mining land value, but for a given purchase price, it will show if there will be sufficient cash flow to meet the annual operating expenses as well as the interest and principal payments. Table 20-2 con-tains a summarized cash flow analysis for the purchase of the example 160-acre tract of land for $1,600 per acre, or a total of $256,000.

The analysis assumes a 35 percent down payment ($89,600) and a 20-year loan of $166,400, with an 8.0 percent annual interest rate and equal annual principal payments. Annual cash receipts and cash expenditures are based on the figures in Table 20-1, with 50 percent of the

TABLE 20-2	Cash Flow Analysis of the Purchase of a 160-Acre Tract at $1,600 per Acre*				
Item	Year 1	Year 2	Year 3	Year 4	Year 5
Cash receipts ($)	47,150	48,565	50,021	51,522	53,068
Cash expenditures ($)					
Seed, fertilizer, pesticides, trucking, drying	16,150	16,635	17,134	17,648	18,177
Machinery	5,750	5,923	6,100	6,283	6,472
Family living	4,000	4,120	4,244	4,371	4,502
Taxes, repairs	3,300	3,399	3,501	3,606	3,714
Annual loan payment ($)					
Principal	8,320	8,320	8,320	8,320	8,320
Interest	13,312	12,646	11,981	11,315	10,650
Total cash outflow ($)	50,832	51,043	51,280	51,543	51,835
Net cash flow ($)	−3,682	−2,478	−1,259	−21	1,233

*Assumes a down payment of $89,600, with the balance financed by a 20-year loan of $166,400 at 8.0% interest, with equal principal payments made annually.

machinery ownership costs estimated to be cash expenses. Although depreciation is a noncash expense, some cash outflow must be included for long-term machinery replacement. Labor costs should reflect the cost of hired labor and a portion of family living costs (estimated at $4,000 in this example), but no opportunity cost for unpaid labor. In the example, cash receipts and expenditures are assumed to inflate at a rate of 3 percent annually. Principal payments and the interest rate are assumed to be fixed for the term of the loan, however.

For the first four years, the cash operating income is not enough to meet both the cash operating expenses and the required principal and interest payments. Because of increasing receipts and the declining interest payment each year, net cash flow is projected to become positive in the fifth year of ownership. In the meantime, the cash flow deficits in the first four years must be met with cash from other sources, such as other land being farmed, livestock, or off-farm income.

The situation shown in Table 20-2 is typical of many land purchases. Land prices include expectations of appreciation in value, an infinite income stream, security, and pride of ownership. These factors cause higher prices but do not increase cash returns. Thus, negative cash flows

from a debt-financed land purchase tend to be the rule rather than the exception. Earnings often will be insufficient to meet both operating expenses and debt repayment. There are several ways to reduce the cash flow deficit, including a lower purchase price, a lower interest rate, a larger down payment, or a longer term on the loan. An amortized loan with equal total payments or a balloon payment at the end would also help reduce the cash outflow during the first few years. A cash flow projection for the entire business should be done to determine whether cash will be available from other parts of the business to help make the loan payments on a new land purchase.

Legal Aspects

Purchasing land is a legal as well as a financial transaction. The legal description of the property should be checked for accuracy, and the title examined for any potential problems, such as unpaid property taxes or mortgages. The buyer should also be aware of any easements for roads, pipelines, or power lines, which might interfere with the use of the land. Zoning and other local land use restrictions should also be checked.

Any water rights and mineral rights passing to the new owner should be carefully identified

---

Box 20-1   **Case Study: A Comparative Sale**

Western Land Management Company has been asked to appraise a 750-acre parcel of land near Rio Frio that is used to grow mostly dryland wheat, milo, and grass hay. Records they have obtained from county courthouses in the area show that several similar properties have sold in the past five years.

The Anderson farm near Habeville was 480 acres, mostly tillable, which sold for $550 per acre three years ago. Because it was a smaller farm, bidding on it was more aggressive than is expected for the Rio Frio farm. Thus, Western decides to discount the sale price of the Anderson farm by 15 percent, to reflect a comparable sale.

$$\$550 \times 85\% = \$467 \text{ per acre}$$

Land values in this part of the state have been increasing gradually in recent years, a total of about 7 percent since the Anderson farm was sold. To put the Anderson sale on a current basis, the Western appraiser adds 7 percent to the adjusted price.

$$\$467 \times 107\% = \$500$$

The final adjusted price of $500 per acre can now be used as a comparable sale to support the appraised value of the Rio Frio land.

---

and understood. Rights to underground minerals do not automatically transfer with the rights to the land surface. Where oil, gas, coal, or other minerals are important, the fraction of the mineral rights being received can have a large impact on land values. In the irrigated areas of the western United States, the right to use water for irrigation purposes is often limited and allocated on an individual farm basis. Obviously, the amount, dependability, and duration of these water rights should be determined carefully. Moreover, an existing but undetected environmental problem could result in anything from difficulty obtaining a loan to a large expense for eliminating the problem.

This is only a partial list of the many ways land buyers can receive less value than they expected because of some unknown restriction or problem. For these and other reasons it is advisable for land buyers to retain the services of an attorney with experience in land transactions before any verbal or written offer is made on a property.

LEASING LAND

Obtaining control of land through leasing has a long history in the United States and other countries. Not all this history has been good. Poor leasing practices have led to exploitation of tenants and sharecroppers and poor land use in some cases. Full landownership has been advocated by some as a means of eliminating these problems. However, it is not likely that we will ever have an agriculture where all farmers own all the land they operate. The capital requirements are too large, and improvements in leasing arrangements have reduced or eliminated many of the past problems and inefficiencies.

Leases on agricultural land are strongly influenced by local custom and tradition. The type, terms, and length of leases tend to be fairly uniform within a given area or community. This reliance on custom and tradition results in fairly stable leasing arrangements over time, which is desirable. However, it also means lease terms are slow to change in response to changing economic conditions and new technology. Inefficient land use can result from outdated leasing arrangements.

A lease is a legal contract whereby the landowner gives the tenant the possession and use of an asset such as land for a period of time in return for a specified payment. The payment may be cash, a share of the production, or a combination of the two. Oral leases are legal in most states but are not recommended. It is too easy for memories to fail, causing disagreements over the terms of the original agreement. Moreover, documentation may be needed for an estate settlement or a tax audit.

A farm lease should contain at least the following information: (1) the legal description of the land, (2) the term of the lease, (3) the amount of rent to be paid and the time and method of payment, (4) the names of the owner (lessor) and the tenant (lessee), and (5) the signatures of all parties to the lease. These are only the minimum requirements of a lease. A good lease will contain other provisions spelling out the rights and obligations of both the landlord and tenant. Many leases also contain a clause that describes the arbitration procedure to be followed in case of unresolved disputes. Dates and procedures for notification of lease renewal and cancellation should also be included, particularly if they differ from state law requirements.

An example of a typical lease form is included at the end of this chapter, but there are many variations. Blank lease forms are available through the Cooperative Extension Service in many states, as well as from attorneys and professional farm managers. It is important to modify the language in a lease form to fit the characteristics of the particular property as well as the laws of the particular state. As with all legal documents, the contracting parties may wish to obtain the advice of an attorney before signing a lease. In the long run, however, a good farm lease is based on mutual trust and communication as well as fair economic terms for both parties.

The three basic types of leases commonly used for renting agricultural land are cash rent, crop share, and livestock share. Each has some advantages and disadvantages to both the owner and the tenant.

Cash Rent

About two-thirds of the farm leases in the United States are cash leases. A cash rent lease specifies that the rent will be a cash payment of either a fixed amount per acre or a fixed lump sum. Rent may be due in advance, at the end of the crop season, or as some combination of these. If the rent is set per acre, the number of acres should be specified in the lease. Some cash leases show rent for land and buildings separately. Under a cash lease, the landlord furnishes the land and buildings, while the tenant receives all the income and typically pays all expenses except property taxes, property insurance, and major repairs to buildings and improvements. The lease may contain restrictions on land use and require the tenant to use certain fertility practices, to control weeds, and to maintain fences, waterways, terraces, and other improvements in their present condition.

The characteristics of a cash lease create both advantages and disadvantages. Some of the more important are:

1. **Simplicity** There is less likelihood of disagreement, because the terms can be easily written out and understood by both parties. A cash lease is also easy for a landlord to supervise, because there are few management decisions to be made. For this reason, landlords who live a long distance from their farm or who have little knowledge of agriculture often prefer a cash lease.

2. **Managerial Freedom** Tenants are free to make their own decisions regarding crops, livestock, and other management decisions. Tenants who are above-average managers often favor a cash lease, because they receive the total benefits from their management decisions.

3. **Risk** A cash lease provides the landlord with a known, steady, and dependable rental income. The tenant stands all risk from yield, price, and cost variability. This avoidance of risk is one reason that cash rents yield lower long-term average returns to the owner than a crop share lease on comparable land.

4. **Capital Requirements** The landlord has fewer capital requirements under a cash lease, because there is no sharing of the annual operating inputs. Conversely, the tenant has a larger capital requirement, including all operating inputs and the cash rent payments.

5. **Land Use** With all income accruing to the tenant, some may be tempted to maximize short-term profits from the land at the expense of long-term productivity, particularly under short-term leases. This can be prevented by including fair and proper land use restrictions in the lease, or negotiating long-term leases.

6. **Improvements** Landlords may be reluctant to invest in buildings and other improvements when using a cash lease, because they do not share in any additional income they generate. Conversely, some owners may expect to receive rent for buildings for which the tenant has no use.

7. **Rigid Terms** Cash rents tend to be inflexible and slow to change. Unless they are renegotiated each year to reflect changes in prices, land values, and technology, inequities can soon develop.

8. **Tax Effects** Some landlords may favor cash leases for tax purposes. Cash rental income where the landlord is not closely involved in the management of the business may qualify as investment income rather than self-employment income. There is no self-employment tax on investment income. The type of lease agreement used can also affect social security payments and estate taxes.

Setting a Fair Cash Rent

The cash rental rates agreed upon for a particular parcel of farmland ultimately depend on the productivity of the land, the value of the crops produced, costs of production, the supply of and demand for farmland in the area, and the bargaining positions of the owner and the operator.

Figure 20-3 shows the average cash rent paid per acre for cropland in different regions across the United States.

Several different approaches can be used to estimate a fair rent. These are illustrated in Table 20-3.

1. **Landowner's Costs** In the long run, the landowner wants to receive at least enough rent to cover both the cash and opportunity costs of owning the land. Assume that the owner of the example tract of land in Table 20-1 feels that it has a current market value of $1,500 per acre, or $240,000, and the opportunity cost rate for similar investments is 5 percent. As shown in Table 20-3, the owner's total costs would be 5 percent of $240,000, or $12,000, for opportunity cost on the investment, $2,800 for property taxes and insurance, $500 for repairs and maintenance, and $800 for depreciation, for a total cost of $16,100, or $101 per acre. This would be the minimum rent needed to pay all the owner's costs.

2. **Tenant's Residual** A second approach is to estimate how much income the tenant will have left after paying all other costs. Using the same example, the tenant would have gross income of $47,150 and total expenses of $29,150. The difference is $18,000, or $113 per acre. Note that the tenant would not include any land ownership costs such as property taxes. If the rented land could be farmed with no extra machinery investment, then machinery ownership costs could be excluded also. This approach estimates the maximum rent the tenant could pay without allowing for any return to risk or profit.

3. **Crop Share Equivalent** A partial budgeting approach can be used to estimate how much cash rent the tenant could pay and receive the same return as from a crop share lease. Suppose that by shifting from a 50/50 crop share to a cash lease, the tenant would receive 100 percent

Figure 20-3   **Average cropland cash rental rates per acre, by region.**
Source: National Agricultural Statistics Service, USDA (2002).

of the crop instead of 50 percent. Using the same example, this would increase the tenant's income by $23,575 (half of $47,150). However, the landowner would no longer pay half of the fertilizer, seed, pesticide, trucking, and drying costs, so the tenant would have $8,075 in additional costs. The net gain would be $15,500, or $97 per acre. This is how much cash rent the tenant could pay and still receive the same net return as under a 50/50 crop share lease.

4. **Share of Gross Income** The last approach estimates a cash rental rate as a percent of the expected gross income from the land. For example, if land costs generally represent about 35 percent of the total cost of production for particular crops in a region, rent could be estimated at 35 percent of the expected gross income. In the example, this would be

equal to $47,150 \times 0.35 = $16,502, or $103 per acre.

These various approaches will not give identical answers. However, they can help define a range within which the owner and operator can negotiate.

Pasture Rent   A fair cash rent for pasture land is more difficult to determine because the potential income is very uncertain. Factors such as the quality of the established pasture, water supply, condition of fences, buildings, and location must all be considered.

Pasture can be rented for a fixed rate per acre or by the month. One common method is to establish a charge per animal unit month (AUM). An AUM is equivalent to one mature cow grazing for one month. With this method, the rent paid is proportional to the stocking capacity of the pasture.

TABLE 20-3	Setting a Fair Cash Rent

1. Landowner's Costs

Opportunity cost on investment (160 acres)	$240,000 × 5% =	$12,000
Property taxes and insurance		2,800
Maintenance		500
Depreciation		800
Total		$16,100
Total cost per acre		$ 101

2. Tenant's Residual

Gross Income	$47,150
Expenses	
fertilizer	6,500
seed	3,350
pesticides	4,500
trucking	500
drying	1,300
labor and management	4,500
machinery	8,500
Total expenses	$29,150
Net income available to pay rent	$18,000
Net income available per acre	$ 113

3. Crop Share Equivalent

Additional income:	$47,150 × 50% =	$23,575
Additional expenses:		
fertilizer	6,500 × 50%	3,250
seed	3,350 × 50%	1,675
pesticides	4,500 × 50%	2,250
trucking	500 × 50%	250
drying	1,300 × 50%	650
Total additional expenses		$ 8,075
Additional net income		$15,500
Additional net income per acre		$ 97

4. Share of Gross Income

Gross income	$47,150
Share of total costs from land	35%
Share of gross income to land	$16,502
Estimated rent per acre	$ 103

Crop Share Leases

Crop share leases are popular in areas where cash grain farms are common. These leases specify that the landlord will receive a certain share of the crops produced, with the proceeds from their sale becoming the rent. Where farms are enrolled in certain USDA farm programs, payments received are usually divided in the same proportion as the crop. The tenant typically provides all the labor and machinery. Fertilizer, seed, pesticides, and irrigation costs may be shared, along with harvesting and other costs in some areas. Along with sharing production and expenses, landlords often participate in management decisions about cropping practices.

The landlord's share of the crop will vary depending on the type of crop, local custom, and how operating costs are shared. Soil productivity and climate are important factors, because many of the tenant's costs, such as labor and machinery, will be nearly the same regardless. Therefore, tenants receive a larger share of the production from less productive areas, because the value of the land contribution is less. In the Midwest the landowner's share may be as high as 50 or 60 percent of the crop, but in more arid regions it may be as low as 25 percent. Landlords whose farms contain poorer soils may find they have to take a smaller share and/or pay more of the variable costs to attract a good tenant.

The advantages and disadvantages of crop share leases can be summarized as follows:

1. **Risk** The crop value will vary with changes in yields and prices, so risk is shared by the tenant and owner. This may be a disadvantage to the owner if the rent is an important part of the owner's total income, which it may be for some retired persons. Variable rent is an advantage to the tenant, as the value of the rental payment varies with the tenant's ability to pay.
2. **Management** Landlords using a crop share lease usually maintain some direct or indirect control over crop selection and other management decisions. This may be

an advantage to an inexperienced tenant and gives the landlord some control over land use.

3. **Capital Requirements** Because some crop production expenses are shared and no cash rent must be paid, the landlord's capital requirements increase, while the tenant's decrease, when compared with a cash lease.
4. **Expense Sharing** One problem with crop share leases is determining a fair and equitable sharing of expenses. As a general rule, variable expenses should be shared in the same proportion as the crop, so as to maintain optimal input levels. The adoption of new technology often creates new problems in determining the proper way to share its costs and benefits. An example of this is presented later in the chapter.
5. **Buildings and Pasture** Because the landlord receives no direct benefits from buildings (except grain storage) and livestock pasture, problems can arise concerning a fair rent for these items. A cash rent supplementing the crop share is often paid by the tenant for use of buildings and livestock pasture.
6. **Marketing** The owner is generally free to sell his/her share of the crops wherever or whenever desired. This requires some additional expertise by the owner. Some owners let the tenant or a professional farm manager make all the marketing decisions.
7. **Material Participation** Landowners are usually more involved in management decisions under a crop share lease. This management participation often qualifies share rent income as self-employment income, which helps build social security earnings. Also, it may possibly qualify the farm for a lower valuation for estate tax purposes.

Livestock Share Leases

A livestock share lease is much like a crop share lease, except livestock income and expenses are also shared between landlord and tenant. The tenant typically furnishes all labor and machinery and a share of the livestock and operating inputs, with the landlord furnishing land, buildings, and the remaining share of livestock and operating inputs. Most livestock share leases are 50/50 shares, although other share arrangements are possible depending on the type of livestock and how expenses are shared.

Livestock share leases can be complex, and there is considerable latitude in the number and type of expenses to be shared. In addition to costs such as feed and health expenses, the landlord may share in the cost of machinery and equipment related to livestock production. Examples include milkers, feed grinders, feeders, waterers, and forage harvesting equipment. The lease should contain a full and detailed list of the expenses to be shared and the portion to be paid by each party.

The advantages, disadvantages, and potential problems with a livestock share lease are similar to those of a crop share lease, with the following additional considerations:

1. **Buildings** A livestock share lease provides the landlord with some return from buildings and pastureland by sharing in the livestock production. Landlords are more likely to furnish and maintain a good set of buildings when they receive part of the income. However, tenants may desire additional building improvements to reduce their labor requirements or improve performance in the shared livestock production.
2. **Records** The sharing of both crop and livestock income and expenses requires good records to ensure a proper division. There should be a periodic accounting of income and expenses with compensating payments made to balance the accounts.
3. **Lease Termination** Terminating a livestock share lease can be complex and time-consuming. All livestock and shared equipment must be divided in a fair and

equitable manner. The lease should contain a method for making the division and a procedure for settling any unresolved disputes.
4. **Management** There is a greater need and opportunity for sharing management decisions under a livestock share lease, and this requires a good working relationship between the landlord and tenant.

Other Types of Leases

Several other types of leases, and variations of the above types, are used.

Labor Share Lease In a labor share lease, the landlord provides all the land, machinery, and other variable inputs. The tenant provides only the labor and receives a share of the production. However, this share would be smaller than with the other share leases. This arrangement works well for a tenant who would like to begin farming but has very limited capital, and for a landlord who has a complete set of farming resources but is ready to retire.

Variable Cash Lease Rigidity of cash rents was identified as one of the disadvantages of cash leases. Variable cash leases are sometimes used to overcome this problem and divide some of the risk between the landlord and tenant. Annual cash rent under a variable cash lease can be tied to the actual yield, the price received, or to both of these factors. For example, the cash rent may increase or decrease by a specified amount for each bushel the actual yield is above or below some base or average yield. The same could be done for price or for gross income. Variable cash leases maintain most of the properties of fixed cash leases but allow the landlord and tenant to share at least part of the price and/or production risk.

One type of variable cash lease sets a rent based on the most likely price and yield, and then adjusts it in proportion to the amount by which the actual price and yield exceed the base values. In the example below, the base rent is $50 per acre for a wheat yield of 60 bushels per acre and a price of $3.20. Assume that the actual yield turns out to be 78 bushels per acre, but the price is only $2.75. The actual rent is then calculated at $55.86 per acre, as shown below:

$$\text{Actual rent} = \$50 \times \frac{\text{actual yield}}{60} \times \frac{\text{actual price}}{\$3.20}$$

$$= \$50 \times \frac{78}{60} \times \frac{\$2.75}{\$3.20} = \$55.86$$

Another common cash rent formula is to simply pay a fixed percent of the actual gross income earned from the crops on the rented land. Since the tenant's ability to pay is affected by both price and yield, a variable cash rent formula that depends on both of these factors provides the most risk reduction. The tenant and landowner should agree in advance on how to determine the actual yield and price to be used for calculating the rent.

Bushel Lease   Another type of lease is the bushel lease, or standing rent lease. With this lease, the landlord typically pays no crop production expenses but receives a specified number of bushels or quantity of the production as rent. Production risk falls entirely on the tenant, because a fixed quantity of the product must be delivered to the landlord regardless of the actual yield obtained. Part of the price risk is shared with the landlord, however, because the actual

rent value will depend on the price received for the product.

Custom Farming   The practice of custom farming is not a true leasing agreement but represents another alternative arrangement between owner and operator. Typically, the operator provides all machinery and labor for the field operations and hauling, in exchange for a fixed payment. There may be a bonus for above-average yields. The owner supplies all the operating inputs, receives all the income, and stands all the price and yield risks. Another variation is for the custom operator to receive a percentage of the production rather than a cash payment.

Table 20-4 summarizes the important characteristics of the lease types discussed in this section.

Efficiency and Equity in Leases

Many leases are based on local custom and tradition, which may result in inefficiencies, poor land use, and a less than equitable division of income and expenses. There have been many critics of a land tenure system based on leasing because of the existence of these problems. It may not be possible to write a perfect lease, but improvements can be made. There are two broad areas of concern in improving the efficiency and equity of a lease. The first is the length of the lease, and the second is the cost-sharing arrangements.

TABLE 20-4	**Comparison of Lease Types**				
	Fixed cash	Variable cash	Fixed bushel	Crop or livestock share	Custom farming
Price risk borne by:	Tenant	Both	Both	Both	Owner
Production risk borne by:	Tenant	Both	Tenant	Both	Owner
Operating capital supplied by:	Tenant	Tenant	Tenant	Both	Owner
Management decisions made by:	Tenant	Tenant	Tenant	Both	Both
Marketing done by:	Tenant	Tenant	Both	Both	Owner
Terms adjust:	Slowly	Quickly	Medium	Quickly	Slowly

Box 20-2   Case Study: Negotiating a Lease

Chad and Maria Grabowski started farming five years ago. They own a tractor and some tillage equipment, but share the use of a planter and harvester with Maria's parents. They borrow enough operating capital from their local bank to finance their crop inputs, with a guarantee from the Farm Service Agency (FSA).

The Grabowskis have the opportunity to cash rent 265 acres a few miles away. The owner, whose parents farmed the land for many years, is an attorney in Atlanta. She is asking $90 per acre cash rent, payable in advance.

Chad and Maria visit with their banker and decide that their equity is too small to risk paying the cash rent. They offer to rent the farm on a 60/40 share arrangement, instead. The owner decides that she doesn't want to get involved with paying part of the input bills and selling her share of the crop. However, she is willing to adjust the cash rent each year based on the actual yields obtained and the selling prices available at harvest.

Finally the Grabowskis reach an agreement whereby they will pay $50 per acre on March 1 to rent the farm, plus a bonus equal to 20 percent of the gross income produced, due by December 1. The total rent cannot exceed $100 per acre, however.

Farm leases are typically written for a term of one year, although some livestock share leases are written for three to five years. Most one-year leases contain a clause providing for automatic renewal on a year-to-year basis if neither party gives notice of termination before a certain date. Many such leases have been in effect for long periods of time, but the possibility always exists that the lease may be canceled on short notice if the landlord sells the farm or finds a better tenant. This places the tenant in a state of insecurity. It also tempts the tenant to use practices and plant crops that maximize immediate profits rather than conserve or build up the property over time.

Short leases also discourage a tenant from making improvements, because the lease may be terminated before the cost can be recovered. These problems can be at least partially solved by longer-term leases and agreements to reimburse the tenant for the unrecovered cost. Improvements at the tenant's expense, such as application of limestone and erection of soil conservation structures, can also be covered under an agreement of this type.

Inefficiencies can also arise from poor cost-sharing arrangements if the costs of inputs that directly affect yields are not shared in the same proportion as the income or production. Seed, fertilizer, pesticides, and irrigation water are examples. In Table 20-5, the profit-maximizing level of fertilizer use is where its marginal value product is equal to its marginal input cost, which occurs at 140 pounds of fertilizer per acre. However, if the tenant receives only one-half the crop but pays all the fertilizer cost, the marginal value product for the tenant is only one-half the total. This is shown in the last column of Table 20-5. Under these conditions the tenant will use only 100 pounds of fertilizer per acre. While this amount will maximize profit for the tenant, it reduces the total profit per acre from what it would be if 140 pounds of fertilizer were applied.

On the other hand, fertilizer has a zero marginal input cost to the landlord. Additional fertilizer use does not increase the landlord's costs but does increase yield, which is shared. The landlord would like to fertilize for maximum yield as the way to maximize profit. This would be at 160 pounds of fertilizer per acre in the example in Table 20-5. These types of conflicts can be eliminated by sharing the cost of yield-determining inputs in the same proportion that

TABLE 20-5	**Example of Inefficient Fertilizer Use Under a Crop Share Lease**
	(Landlord Receives One-Half of Crop but Pays No Fertilizer Cost)

Fertilizer (lb)	Yield (bu)	Marginal input cost at $0.20/lb ($)	Total marginal value product, corn at $2.20 per bu ($)	Tenant's marginal value product ($)
60	95			
80	100	4.00	11.00	5.50
100	104	4.00	8.80	4.40
120	107	4.00	6.60	3.30
140	109	4.00	4.40	2.20
160	110	4.00	2.20	1.10

the production is shared. If fertilizer costs were shared equally in this example, both parties would pay one-half the marginal input cost and receive one-half the marginal value product. They would both agree on 140 pounds of fertilizer as the profit-maximizing level. Tenants are understandably reluctant to adopt any new yield-increasing technique or technology if they must pay all the additional cost but receive only a share of the yield increase.

Similar problems can arise when changing technology allows the substitution of shared inputs for nonshared inputs. An example is the use of herbicides in place of mechanical or hand weed control. If the tenant pays only one-half the cost of the herbicide and all the cost of labor and machinery, the profit-maximizing combination of inputs for the tenant will include more herbicide and less labor and machinery than in a cash lease or owner-operator situation.

Determining Lease Shares

The objective of any lease should be to provide a fair and equitable return to both parties for the inputs they contribute to the total farm operation. A fair and equitable sharing arrangement exists when both parties are paid for the use of their inputs according to the contribution those inputs make toward generating income. Application of this principle requires the identification and valuation of all resources contributed by both the landlord and tenant.

One method for determining the proper shares for a crop share lease is shown in Table 20-6. The example starts from an assumed 50/50 sharing of expenses for seed, fertilizer, pesticides, and drying fuel. Other costs are attributed to the landlord or tenant on the basis of who owns the asset or provides the service.

The most difficult part of this procedure is placing a value on the services of fixed assets such as labor, land, and machinery. The same procedures that are used for estimating cash and opportunity costs for farm budgets should be applied. Land costs can include both cash costs and an opportunity interest charge, or they can be estimated from current cash rental rates. Do not use principal and interest payments, however, because they represent debt repayment rather than economic costs.

This example has a cost sharing proportion of nearly 50/50. It may be so close that both parties feel that a 50/50 division of production is fair without any adjustments. If the cost shares are substantially different, one of two methods can be used to make adjustments. First, the party contributing the smaller part of the costs can

TABLE 20-6	Determining Income Shares Under a Crop Share Lease		
Cost Item	**Whole Farm**	**Owner**	**Tenant**
Fertilizer	$ 6,500	$ 3,250	$ 3,250
Seed	3,350	1,675	1,675
Pesticides	4,500	2,250	2,250
Trucking	500	0	500
Drying fuel	1,300	650	650
Labor, management	4,500	0	4,500
Machinery ownership	5,500	0	5,500
Machinery operating	3,000	0	3,000
Property taxes, insurance	2,800	2,800	0
Repairs and maintenance	500	500	0
Depreciation of buildings, fences	800	800	0
Opportunity cost for land	12,000	12,000	0
Total costs	$45,250	$23,925	$21,325
Percent contributed		53%	47%

agree to furnish more of the capital items or pay for more of the variable costs, to make the cost sharing closer to 50/50. The second method would be to change the production shares to 60/40 or some other division to more nearly match the cost shares. This should include changing the cost sharing on the variable inputs to the same ratio, to prevent the inefficiencies discussed earlier. Whichever alternative is used to match the cost and production shares, the final result should have the tenant and landlord sharing farm income in the same proportion as they contribute to the total cost of production.

CONSERVATION AND ENVIRONMENTAL CONCERNS

Conservation can be defined as the use of farming practices that will maximize the present value of the long-run social and economic benefits from land use. This definition does not prevent land from being used but does require following practices that will maintain soil productivity and water quality over time. Farming

systems that accomplish these goals are sometimes called *sustainable agriculture.*

Ordinary budgeting techniques are often inadequate for deciding how best to achieve the goals of conservation. Nevertheless, farm and ranch managers need to be conscious of how their decisions affect the long-term quality of life for themselves and society in general. There are three general areas in which sustainable agriculture considerations go beyond conventional budgeting analysis.

Long-Run versus Short-Run Consequences

Narrow profit margins and tight cash flows make it tempting to "mine" the soil, to maximize short-run returns. Most conservation practices require some extra cash expenditures. They may also reduce crop yields temporarily as soil and cropping patterns are disturbed. This short-run reduction in profit may be necessary to achieve higher profits in the future or to prevent a long-run decline in production if no conservation practices are used. The long-term effects of soil depletion and erosion on productivity are not always well known or understood. Likewise, it takes years of study to determine the long-term effects of continued use of high rates of fertilizer and pesticides on soil, water, wildlife, and humans.

Conservation practices such as terraces, drainage ditches, and diversion structures require large initial investments. The present value methods discussed in Chapter 17 can be used to evaluate the profitability of a particular practice. The opportunity cost of capital and the planning horizon of the landowner become very important in the analysis. Higher discount rates reduce the present value of the larger future incomes resulting from conservation. Shorter planning horizons also discourage conservation. For this reason, many farmers look at changes in tillage practices and crop rotations as lower-cost alternatives to achieve conservation goals.

Leasing arrangements also affect the type of conservation practices followed. One-year leases discourage tenants from considering the long-term effects of their farming practices. On the other hand, landlords may be reluctant to make large conservation investments if they feel the tenant will receive all or a large share of the benefits. Share leases should attempt to divide the costs and benefits of such practices whenever possible. Tenants and landlords should fully discuss the farming practices needed to meet long-run conservation and environmental considerations.

Farming Systems Analysis

Because most farms and ranches carry out more than one type of crop or livestock activity, it is important to understand how they affect each other. This is called *farming systems analysis.* For example, crops grown in rotation or in combination with other crops have different fertility and pest control needs than the same crops grown alone or continuously. Tillage practices and soil and water runoff also differ. Livestock enterprises can complement crop enterprises by

efficiently utilizing feedstuffs with a low market value and returning fertility to the land through manure disposal. There can be interactions among production practices used on a single crop, such as the placement of fertilizer and pesticides and the type of tillage practices followed. These interactions are difficult to quantify and difficult to incorporate into standard enterprise budgets. However, they should be considered in whole-farm planning and budgeting.

Off-Farm Effects

Many of the decisions farmers make with regard to land use and production practices have consequences that go far beyond the boundaries of the farm. Buildup of silt in rivers and lakes, pollution and contamination of groundwater, destruction of wildlife habitat, and the presence of chemical residues in livestock products are just a few examples. These effects are very difficult to evaluate in monetary terms, and are often difficult to relate back to specific agricultural practices and sources. Nevertheless, research into causes and effects is continuing. Agriculture must consider more than just farm input costs when making decisions about input use. The total societal costs of using various technologies is quickly becoming an important factor in choosing production practices.

Regulations and Incentives

Both the federal and state governments have enacted laws to promote and sometimes require land use and livestock production practices that preserve and enhance soil, water, and air resources. As more and more of these regulations are enacted, future conservation efforts may increasingly become a matter of selecting the least-cost combination of practices to meet the relevant requirements.

Long-term land retirement programs, such as the Conservation Reserve Program, offer guaranteed annual payments in return for taking highly erodible land out of production. The opportunity cost of the lost production must be evaluated against the incentive payments and the long-run conservation benefits. Some programs pay landowners in return for conservation easements that control land use. Other regulations, such as "swampbuster" and "sodbuster" rules, have restricted farmers from cropping certain areas with a history of pasture or wetland use. Farmers are also required to develop and follow an approved conservation plan in order to participate in some government farm programs.

Restrictions on the use of pesticides, some chemical fertilizers, and animal manure vary among states. Regular soil testing and careful scouting for pest problems will ensure that such products are used only at environmentally safe and economically profitable levels. The principles of marginal cost and marginal revenue can be used to determine the economic threshold at which the potential losses from a pest problem justify the cost of treating it.

Some farms have been found to have environmental hazards such as leaking underground fuel storage tanks or accumulations of discarded pesticide containers. The costs of cleaning up such problems can significantly reduce the value of a farm. Before closing a sale of a farm or ranch property, a prudent buyer will arrange for an environmental audit to identify any potential problems of this type and the estimated costs of correcting them.

Several state and federal laws regulate the discharge of pollutants into streams and lakes. Most agricultural runoff is considered to be "nonpoint source" pollution. However, confined animal feeding operations (CAFOs) are considered to be point sources, and are more tightly controlled.

A high opportunity cost of capital, short planning horizons, and limited direct evidence of environmental effects combine to explain why some landowners are reluctant to adopt sound environmental practices. However, society has an interest in conservation in order to maintain and expand the nation's long-run food production potential, as well as to ensure food safety. Societal planning horizons are typically longer and broader than those for an individual farmer. In recognition of this and the limited capital position of many farmers, society has devised means to encourage more sustainable practices. Technical assistance is available at no cost through local offices of the Natural Resources Conservation Service. Financial assistance to pay for part of the costs of certain conservation practices has been made available through the U.S. Department of Agriculture and some state programs. And, USDA program payments are available to producers who follow certain conservation practices.

Today's farmers and ranchers must think beyond simply complying with regulations and maximizing current profits. The ethic of conservation tells us that owning or using land for agricultural purposes carries with it the responsibility to follow practices that will support society well into the future.

SUMMARY

Land is an essential resource for agricultural production. It is a permanent resource that is fixed in supply and location. The decision to buy or lease land is an important one that will affect the production capacity and financial condition of the business for many years. Land ownership has many advantages. However, purchasing land requires a strong capital position and adequate cash flow potential if credit is used. Land can be valued based on its net earnings or by comparing sale prices of similar land.

A mixture of owned and leased land exists in many farm and ranch businesses. Cash rent, crop share, and livestock share are the most common types of lease arrangements. Each lease type has advantages and disadvantages to both the tenant and landlord in terms of the capital contribution required, the amount of price and yield risk to be shared, and the making of management decisions. Share leases should provide for the sharing of income in the same proportion as each party contributes to the total cost of production. Variable inputs should also be shared in this proportion in order to allocate resources efficiently.

Land-use decisions need to consider long-run environmental effects, interactions among enterprises, and consequences that occur beyond the borders of the farm in order to conserve resources and sustain agriculture into the future.

QUESTIONS FOR REVIEW AND FURTHER THOUGHT

1. List as many reasons as you can to explain why people buy farmland. How would your list, or the importance of each reason, be different for an operating farmer than for a nonfarm investor who lives in another state?

2. What are the advantages and disadvantages of owning land compared to renting land?

3. Using the capitalization method of valuation and a 6 percent discount rate, how much could you afford to pay for an acre of land that is expected to have a net return of $81 per year? What would the answer be with a discount rate of 4 percent?

4. Why is a cash flow analysis important in a land purchase decision?

5. List three advantages and disadvantages of cash and crop share leases for both the tenant and the landlord.

6. What are the typical terms for crop share leases in your community? Analyze them using Table 20-6 as a guide. Are the terms fair? What problems did you encounter in the analysis?

7. Develop a flexible cash lease for your area using both price and yield as the variable factors upon which the rent is based. Show how the rent varies for different combinations of yield and price.

8. What environmental concerns are associated with agriculture in your area? Give two examples.

Cash Farm Lease
with Flexible Provisions

This form can provide the landlord and tenant with a guide for developing an agreement to fit their individual situation. This form is not intended to take the place of legal advice pertaining to contractual relationships between the two parties. Because of the possibility that an operating agreement may be legally considered a partnership under certain conditions, seeking proper legal advice is recommended when developing such an agreement.

This lease entered into this _____ day of _____ ,20 _____ , between

_____ , landlord, of _____
                                                          address
                                                        _____

_____ , spouse, of _____
                                                          address
                                                        _____

hereafter known as "the landlord," and

_____ , tenant, of _____
                                                          address
                                                        _____

_____ , spouse, of _____
                                                          address

hereafter known as "the tenant."

I. Property Description

The landlord hereby leases to the tenant, to occupy and use for agriculture and related purposes, the following described property:

_____

_____

consisting of approximately _____ acres situated in _____ County (Counties),

_____ (state) with all improvements thereon except as follows:_____

_____

II. General Terms of Lease

A. Time period covered. The provisions of this agreement shall be in effect for _____ year(s), commencing on the _____ day of _____ , 20 _____ . This lease shall continue in effect from year to year thereafter unless written notice of termination is given by either party to the other at least _____ days prior to expiration of this lease or the end of any year of continuation.

B. Review of lease. A written request is required for general review of the lease or for consideration of proposed changes by either party, at least _____ days prior to the final date for giving notice to terminate the lease as specified in **II - A.**

C. Amendments and alterations. Amendments and alterations to this lease shall be in writing and shall be signed by both the landlord and tenant.

D. No partnership intended. It is particularly understood and agreed that this lease shall not be deemed to be, nor intended to give rise to, a partnership relation.

E. Transfer of property. If the landlord should sell or otherwise transfer title to the farm, such action will be done subject to the provisions of this lease.

F. Right of entry. The landlord, as well as agents and employees of the landlord, reserve the right to enter the farm at any reasonable time to (a) consult with the tenant; (b) make repairs, improvements, and inspections; and (c) (after notice of termination of the lease is given) do tilling, seeding, fertilizing, and any other customary seasonal work, none of which is to interfere with the tenant in carrying out regular operations.

G. No right to sublease. The landlord does not convey to the tenant the right to lease or sublet any part of the farm or to assign the lease to any person or persons whomsoever.

H. Binding on heirs. The provisions of this lease shall be binding upon the heirs, executors, administrators, and successors of both landlord and tenant in like manner as upon the original parties, except as provided by mutual written agreement.

I. Additional agreements regarding terms of lease:

_____

_____

_____

III. Land Use

A. General provisions. The land described in Section I will be used in approximately the following manner. If it is impractical in any year to follow such a land-use plan, appropriate adjustments will be made by mutual written agreement between the parties.

1. **Cropland** a) Row crops _____ _____ Acres

    b) Small grains _____ _____ Acres

    c) Legumes _____ _____ Acres

    d) Rotation pasture _____ _____ Acres

2. **Permanent pasture** _____ _____ Acres

3. **Other:** _____ _____ Acres

    _____ _____ Acres

    TOTAL Acres _____ _____ Acres

B. Restrictions. The maximum acres harvested as silage shall be _____ acres unless it is mutually decided otherwise. The pasture stocking rate shall not exceed:

Pasture Identifications	Animal Units / Acre
_____	_____
_____	_____
_____	_____

(1000-pound mature cow is equivalent to one animal unit.)
Other Restrictions:

_____

_____

C. Government Programs. The extent of participation in government programs will be discussed and decided on an annual basis. The course of action agreed upon should be placed in writing and be signed by both parties. A copy of the course of action so agreed upon shall be made available to each party.

IV. Amount and Payment of Rent

If a flexible cash rental arrangement is desired, use material on the last page of this form and omit section A below.

A. Cash rental rates. The tenant agrees to pay as cash rent the amount as calculated below for each kind of land; or, one total may be entered for **Entire Farm** unit.

Amount of Cash Rent

Kind of Land or Improvements	Acres	Rate per Acre	Amount
Row crops	____	$ ____	$ ____
Small grains	____	$ ____	$ ____
Legumes	____	$ ____	$ ____
Permanent pasture	____	$ ____	$ ____
Timber	____	$ ____	$ ____
Waste	____	$ ____	$ ____
Farm buildings			$ ____
Dwelling			$ ____
Other	____	$ ____	$ ____
Entire Farm	____		$ ____

B. Rental payment. The annual cash rent shall be paid as follows:

$_____ on or before _____ day of _____ (month)

$_____ on or before _____ day of _____ (month)

$_____ on or before _____ day of _____ (month)

$_____ on or before _____ day of _____ (month)

Rental adjustment. Additional rental payment agreements:

_____

_____

_____

V. Operation and Maintenance of Farm

In order to operate this farm efficiently and to maintain it in a high state of productivity, the parties agree as follows:

A. The tenant agrees:

1. **General maintenance:** To provide the labor necessary to maintain the farm and its improvements during the rental period in as good condition as it was at the beginning. Normal wear and depreciation and damage from causes beyond the tenant's control are excepted.

2. **Land use.** Not to: a) plow pasture or meadowland, b) cut live trees for sale or personal use, or c) pasture new seedlings of legumes and grasses in the year they are seeded without consent of the landlord.

3. **Insurance.** Not to house automobiles, trucks , or tractors in barns, or otherwise violate restrictions in the landlord's insurance policies without written consent from the landlord. Restrictions to be observed are as follows: _____

_____

_____

4. **Noxious weeds.** To use diligence to prevent noxious weeds from going to seed on the farm. Treatment of the noxious weed infestation and cost thereof shall be handled as follows: _____

_____

5. **Addition of improvements.** Not to: a) erect or permit to be erected on the farm any nonremovable structure or building, b) incur any expense to the landlord for such purposes, or c) add electrical wiring, plumbing, or heating to any building without written consent of the landlord.

6. **Conservation.** Control soil erosion according to an approved conservation plan; keep in good repair all terraces, open ditches, inlets and outlets of tile drains; preserve all established watercourses or ditches including grassed waterways; and refrain from any operation or practice that will injure such structures.

7. **Damage.** When leaving the farm, to pay the landlord reasonable compensation for any damages to the farm for which the tenant is responsible. Any decrease in value due to ordinary wear and depreciation or damages outside the control of the tenant are excepted.

8. **Costs of operation.** To pay all costs of operation except those specifically referred to in Sections V-A-4 and V-B.

9. **Repairs.** Not to buy materials for maintenance and repairs in an amount in excess of $_____ within a single year without written consent of the landlord.

B. The landlord agrees:

1. Loss replacement. To replace or repair as promptly as possible the dwelling of any other building or equipment regularly used by the tenant that may be destroyed or damaged by fire, flood, or other cause beyond the control of the tenant or to make rental adjustments in lieu of replacements.

2. Materials for repair. To furnish all material needed for normal maintenance and repairs.

3. Skilled labor. To furnish any skilled labor tasks that the tenant is unable to perform satisfactorily. Additional agreements regarding materials and labor are: _____

_____

4. Reimbursement. To pay for materials purchased by the tenant for purposes of repair and maintenance in an amount not to exceed $_____ in any one year, except as otherwise agreed upon. Reimbursement shall be made within _____ days after the tenant submits the bill.

5. Removable improvements. Let the tenant make minor improvements of a temporary of removable nature, which do not mar the condition or appearance of the farm, at the tenant's expense. The landlord further agrees to let the tenant remove such improvements even though they are legally fixtures at any time this lease is in effect or within _____ days thereafter, provided the tenant leaves in good condition that part of the farm from which such improvements are removed. The tenant shall have no right to compensation for improvements that are not removed except as mutually agreed.

6. Compensation for crop expenses. To reimburse the tenant at the termination of this lease for field work done and for other crop costs incurred for crops to be harvested during the following year. Unless otherwise agreed, current custom rates for the operations involved will be used as a basis of settlement.

C. Both agree:

1. Not to obligate other party. Neither party hereto shall pledge the credit of the other party hereto for any purpose whatsoever without the consent of the other party. Neither party shall be responsible for debts or liabilities incurred, or for damages caused by the other party.

2. Capital improvements. Costs of establishing hay or pasture seedings, new conservation structures, improvements (except as provided in Section V-B-5), or of applying lime and other long-lived fertilizers shall be divided between landlord and tenant as set forth in the following table. The tenant will be reimbursed by the landlord either when the improvement is completed, or the tenant will be compensated for the share of the depreciated cost of the tenant's contribution when the lease ends based on the value of the tenant's contribution and depreciation rate shown in the "Compensation for Improvements" table. (Cross out the portion of the preceding sentence which does not apply.) Rates for labor, power and machinery contributed by the tenant shall be agreed upon before construction is started.

Table 1. Compensation for improvements.

| Type of Improvement | Date of Completion | Estimated Total Dollar Cost | Percent Contributed by Tenant | | | TOTAL Dollar Value of Tenant's Contribution* | Percent Rate of Annual Depreciation |
			Material	Unskilled Labor	Machinery		
		$	%	%	%	$	%
		$	%	%	%	$	%
		$	%	%	%	$	%
		$	%	%	%	$	%
		$	%	%	%	$	%
		$	%	%	%	$	%
		$	%	%	%	$	%
		$	%	%	%	$	%
		$	%	%	%	$	%
		$	%	%	%	$	%

* To be recorded when improvement is completed.

VI. Arbitration of Differences

Any differences between the parties as to their several rights or obligations under this lease that are not settled by mutual agreement after thorough discussion, shall be submitted for arbitration to a committee of three disinterested persons, one selected by each party hereto and to the third by the two thus selected. The committee's decision shall be accepted by both parties.

Amount of Rent to be paid when Cropland is rented on a Flexible Basis

A.   **Cash rent for inflexible items.** *(Complete at beginning of lease period.)*

    a)   Pasture $ _____

    b)   Hayland $ _____

    c)   Other inflexible cropland $ _____

    d)   Timber, wasteland $ _____

    e)   Farmstead $ _____

    TOTAL Inflexible Rent $ _____

B.   **Flexible cropland rent.** *(From Method I, II, or III below.)* $ _____

C.   **TOTAL Rent for Year** $ _____

D.   **Flexible cropland rent.** *(Use Method I, II, or III.)*

    1. Basic information to be used in Methods I and II

Crop(s)	Base Cash Rent (per acre)	Base Yield (bushel or ton per acre)	Base Price (per bushel or ton)	Minimum Cash Rent (per acre)	Maximum Cash Rent (per acre)
_____	$ _____	_____	$ _____	$ _____	$ _____
_____	$ _____	_____	$ _____	$ _____	$ _____
_____	$ _____	_____	$ _____	$ _____	$ _____

    2. The current price for the current year shall be Average Price at close of day based on the following time periods(s) and location(s).

Crop(s)	Day	Month	through	Day	Month	at	Price Source
_____	Day	Month	through	Day	Month	at	Price Source
_____	Day	Month	through	Day	Month	at	Price Source
_____	Day	Month	through	Day	Month	at	Price Source

For each year of this lease, the Per-acre Base Cash Rent for each crop shall be adjusted at the close of the cropping season by one of the following methods:

Method I - Flexing for Price Only

Crop(s)	Base Rent	x	(Current Price ÷ Base Price)	=	Rent per acre[1]	x	Acres Grown	=	Adjusted Rent for the Year
_____	_____	x	_____	=	_____	x	_____	=	_____
_____	_____	x	_____	=	_____	x	_____	=	_____
_____	_____	x	_____	=	_____	x	_____	=	_____

Method II - Flexing for Price and Yield

Crop(s)	Base Rent	x	(Current Price ÷ Base Price)	x	(Current Yield ÷ Base Yield)[2]	=	Rent per acre[1]	x	Acres Grown	=	Adjusted Rent for the Year
_____	_____	x	_____	x	_____	=	_____	x	_____	=	_____
_____	_____	x	_____	x	_____	=	_____	x	_____	=	_____
_____	_____	x	_____	x	_____	=	_____	x	_____	=	_____

Method III - Work Out and Record Procedure to be used.

[1] If calculated figure is less that "minimum cash rent" in D-1, use the set minimum. If calculated figure is more than "Maximum Cash Rent" in D-1, use the set maximum.
[2] The current yield shall be the "farm" yield for the current lease year.

Executed in duplicate on the date first above written:

_____   _____
tenant   *landlord*

_____   _____
tenant's spouse   *landlord's spouse*

State of _____

County of _____   }   SS:

On this _____ day of _____ , A.D. 20 _____ , before me, the undersigned, a Notary Public in said State, personally appeared _____ , _____ , _____ , and _____ to me known to be the identical persons named in and who executed the foregoing instrument, and acknowledged that they executed the same as their voluntary act and deed.

_____
Notary Public

NCR-76   Revised First Edition, 2nd Printing, 1M, 1998

Copyright © 1998, MidWest Plan Service, Iowa State University, Ames, Iowa 50011-3080 (515-294-4337)

For additional copies of this publication and a FREE Catalog of other agricultural publications contact:

MidWest Plan Service (MWPS), 122 Davidson Hall, Iowa State University, Ames, Iowa 50011-3080 or CALL: 1-800-562-3618

Printed on Recycled Paper

MWPS
MidWest Plan Service
A Foundation of Knowledge
Publisher and Distributor

HUMAN RESOURCE MANAGEMENT

CHAPTER OUTLINE

Characteristics of Agricultural Labor

Planning Farm Labor Resources

Measuring the Efficiency of Labor

Improving Labor Efficiency

Improving Managerial Capacity

Obtaining and Managing Farm Employees

Agricultural Labor Regulations

Summary

Questions for Review and Further Thought

CHAPTER OBJECTIVES

1. To describe trends in the use of human resources in agriculture

2. To illustrate how to plan for the quantity and quality of human resources needed for different farm and ranch situations

3. To outline methods for measuring and improving labor efficiency

4. To suggest ways to improve the management of agricultural employees, including selection, compensation, and motivation

5. To summarize laws regulating agricultural workers and employers

Human labor is one of the few inputs in agriculture whose use has diminished substantially over time. The decline has been especially dramatic since 1950, as shown in Table 21-1, although the total number of workers seems to have stabilized in the last decade. The introduction of mechanization and other labor-saving technology has allowed agricultural production to increase in spite of the decrease in labor use, however. Energy in the form of electrical and mechanical devices has replaced much of the physical energy provided by humans and draft animals in the past. More of the labor input on today's farm is spent operating, supervising, and monitoring these mechanical activities, and less is expended on physical effort. Changes in the tasks performed by agricultural labor have required both employees and managers to increase their education, skills, and training.

The mere availability of new technology such as larger machinery, mechanical feed and manure handling systems, and computers does not alone explain their rapid and widespread adoption. There has to be an economic justification before farmers will use any new technology, or it will simply "sit on the shelf." Most labor-saving technology has been adopted for one or more of the following reasons:

1. It is less expensive than the labor it replaces.
2. It allows farmers to increase their volume of production and total profit.
3. It makes work easier and more pleasant.
4. It allows certain operations such as planting and harvesting to be completed on time, even when weather is unfavorable or labor is in short supply.
5. It does a better job than could be accomplished manually.

Input substitution occurs because of a change in the marginal physical rate of substitution and/or a change in the relative prices of inputs, as was discussed in Chapter 8. Both factors have been important in the substitution of capital-intensive technology for labor in agriculture. Marginal rates of substitution have changed as new technology altered the shape of the relevant isoquants, making it profitable to use less labor and more capital.

Both interest rates and wage rates have increased since 1950, but wage rates have increased by a higher percentage, making labor relatively more expensive than capital. This change affects the price ratio in the substitution problem, also making it profitable to use more capital and less labor. The increased amount of capital per worker in agriculture has caused a substantial increase in the productivity of agricultural labor, making it both feasible and necessary to pay higher wages. This increased productivity has allowed farmers and ranchers to enjoy a standard of living comparable to that of nonfarm families.

CHARACTERISTICS OF AGRICULTURAL LABOR

A discussion of agricultural labor must recognize the unique characteristics that affect its use and management on farms. Labor is a continuous-flow input, meaning the service it provides is available hour by hour and day by day. It cannot be stored for later use; it must be used as it becomes available or it is lost.

TABLE 21-1	Workers on United States Farms (1950–2000)		
Year	Operators and unpaid workers	Hired workers	Total workers
1950	7,597,000	2,329,000	9,926,000
1960	5,172,000	1,885,000	7,057,000
1970	3,348,000	1,175,000	4,523,000
1980	2,401,000	1,298,000	3,699,000
1990	1,999,000	892,000	2,891,000
2000	2,062,000	890,000	2,952,000

Source: National Agricultural Statistics Service, U.S. Department of Agriculture, 2001.

Full-time labor is also a "lumpy" input, meaning it is available only in whole, indivisible units. Part-time and hourly labor is often used also, but the majority of agricultural labor is provided by full-time, year-round employees. Typically, only a few employees are hired on each farm (Table 21-2). If labor is available only on a full-time basis, the addition or loss of an employee constitutes a major change in the labor supply of the farm. For example, a sole proprietor hiring the first employee is increasing the labor supply by 100 percent, and the second employee represents a 50 percent increase. A problem facing a growing business is when and how to acquire the additional resources needed to keep a new worker fully employed. When other resources such as land and machinery also come in "lumpy" units, it becomes very difficult to avoid a shortage or excess of one or more resources.

The operator and other family members provide all or a large part of the labor used on most farms and ranches. This labor does not generally receive a direct cash wage, so its cost and value can be easily overlooked or ignored. However, as with all resources, there is an opportunity cost for operator and family labor, which can be a large part of the farm's total noncash fixed costs. Compensation for operator and family labor is received indirectly through expenditures for family living expenses and other cash withdrawals. This indirect wage or salary may vary widely, particularly for nonessential items, as net farm income varies from year to year. Cash farm expenses have a high-priority claim on any cash income, causing nonessential living expenses to fluctuate with farm income.

The human factor is another characteristic that distinguishes labor from other resources. If an individual is treated as an inanimate object, productivity and efficiency suffer. The hopes, fears, ambitions, likes, dislikes, worries, and personal problems of both the operator and employees must be considered in any labor management plan.

PLANNING FARM LABOR RESOURCES

Planning the farm's labor resources carefully will help avoid costly and painful mistakes. Figure 21-1 illustrates the process. The first step is to assess the farm's labor needs, both the quantity and quality, and the conditions under which workers will function.

Quantity of Labor Needed

Most farm managers judge the quantity of labor needed by observation and experience. When new enterprises are being introduced, typical labor requirements from published enterprise budgets can be used. A worksheet such as the one illustrated in Table 21-3 is useful for summarizing labor needs.

The seasonality of labor needs also must be considered. For example, labor requirements may exceed available labor during the months when planting, harvesting, calving, and farrowing take place. Figure 21-2 shows an example of the total monthly labor requirements for all farm enterprises and the monthly labor provided by the farm operator and a full-time employee. The farm operator in this example has a problem common to many farms and ranches. Operator

TABLE 21-2	Number of Farm Employees on U.S. Farms, 2002, by Number per Farm

Number of employees per farm	Percent of all employees
1	9
2	10
3–6	20
7–10	12
11 or more	49

Source: National Agricultural Statistics Service, USDA.

Step 1—Assess your situation

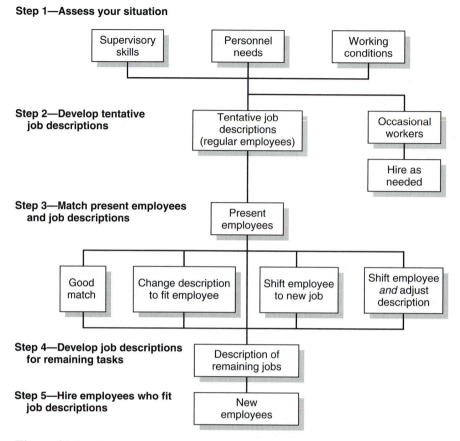

**Step 2—Develop tentative
job descriptions**

**Step 3—Match present employees
and job descriptions**

**Step 4—Develop job descriptions
for remaining tasks**

**Step 5—Hire employees who fit
job descriptions**

Figure 21-1   Flow chart of the farm labor planning process.
(*Source:* Thomas, Kenneth H., and Bernard L. Erven, *Farm Personnel Management,* North Central Regional
Extension Publication 329).

labor will not meet the requirements in some months, but the addition of a full-time employee results in large amounts of excess labor during other months. Longer workdays, temporary help, or hiring a custom operator may be necessary to perform the required tasks on time. A more permanent solution may be to increase the capacity of field machinery or processing equipment, or to shift to different enterprises. How much labor should be utilized to maximize profits depends on its availability, its cost, and whether it is a fixed or variable input.

Labor as a Fixed Cost   The total labor supply of the operator and/or full-time employees may

be fixed but not fully utilized. If this labor is being paid a fixed salary regardless of the hours worked, there is no variable or marginal cost for utilizing another hour. In this situation, labor can be treated as a fixed cost. In an enterprise budget labor costs do not affect the gross margin, nor the choice of enterprises. In a partial budget, labor costs are not included in added costs nor in reduced costs.

Labor as a Variable Cost   However, even permanent labor may have an opportunity cost greater than zero, from either leisure or off-farm employment. This becomes the minimum acceptable earning rate, below which the individual

TABLE 21-3 | **Labor Estimate Worksheet**

(1)	Acres / No. Units	Hr/Acre / Hr/Unit	Total hours for year (2)	December–March (3)	April–June (4)	July–August (5)	September–November (6)
					Distribution of hours		
1 Operator (or Partner No. 1)			2,900	800	750	600	750
2 Partner No. 2							
3							
4 Family labor			1,300	200	300	500	300
5 Hired labor							
6 Custom machine operators							
7 Total labor hours available			4,200	1,000	1,050	1,100	1,050
Direct labor hours needed by crop and animal enterprises							
Crop enterprises	Acres	Hr/Acre					
8 Wheat	700	1.8	1,260	10	100	750	400
9 Sorghum	300	2.1	630	0	400	0	230
10 Alfalfa	200	6.2	1,240	0	400	700	140
11							
12							
13							
14							
15							
16 Total labor hours needed for crops			3,130	10	900	1,450	770
Animal enterprises	No. Units	Hr/Unit					
17 Beef cows	250	6.0	1,500	600	500	200	200
18 Background	400	.25	100	50	0	0	50
19							
20							
21 Total labor hours needed for animals			1,600	650	500	200	250
22 Total hours needed for crops and animals			4,730	660	1,400	1,650	1,020
23 Total hours of indirect labor needed			600	200	150	100	150
24 Total labor hours needed			5,330	860	1,550	1,750	1,170
25 Total available (line 7)			4,200	1,000	1,050	1,100	1,050
26 Additional labor hours required (line 24 minus line 25)					500	650	120
27 Excess labor hours available (line 25 minus line 24)				140			

Source: Missouri Farm Planning Handbook, Manual 75: Feb. 1986. Department of Agricultural Economics, University of Missouri, Columbia, MO.

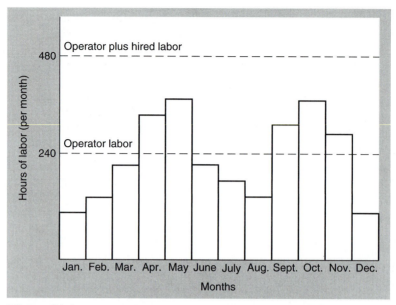

Figure 21-2   **Profile of labor requirements and availability.**

will choose not to work additional hours. Labor has a variable opportunity cost when increasing or decreasing the size of one enterprise affects the possible scale of other enterprises.

When labor is hired on a part-time or "as-needed" basis, it should always be treated as a variable cost. The cost of such labor, including benefits and payroll taxes, should be included in any budgeting decisions as well as marginal cost, marginal revenue analyses.

Quality of Labor Needed

Not all farm labor is equal in training, ability, and experience. New agricultural technology requires more specialized and sophisticated skills. Some activities, such as application of certain pesticides, may even require special training and certification. An assessment of labor needs requires identification of special capabilities needed, such as operating certain types of machines, performing livestock health chores, balancing feed rations, using computers or electronic control devices, or performing mechanical

repairs and maintenance. If the operator or family members do not possess all the special skills needed, then employees must be hired who do, or certain jobs may be contracted to outside consultants, repair shops, or custom operators. Training programs may also be available to help farm workers acquire new skills.

Management Style

Some farm and ranch managers utilize hired personnel more effectively than others. Many operators are accustomed to working alone, and prefer employees who can work independently with a minimum amount of supervision and instruction. Other employers prefer to work closely with workers, and give specific instructions about how a job is to be performed. Both management styles can be effective, but a good manager recognizes his or her own style and seeks out workers who can function effectively under it.

Once the quantity and quality of labor needed by the farm or ranch operation have

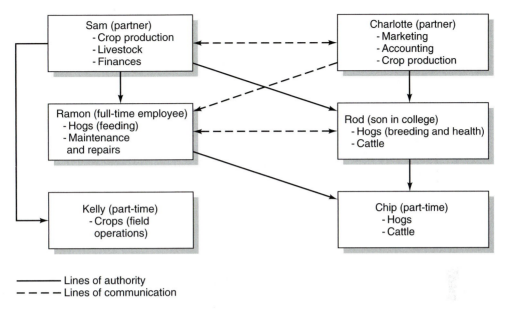

Figure 21-3   Organizational chart for a farm business.

been analyzed, tentative job descriptions should be developed. Larger operations will allow for much more specialization of duties than smaller ones, of course. Then the skills of the workers currently available should be compared to the job descriptions. Some duties may have to be rearranged to match the skills and interests of certain employees, family members, or partners. Needs that cannot be met by the current work force will have to be filled by providing job training, securing additional workers, or contracting with outside services.

Developing an organizational chart is especially useful where several employees and managers are involved. In particular, it should be made clear whether some employees are expected to take directions from other employees, or from members of the operator's family. Figure 21-3 shows an example of an organizational chart for a medium-size operation. Besides indicating which members of the labor force supervise other members, an organizational chart should show the lines of communication that

need to be maintained among managers and employees.

MEASURING THE EFFICIENCY OF LABOR

Labor efficiency depends not only on the skills and training of the labor used but also on the size of the business, enterprises, degree of mechanization, type of organization, and many other factors. Measures of labor efficiency should be used to compare and evaluate results only on farm businesses of approximately the same size and type.

Measures of labor efficiency often use the concept of person-years or full-time equivalents of labor employed. This is a procedure for combining operator, family, and hired labor into a total labor figure which is comparable across farms. The example on the next page shows 21 months of labor provided from three sources. Dividing this total by 12 converts it into 1.75

person-years, or the equivalent of 1.75 persons working full time during the year.

Operator labor	12 months
Family labor	4 months
Hired labor	5 months
Total	21 months

21 ÷ 12 = 1.75 person-years equivalent

Measures of labor efficiency convert some physical output, cost, or income total into a value per person-year. The following measures are commonly used.

Value of Farm Production per Person   This measures the total value of agricultural products generated by the farm per person-year equivalent. It is affected by business size, type of enterprise, and the amount of machinery and other labor-saving equipment used.

Labor Cost per Crop Acre   The cost of labor per crop acre is found by dividing the total crop labor cost for the year by the number of acres in crop and fallow. The opportunity cost of operator and family labor is included in total labor cost. Values will be affected by machinery size, type of crops grown, and whether or not custom operators are used.

Crop Acres per Person   The number of crop acres per person is found by dividing total crop acres by the number of person-years of labor used for crop-related activities.

Cows Milked per Person   The total number of productive cows in the dairy herd is divided by the number of person-years of labor. Other livestock enterprises use similar measures of labor efficiency.

Table 21-4 contains some labor efficiency values for farms in Iowa. The data show that productivity per person usually increases as farm size increases, due mostly to more investment in larger machinery and equipment, and spreading labor overhead over more units of production.

IMPROVING LABOR EFFICIENCY

Labor efficiency can be improved by more capital investment per worker, the use of larger-scale machinery, or adopting less labor intensive technology. However, if the objective is to maximize profit and not just to increase labor efficiency at any cost, then marginal rates of substitution and the prices of labor and capital must determine the proper combination. Increasing the capital investment per worker will increase profit only if: (1) total cost is reduced, while revenue increases, remains constant, or at least decreases less than costs; or (2) the labor that is saved can be used to increase output value elsewhere by more than the cost of the investment.

TABLE 21-4	Labor Efficiency by Farm Size for Iowa Farms, 2001		
	Farm size, annual sales		
Item	**$40,000 to $99,999**	**$100,000 to $249,999**	**$250,000 or more**
Months of labor per farm	8.0	12.5	20.7
Value of farm production per person	$142,589	$160,291	$248,803
Labor cost per crop acre	$ 52.80	$ 41.99	$ 39.45
Crop acres per person	477	486	608

Source: 2001 Iowa Farm Costs and Returns, Iowa State University Extension Publication FM-1789, 2002.

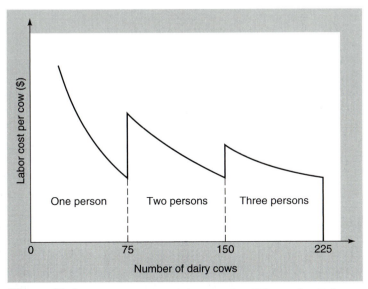

Figure 21-4   **Labor cost per dairy cow for different herd sizes.**

When the labor supply is increased by adding a full-time worker, some additional units of an enterprise may be needed to fully utilize the available labor. Figure 21-4 shows how costs per cow may go up when a dairy farm adds employees, if the number of cows is not increased at the same time. This illustrates the "lumpy" nature of full-time labor as an input.

Simplifying working procedures and routines can have a high payoff in increased labor efficiency. Considerable time can be saved by having all necessary tools and other supplies at the work area, not having to stop to open and close gates, keeping equipment well maintained, and having spare parts on hand. Making changes in the farmstead layout, building designs, field size and shape, and the storage sites of materials relative to where they will be used can also save time and increase labor efficiency. When materials must be moved, consider using conveyers, carts, small vehicles, augers, and other labor-saving devices. Automatic switches and timers can eliminate time spent waiting. Carrying radios or portable telephones in vehicles or tractors makes timely communication possible,

allows coordination of activities over a wide geographic area, and reduces trips. Production data can be entered directly into small notebook or hand-held computers, rather than transferred from paper records. As always, the additional cost of any changes must be weighed against the value of the labor saved.

Labor efficiency can also be improved by making sure workers have safe and comfortable working conditions whenever possible. Although most agricultural work is still performed outside, modern machinery cabs and vehicles help reduce fatigue from heat, cold, dirt, and noise. Making sure that machinery guards are in place, livestock buildings are well ventilated, and all safety measures are followed will prevent lost work time from injuries or illness. Workers should be provided with suitable clothing and other safety equipment when working with agricultural chemicals or performing other hazardous jobs.

The simple procedure of planning and scheduling work in advance will help reduce wasted time. Tasks that must be done at a specific time should be scheduled first, and those that are not time-specific, such as building repairs or other maintenance, can be planned for months of slack labor. Keep a list of jobs to be done, and assign a priority and deadline to each one. This list should be where all workers can see it, add to it, and cross off tasks that have been completed. From this, a daily work schedule can be planned in a few minutes each morning or evening. Time used to organize the next day's tasks and their order of importance is time well spent.

IMPROVING MANAGERIAL CAPACITY

The most important labor resources in any business are the managers themselves. Although a manager may begin with a wealth of experience and education, this will not be adequate for an entire career. Management skills must continually be reinforced and upgraded.

Agricultural technology is continually evolving. Simply repeating prescribed techniques is not enough. A successful ranch or farm manager must understand the principles behind the technology, such as animal nutrition, plant physiology, genetics, and farm mechanics. New applications of these principles then will be easier to master. Farm magazines, Extension seminars, continuing education courses and field days are all excellent sources of information about new technology. Ultimately, though, the manager may realize that it is not possible for one person to master everything, and that delegating responsibility to others in the business or hiring outside consultants is a useful way of accessing additional expertise.

Developing an efficient office or business center is also a sign of a good manager. The physical facilities should be comfortable and functional, and allow the use of up-to-date communications and information management technology. The flow of bills, receipts, reports and correspondence should move quickly and frequently. Records should be sorted and filed for easy retrieval.

Farm operators are seldom underemployed. The work to be done always expands to fill the time available, so setting priorities is important. It is easy to simply look around and do whichever task comes into view first. However, the principles of *immediacy* and *impact* must be applied to decide what jobs will have the most effect on the business and/or need to be completed first. Making monthly and weekly lists helps identify and prioritize tasks to be done.

Many successful farm managers are active in professional and volunteer organizations outside their own businesses. Farm policy and commodity organizations, school boards, churches, cooperatives and marketing clubs all offer chances to develop organizational skills and personal relationships. Not only is participation in such groups a way to contribute to a community, but some of the best ideas and philosophies come from interacting with other highly capable people.

Finally, an agricultural manager needs to develop a "world view." Over 25 percent of U.S. agricultural products are consumed in other countries, so it is important to understand tastes and customs from other cultures. Farmers in other countries also offer significant competition in the marketplace. Traveling abroad, especially with tours designed to learn more about agricultural production, trade, and consumer preferences is a very effective way of learning about the rest of the world. Hosting international visitors and attending programs or reading about other cultures also are valuable ways to increase international understanding.

OBTAINING AND MANAGING FARM EMPLOYEES

Paid employees account for almost 30 percent of the total labor supply on farms and ranches in the United States, as was shown in Table 21-1. Acquisition, training, and retention of hired workers are common subjects of conversation whenever managers of larger farms get together. Farm managers are finding that skills in human relations and personnel management are valuable assets.

Box 21-1 | **Three Ads for the Same Job: Which One Would You Apply for?**

Wanted: Hired Man. Phone 123-456-7890

Wanted: Person for general farm work, full-time, on a beef and crop farm. Requires two years farm experience or equivalent and/or two years ag education beyond high school. Good wage and benefit package. Call: 123-456-7890

Source: Shapley, A. E.: Farm Employer's Handbook, Michigan State University.

Maple Grove Farms has an opening for a full-time person to assist the owner/operator in management and production of beef and crops. The position offers considerable variety and opportunity for growth. At least two years farm experience and/or ag education beyond high school required. Pay and benefit package based on applicant's experience and training. Write for application to Maple Grove Farms (address).

Recruiting

The process of hiring an employee starts with recruiting, including announcing the job opening, publicizing it, and receiving applications. Placing newspaper ads, talking with other farmers, relatives, agribusiness people, or farm consultants in the community, or contacting college placement offices and employment agencies are ways to inform individuals about a job opening and identify potential candidates to fill it. In some areas, the employer may have to negotiate with a labor contractor to supply a large number of temporary workers for harvesting or other labor-intensive activities.

The job announcement should clearly state the skills and experience desired. In addition, it should make the position sound desirable. Emphasize reasons why the applicant would want to work for this operation instead of the one down the road.

It is generally helpful to provide an application form to each applicant. Basic information should be obtained about the applicant's background, work experience, training, personal goals, references, and other factors. Remember that questions about race, religion, age, or marital status should be avoided.

Interviewing and Selecting

Completed application forms can be used to select a small number of candidates for the next step, interviewing. The interview should be planned carefully to efficiently acquire more information. Sufficient time and opportunity should be allowed for the applicant to ask questions about the job, its duties, and responsibilities. Interviewing involves not only obtaining information about applicants but also providing information to them so they can assess their own interest in and qualifications for the job. A skills test may be needed for certain technical jobs. A tour of the work location should be provided, and the candidate should have a chance to visit with other employees.

The information obtained about each job applicant through the application form, the interview, and references must now be evaluated. Many factors must be considered in selecting a candidate, including personal compatibility. Farm employers often work more closely with their employees on a day-to-day basis than other employers do, sometimes under stressful conditions. This close working relationship increases the chance for friction if the individuals are not compatible.

The Employment Agreement

Once a job offer is made and accepted, a written employment agreement should be developed, as shown in the example in Box 21-2. The purpose of the agreement is to clarify the work expectations of both the employer and employee, and to serve as a reference for evaluating performance later on.

The employment agreement should start with the job description, including duties and responsibilities, lines of authority, and the job title. Other important information includes wages and benefits, working hours and days, vacation, sick leave and personal leave policy, safety rules, allowable uses of farm property, training opportunities, bonus or incentive plans, and procedures for evaluation and promotion or termination. A review of the employment agreement should be made once or twice a year as part of the evaluation process.

Compensation

A competitive compensation package is essential in a successful labor hiring program. The total value of take-home pay, benefits, and bonuses should be compared against comparable figures in other employment.

Wages and Salaries   The actual cash wage or salary paid is the most important item. Whether to pay a fixed or variable wage is the first decision. Positions in which the duties and hours worked will be fairly constant throughout the year usually receive a fixed weekly or monthly salary. Other positions with highly variable hours, such as those found on a fruit and vegetable or cash grain farm, are often paid by the hour, as are most part-time positions. Workers who are employed in harvesting activities are sometimes paid on a piece-rate basis.

Salaries depend on the position and the particular skills of the employee filling it. The size of the farming operation, the duties performed, and the number of years the employee has worked on the farm are other factors that influence the level of compensation.

Skill-based pay is an approach that sets compensation based on the level of responsibility each worker has rather than the specific duties. Characteristics such as decision-making authority, supervision of other employees, and

specialized skills needed are used to rate each position and assign it a wage range.

Fringe Benefits   Fringe benefits are often a large part of the total compensation for farm employees. Table 21-5 shows the employee benefits that were reported in a recent study of employees on swine farms. They accounted for 12 percent of the average employee's compensation.

Prospective employees should be made aware of the value of their benefits so they can fairly evaluate a job offer. Fringe benefits such as retirement plans, insurance, housing, utilities, farm produce, and the use of a vehicle will make a lower cash wage for farm work competitive with nonfarm employment that has a higher cash salary but fewer benefits.

Fringe benefits are most useful when the employer can provide them for less than it would cost the employee to obtain them elsewhere. Examples include the use of existing housing, vehicles, or livestock facilities. Some benefits such as health insurance are tax deductible for the employer but not taxable to the employee.

Incentive Programs and Bonuses   Bonuses are often used to supplement base wages, improve labor productivity, and increase employee retention. However, bonuses help increase labor efficiency only if they are tied closely to performance. If not, employees soon come to expect the bonus and consider it part of their basic cash salary. When the size of the bonus is tied to annual profit, an employer may find it difficult to decrease the bonus in a poor year after employees have experienced several years with a larger bonus.

Most bonus plans are based on one or more factors: volume, performance, longevity, or profitability.

1. *Volume* can be measured by the total number of pigs weaned, calves weaned, or bushels harvested. The employee's wages increase when the work load increases, and a modest incentive for efficiency is

TABLE 21-5	**Benefits Received by Employees on Swine Farms (U.S., 1995)**

Benefit	Percent receiving
Major medical coverage	80.2
Paid vacation	78.6
Life insurance	65.5
Workers compensation	65.4
Paid holidays	63.4
Disability insurance	54.6
Paid sick leave	51.9
Dental coverage	46.4
Processed meat	39.0
Housing	37.9
Pension/retirement plan	35.8
Unemployment insurance	34.3
Continuing education	26.3
Vehicle	23.4
Paid utilities	22.9
Profit-sharing plan	18.1
Other	8.6

Source: Employee Management, National Hog Farmer, 1996.

provided. Care should be taken that higher costs are not incurred simply to increase production, however.

2. *Performance* can be measured by the number of pigs weaned per sow, calving percentage, milk production per cow, or crop yields per acre. The bonus is often based on how much the actual performance exceeds a certain base level. This type of bonus can be an effective work incentive as long as it is based on factors over which the employee has at least some control.

3. *Longevity* is rewarded by paying a bonus based on the number of years the employee has worked for the business. This recognizes the value of experience and worker continuity to the employer, and rewards loyalty.

4. *Profitability* bonuses are usually based on a percent of the gross or net income of the business. They allow the employee to share in the risks and rewards of the business, but may depend on many factors not under the control of the employee. They also require the employer to divulge some financial information to the employee.

Following several basic principles will increase the effectiveness of any incentive program.

- The program should be simple and easily understood by the employee.
- The program should be based on factors largely within the control of the employee.
- The program should reward work that is in the best interest of the employer.
- The program should provide a cash return large enough to motivate improved performance.
- The incentive payment should be made promptly or as soon after the completion of work as possible.
- An example of how the bonus will be computed should be provided in writing, using typical performance levels.
- The incentive payment should not be considered as a substitute for a competitive base wage and good labor relations.

A good bonus plan not only rewards employees financially, but also provides recognition for their accomplishments.

Total compensation packages offered to employees should have *internal consistency,* that is, they should be fair among employees of the same business. They should also have *external consistency* when compared to workers doing similar jobs on other farms and ranches.

Training Hired Labor

Farm and ranch managers sometimes hire unskilled workers and then expect them to perform highly skilled tasks in livestock management or with expensive machinery. They may also expect employees to automatically do things exactly the way they themselves would do them. The result is disappointment, frustration, high

Box 21-2   Employment Agreement Example

Farm Employer-Employee Agreement

Beau Valley Farms (employer) of Greenville, Oregon, agrees to employ _____ (employee), to perform agricultural work in and around Jefferson County, Oregon, beginning _____ (date) and continuing until such time as either wishes to terminate this agreement by a 30-day notice. The employer and the employee agree to comply with the following conditions and actions:

1. To pay the employee $ _____ per hour, from which the employee's income tax and social security taxes will be withheld, as required by law. Wages will be paid on the 1st and 15th day of each month.

2. To provide a 3-bedroom house, with utilities to be paid by the employee. Routine maintenance is to be done by the employee and materials paid for by the employer. Any other agreements pertaining to the employee's housing will be noted at the end of the agreement.

3. The normal work week is Monday through Saturday. The normal paid working hours are from 7:00 A.M. to 6:00 P.M. (12:00 noon on Saturdays), with two 30-minute rest breaks and one hour off for lunch. Longer hours may be required to complete jobs in progress. Overtime will be paid for any work done after normal working hours at the rate of 150% of the normal wage rate.

4. Time off shall be given every Sunday and holiday, and one Saturday per month. The holidays for purposes of this agreement are New Year's Day, Easter, Memorial Day, Independence Day, Labor Day, Thanksgiving Day, and Christmas Day.

5. The employee is entitled to 10 days of vacation with pay annually, which shall be taken during the nonheavy work season and agreed upon with the employer 30 days in advance.

6. The employee is entitled to 5 days sick leave with pay annually for the time off due to actual illness, and 1 day of personal emergency leave with pay.

7. The employee is entitled to drive a vehicle provided by Beau Valley Farms for purposes related to employment, including commuting to work and for personal use within Jefferson County.

8. The following insurance plans will be carried on the employee:

   • comprehensive medical and health

   • term life insurance coverage up to $100,000

9. A performance review and written performance evaluation will be carried out once during the first 6 months of employment and once each 12 months thereafter.

10. A bonus or incentive plan (is, is not) included. If included, the provisions are noted on the attached appendix.

11. Other provisions not included above are listed on the appendix to this form.

Employer Representative
   Signature _____ Date _____
Social Security No. _____
Employee
   Signature _____ Date _____
Social Security No. _____

Adapted from: Thomas, Kenneth H., and Bernard L. Erven: *Farm Employee Management.*

repair bills, poor labor productivity, and employee dissatisfaction.

Studies of employment practices on ranches and farms show little evidence of formalized training programs for new employees. Even a skilled employee will need some instruction on the practices and routines to be followed for a particular operation, and the meaning of specialized terminology. Employees with lesser skills should receive complete instructions and proper supervision during a training period. Employers need to develop the patience, understanding, and time necessary to train and supervise new employees. Unfortunately, in production agriculture, the training period may need to last as long as a year, or until the new employee has had a chance to perform all the tasks in a complete cycle of crop and livestock production.

Periodic retraining may be necessary for even long-time employees. The adoption of new technology in the form of different machinery, new pesticides, feed additives, seed varieties and electronic data gathering, or the introduction of a new enterprise may require additional training for all employees. Extension short courses, bulletins, farm magazines, video tapes, internet sites, field days, and seminars can be used for employee training. Participation in these activities will improve not only employees' skills but also their self-esteem. The cost of training should be borne partially or wholly by the employer, as an investment in future productivity.

Motivation and Communication

Hiring and training new employees is costly in terms of both time and money. If labor turnover is high, these costs can become excessive, and labor efficiency will be low. Employers should be aware of the reasons for poor employee retention and take action to improve the situation.

Farm employees often report that they are attracted to farm work by previous farm experience, the chance to work outdoors, and an interest in working with crops and livestock. Disadvantages cited are long hours, little time off, early-morning or late-evening work, uncomfortable working conditions, and poor personal relationships with their employer. Low pay is seldom at the top of the list of disadvantages, indicating that they have personal goals other than obtaining the highest pay.

A set policy for adequate vacations and time off is important to employees, as is the opportunity to work with good buildings, livestock, and equipment. Job titles can also be important. The title "hired hand" contributes little to an employee's personal satisfaction and self-image. Herd superintendent, crop manager, group leader, and machine operator are examples of titles most people associate with a higher status than that of a hired hand.

Good human relations is the most important factor in labor management. This includes such things as a friendly attitude, loyalty, trust, mutual respect, the ability to delegate authority, and the willingness to listen to employee suggestions and complaints. Work instructions should be given in sufficient detail that both parties know what is to be done, when, and how. Everyone responds positively to public praise for a job well done, but criticism and suggestions for improvement should be communicated in private. Unpleasant jobs should be shared by all employees, and everyone should be treated with equal respect. Employers who reserve the larger, air-conditioned tractor for their own use, while employees drive smaller tractors without cabs, for example, are more likely to have labor problems.

As employees grow in their abilities and experience, they should be given added responsibilities, along with the chance to make more decisions. Likewise, the employer must be willing to live with the results of those decisions or tactfully suggest changes.

Working environment also has a major impact on job satisfaction. A national survey of employees on swine farms found that when gas levels in hog facilities were low, 47 percent of workers were very satisfied with their job and 38 percent were satisfied. Conversely, when gas levels in facilities were high, only 10 percent and 26 percent of workers, respectively, were very satisfied or satisfied. Results were similar for facilities with low or high dust levels.

Bridging Cultural Barriers

More and more agricultural employees are either recent immigrants to the United States or children of immigrants. A recent study of seasonal

Box 21-3	**Case Example: Understanding Cultural Differences**

Haitian farm workers in an upstate apple growing enterprise had been told repeatedly to leave the apples that had fallen on the ground alone, and not put them with those that were picked from the tree. Problems also occurred when different varieties of apples were placed in the same containers. The supervisor had even told a few people not to return for work because of their insubordination. Discussions with the farm workers explained their actions. In Haiti food was in short supply. Leaving what appeared to be good apples on the ground was a waste of food that they could not understand. Mixing varieties was understandable when we consider that only the rich can afford apples in Haiti. The numerous varieties of apples were as unknown to them as the numerous varieties of bananas would be to U.S. born workers.

Source: Human Resource Management on the Farm, by Kay Embrey, Cornell University Cooperative Extension Service.

agricultural workers in the U.S. revealed that 84 percent of them were foreign born. When employees and employers do not share a common language and culture, communication and motivation can break down quickly.

Immigrant workers often develop extensive networks of relatives and acquaintances from their native communities. Loyalties are very strong, and may supercede loyalty to a job or employer. Such networks can be used effectively for recruiting new employees who will be accepted by the existing workforce. Conversely, mixing workers from different countries or regions may create conflicts, even though the cultural differences are not apparent to the employer.

Training programs are much more critical for employees who have not grown up in the local area. Technical knowledge and habits cannot be taken for granted. Showing non-native speakers how to perform tasks is much more effective than telling them. Fortunately, many technical manuals, regulations, and employee agreements are available in multiple languages today, especially Spanish.

Cultural attitudes affect employer-employee relationships. In many countries workplaces have very hierarchical structures. Supervisors demand respect from employees based on their position or family ties, not on their abilities or personalities. Workers may be uncomfortable having their employer working side-by-side with them, even though this is common on small farms. They also may feel that asking questions about their duties is a sign of ignorance or disrespect. When something goes wrong, it is often attributed to fate, not to the employee's lack of diligence. Punctuality and timeliness may be less important than personal relationships and family responsibilities. Religious or cultural holidays that are observed by some ethnic groups may be different than those of the employer, and need to be considered when the employment agreement or work calendar is developed.

Evaluation

All employers are constantly evaluating their employees' performances. Too often, however, communication takes place only when serious problems arise. Regular times for communication and coordination should be scheduled. If meetings are held only when there is a problem, the employee will immediately be put on the defensive when one is called. Some managers take their employees to breakfast once every week or month to discuss plans and problems. The employer should listen carefully to each employee's concerns and ideas, even if not all of them can be acted upon.

Operations with a sizable work force should use written evaluations, with performance measured against job descriptions. Employees should be warned if their performance is unsatisfactory, first orally and then in writing, and given a chance to improve. If termination is necessary, document the reasons carefully and give written notice in advance. All this takes time, but it can prevent costly complaints or lawsuits later. If an employee quits, hold an exit interview. Discuss the reasons for leaving, and decide whether changes in employee recruitment or management practices are needed.

AGRICULTURAL LABOR REGULATIONS

Federal and state regulations that affect the employment of agricultural labor have become an important factor in labor management in the United States. Their thrust is to extend to farm workers much of the same protection enjoyed by nonfarm workers. For the employer, the effects of these regulations may be increased costs for higher wages and more benefits, more labor records to keep, and additional investments for safety and environmental protection, but also a more satisfied and productive work force.

It is not possible to list and describe all the federal and state regulations pertaining to agricultural labor in a few pages. Only a brief discussion of some of the more important and general regulations will be included here.

Minimum Wage Law   Agricultural employers who used more than 500 person-days of labor in any quarter of the preceding calendar year must pay at least the federal minimum wage to all employees except immediate family members, certain piece-rate hand harvesters, and cowboys and shepherds employed in range production of livestock. The family member exemption does not apply to partnerships and corporations. The minimum wage is subject to increases over time. This law also requires employers to keep detailed payroll records to prove compliance with the minimum wage legislation. Agricultural employers are not required to pay overtime wages to employees who work more than 40 hours per week, however.

Social Security   Employers must withhold social security, or FICA (Federal Insurance Contributions Act), taxes from an employee's cash wages and match the employee tax with an equal amount. This regulation applies to farmers with one or more employees, if the employee was paid $150 or more in cash wages during the year or the employer paid $2,500 or more in total wages to all employees. Social security taxes are not withheld for children under 18 years of age employed by their father or mother. All other employees, including relatives, are subject to FICA withholding rules.

The tax rate, and the maximum earnings subject to this tax, increase from year to year. Employers should contact the local Social Security Administration office or obtain Internal Revenue Service (IRS) Circular E, Employer's Tax Guide, for the current rates. It is important to remember that the tax rate and maximum tax figures represent the amounts withheld from the employee's pay. The employer must match them with an equal amount for remittance to the Internal Revenue Service.

Federal Income Tax Withholding   Farmers are required to withhold federal income taxes from wages paid to agricultural workers if the employer and employee meet the same criteria described for social security withholding. Circular E or Circular A, Agricultural Employee's Tax Guide, provides details. Records should be kept of each employee's name, age, social security number, cash and noncash payments received, and amounts withheld or deducted.

Workers' Compensation   This is an insurance system designed to protect workers who suffer job-related injuries, illnesses, disability, or death. It also frees the employer from additional liability for such injuries. A fixed amount of compensation is paid for each illness or injury.

Laws regarding workers' compensation insurance for farm employees vary from state to state, but most states require it of any farm operator with one or more employees. Where required, the employer pays a premium based on a percentage of the total employee payroll.

Unemployment Insurance   This insurance temporarily replaces part of a person's income lost due to unemployment. Benefits are financed by an employer-paid payroll tax. Farm employers must contribute if, during the current or previous calendar year, they: (1) employed 10 or more workers in each of 20 or more weeks, or (2) paid $20,000 or more in cash wages in any calendar quarter.

Child Labor Regulations   These regulations, established under the Fair Labor Standards Act, require an employee to be at least 16 years of age to be employed in agricultural work during school hours. The minimum age is 14 for employment after school hours, with two exceptions: children 12 or 13 can be employed with written parental consent or when a parent is employed on the same farm, and children under 12 can work on their parents' farm.

Age restrictions also apply to certain jobs that are classified as hazardous. Examples of these jobs are working with certain farm chemicals, operating most farm machinery, working inside storage structures, and working with breeding livestock. Employees under the age of 16 cannot be employed in these hazardous jobs, except that 14- or 15-year-old employees can be certified to operate farm machinery by taking qualified safety courses. Child Labor Bulletin No. 102 of the U.S. Department of Labor contains details.

Occupational Safety and Health Act (OSHA)
OSHA is a federal law passed to ensure the health and safety of employees by providing safe working conditions. Examples are the requirements for slow-moving farm equipment using public roads to display a slow-moving vehicle (SMV) sign on the rear, and for roll bars

to be present on most farm tractors. Employers of seven or more workers must keep records of work-related accidents. Farm employers should check the latest OSHA regulations to be sure they are meeting all current requirements.

The Environmental Protection Agency (EPA) regulates the handling of farm pesticides. All farm workers must complete a pesticide applicator training course to handle both general and restricted-use pesticides.

Immigration Reform and Control Act   This act requires employers to check documents that establish the identity and eligibility of all workers hired after November 6, 1986. Both the employer and employee must complete and retain a Form I-9, available from the Immigration and Naturalization Service or Department of Labor, which certifies that the employee is a U.S. citizen, a permanent alien, or an alien with permission to work in the United States.

Migrant Labor Laws   The U.S. Migrant and Seasonal Agricultural Worker Protection Act prescribes housing, safety, and record-keeping requirements for employers of migrant or seasonal agricultural workers. State laws may also stipulate minimum housing, safety, and health standards.

Civil Rights   Federal civil rights legislation prohibits employment discrimination on the basis of race, color, religion, sex, national origin, handicap, ancestry, or age. These rules also apply to recruiting and hiring practices.

Disability   The Americans with Disabilities Act of 1990 (ADA) prohibits discrimination against employees or potential employees based on physical or mental handicaps. If there are aspects to a job that could not be carried out by a person with certain disabilities, these should be included in a written job description before applicants are interviewed. The ADA applies to businesses that employ 15 or more workers for 20 or more weeks out of the year.

SUMMARY

Although the number of people employed in agriculture has fallen dramatically in the past 50 years, their productivity and skill level have increased even more rapidly. Planning farm personnel needs involves assessing the quantity, quality, and seasonality of labor needed. Whether the cost of farm labor is considered to be fixed or variable depends on its availability and opportunity costs. Good labor management techniques can improve labor efficiency and satisfaction. Farm managers themselves need to continually upgrade their skills.

Effective management of hired employees starts with the recruiting, interviewing, and selection process. An employment agreement should be drawn up that specifies work rules and compensation, including salary, benefits, and bonuses. Bonuses should be planned carefully to provide the desired work incentives. Farm employees also must be trained, motivated, and evaluated. Where cultural differences exist between managers and employees, special care should be taken to ensure communication is effective.

A manager of farm labor must be familiar with and comply with the various state and federal laws that protect and regulate agricultural workers.

QUESTIONS FOR REVIEW AND FURTHER THOUGHT

1. Why has the amount of human resources used in agriculture decreased?

2. Why are the training and skills needed by agricultural workers greater today than in the past?

3. Under what conditions is it profitable to substitute labor for capital?

4. Why are there wide variations in the monthly labor requirements on a crop farm, but less variation on a dairy farm?

5. Why is the labor cost per acre or per head generally less for larger farms than for smaller ones?

6. Observe an actual routine farm chore such as milking or feeding livestock. What suggestions would you have for simplifying the chore to save labor? Would your suggestions increase costs? Would they be profitable?

7. Write an advertisement to place in a newspaper or magazine for recruiting a livestock manager for a large farm.

8. Write a detailed job description for the same position.

9. Discuss what you would include in a training program for a new farm employee.

10. Discuss some procedures that could be used to improve personal relationships between a farm employer or supervisor and farm employees who grew up in a different culture.

11. What type of incentive program would be most effective on a dairy farm if the objective was to increase milk production per cow? If the objective was to retain the better employees for a longer time?

12. Which labor laws and regulations have to do with withholding taxes? Safety and health? Employing minors?

22

MACHINERY MANAGEMENT

CHAPTER OUTLINE

Estimating Machinery Costs

Examples of Machinery Cost Calculations

Factors in Machinery Selection

Alternatives for Acquiring Machinery

Improving Machinery Efficiency

Summary

Questions for Review and Further Thought

CHAPTER OBJECTIVES

1. To illustrate the importance of good machinery management on farms and ranches

2. To identify the costs associated with owning and operating machinery

3. To demonstrate procedures for calculating machinery costs

4. To discuss important factors in machinery selection including size, total costs, and timeliness of operations

5. To compare owning, renting, leasing, and custom hiring as different means of acquiring the use of machinery

6. To present methods for increasing the efficiency of machinery use

7. To introduce factors that influence when machinery should be replaced

Mechanization has had a dramatic effect on production costs, efficiency levels, energy use, labor requirements, and product quality in agriculture all over the world. The decline in farm labor required and the increase in mechanization were discussed in Chapter 21. As tractor sizes have increased, the sizes of other machinery have also increased to keep pace. These increases have contributed to higher machinery investment per farm and more efficient use of labor, and have made it possible for an individual operator to farm many acres.

The quality of work performed by farm machinery has also risen dramatically. Field losses during harvesting have been greatly reduced. Improved seed and fertilizer placement have made it possible to reduce the amount of tillage performed. Sensitive yield monitors, more accurate sprayers and applicators, and the use of satellite positioning technology have spawned a new set of practices known as *precision agriculture.*

Discussions on the increased use of mechanization in agriculture often emphasize tractors and other crop production machinery. Perhaps no less dramatic has been the rise in the use of power and equipment in livestock production and materials handling. Physical labor requirements have been reduced in many areas by the use of small engines and electric motors, conveyors, computers, sensors, and timers. Grain handling, manure collection and disposal, livestock feeding, feed grinding and mixing, hay handling, accounting, and data collection have all been greatly automated to reduce labor requirements and costs, and improve performance.

Good machinery management strives to provide reliable service to the various crop and livestock enterprises at minimum cost. Sometimes machinery is a profit center itself, such as for a custom operator. More often though, it is simply an input for other enterprises. This makes it difficult to separate machinery management completely from an analysis of the total business. The remainder of this chapter will concentrate on the use of economic principles and budgeting in machinery management, as well as the relations among machinery, labor, capital, and the production enterprises.

ESTIMATING MACHINERY COSTS

Machinery is costly to purchase, own, and operate. A single tractor can cost well over $100,000, and a cotton picker over $300,000. A manager must be aware of the costs of owning and operating a machine and understand how they are related to machinery use, interest rates, useful life, and other factors. It is easy to underestimate machinery costs, because many of them involve infrequent, though large, cash outlays, or they are noncash costs.

Machinery costs can be divided into ownership and operating costs. Ownership costs are also called overhead, indirect, or fixed costs, because they are basically fixed with respect to the amount of annual use. Operating costs are also referred to as variable or direct costs, because they vary directly with the amount of machine use. Fixed and variable costs were discussed in Chapter 9 but will be reviewed in the following sections as they apply to farm machinery.

Ownership Costs

Ownership or fixed costs begin with the purchase of the machine, continue for as long as it is owned, and cannot be avoided by the manager except by selling the machine. For this reason, it is important to estimate what the resulting ownership costs will be before a machine is purchased.

Ownership costs can be as much as 60 to 80 percent of the total annual costs for a nonpowered implement. As a rule of thumb, yearly ownership costs will be about 15 to 20 percent of the original cost of the machine, depending on the type of machine, its age, expected useful life, and the cost of capital.

Depreciation

Depreciation is a noncash expense that reflects a loss in value of machinery due to age, wear, and obsolescence. It is also an accounting procedure to recover the initial purchase cost of an asset by spreading this cost over its entire ownership period. Because most depreciation is caused by age and obsolescence, it is considered a fixed cost once the machine is purchased. Recent research, however, has found that the actual decrease in the market value of machinery can vary by as much as 10 percent, depending on the amount of annual use.

Annual depreciation can be estimated using the straight-line, declining balance, or sum-of-the-years' digits methods discussed in Chapter 4. However, if only the *average* annual depreciation is needed, it can be found from the equation for straight-line depreciation:

$$\text{Depreciation} = \frac{\text{cost} - \text{salvage value}}{\text{ownership life}}$$

The salvage value at various ages can be estimated as a percent of the new list price of a similar machine. Table 22-1 shows values that can be used to estimate the salvage value of several common types of machines.

Internal Revenue Service regulations specify certain methods that can be used to calculate machinery depreciation for tax purposes. However, the results may not reflect the actual annual depreciation of the machine. Ownership costs should be based on the concept of economic depreciation, or the actual decline in value, not on income tax depreciation rates. The commonly used tax depreciation methods discussed in Chapter 16 result in high depreciation in the early years of ownership and little or no depreciation in later years.

Interest

Investing in a machine ties up capital and prevents it from being used for an alternative investment. Equity capital has an opportunity cost, and this cost is part of the true cost of machine ownership. The opportunity cost used for machinery capital should reflect the expected return from investing the capital in its next best alternative use. When borrowed capital is used to purchase machinery, the interest cost is based on the loan interest rate. Depending on the source of capital, the proper interest rate to use is the rate of return on alternative investments, the expected interest rate on borrowed capital, or a weighted average of the two.

TABLE 22-1	Estimated Salvage Value as a Percentage of the New List Price for a Similar Machine					
Age of machine in years	Tractor 80-149 HP	Tractor 150+ HP	Combine	Baler	Tillage	Planter
1	68%	67%	70%	56%	61%	65%
2	62	59	59	50	54	60
3	57	54	51	46	49	56
4	53	49	45	42	45	53
5	49	45	40	39	42	50
6	46	42	36	37	39	48
7	44	39	32	34	36	46
8	41	36	29	32	34	44
9	39	34	26	30	31	42
10	37	32	22	28	30	40
11	35	30	21	27	28	39
12	34	28	18	25	26	38
Assumed hours of annual use	400	400	275	—	—	—

Source: ASAE Standards 2001. American Society of Agricultural Engineers, St. Joseph, MI, 2001.

The interest component of average annual fixed costs is calculated from the equations

$$\text{Average value} = \frac{\text{cost} + \text{salvage value}}{2}$$

$$\text{Interest} = \text{average value} \times \text{interest rate}$$

The first equation gives the *average value* of the machine over its ownership period, or its value at mid-life. Since the machine is declining in value over time, its average value is used to determine the average annual interest charge. This procedure assumes that capital tied up in the machinery investment decreases over the life of the machine, just as the balance owed on a loan used to purchase a machine declines as it is paid off.

Taxes

A few states levy property taxes on farm machinery. The actual charge will depend on the valuation procedure and tax rate in a particular location. Machinery cost studies often use a charge of about 1 percent of the average machine value as an estimate of annual property taxes. A higher rate should be used for pickups and trucks to cover the cost of the license and any other road-use fees.

Some states also levy sales taxes on the purchase of farm machinery. Because this is a one-time cost, it should simply be added to the initial purchase cost.

Insurance

Another ownership cost is the annual charge for insurance to cover damage to the machine from collision, fire, theft, hail, or wind, and for any liability coverage. A charge for insurance should be included in ownership costs, even if the owner carries no formal insurance and personally stands the risk, because some losses can be expected over time. The proper charge for insurance will depend on the amount and type of coverage, and insurance rates for a given location. Machinery cost studies assume a typical annual charge equal to about 0.5 percent of the machine's average value. This charge will need to be higher for on-road vehicles because of their higher premiums for property damage, collision, and liability coverage.

Housing

Most machinery cost estimates include an annual cost for housing the machine. Studies have found annual housing costs to be about 0.5 to 1.5 percent of the average value of the machine, so a value of 1.0 percent is often used to estimate machinery housing costs. A housing charge can also be estimated by calculating the annual cost per square foot for the machine shed (possibly including a prorated charge for a shop area) and multiplying this by the number of square feet the machine occupies. Even if a machine is not housed, a charge should be included to reflect additional wear and tear. A national study found that tractors that were left outside had a 16 percent lower trade-in value after 10 years than those that were kept inside.

Leasing

Some machinery is acquired under a long-term leasing agreement. Although the machine is not actually owned, the annual lease payment should be included with other fixed or ownership costs. Typically, the operator also pays the insurance premiums for a leased machine and provides housing for it. Lease payments are often lower than interest and depreciation costs, because the operator is not acquiring any equity position in the machine. Advantages and disadvantages of leasing machinery will be discussed later in this chapter.

Operating Costs

Operating costs are directly related to use. They are zero if the machine is not used, but increase directly with the amount of annual use. Unlike ownership costs, they can be controlled by varying the amount of annual use, by improving efficiency, and by following a proper maintenance program.

Repairs

Annual repair costs will vary with use, machine type, age, preventive maintenance programs,

---

Box 22-1   Capital Recovery

An alternative to calculating depreciation and interest is to compute the annual capital recovery charge, as discussed in Chapter 17. The capital recovery charge includes both annual depreciation and interest costs. It can be found using the equation below.

The amortization factor is the value from Appendix Table 1 that corresponds to the ownership life and the interest rate for the machine. The capital recovery amount is the annual payment that will recover the initial investment lost through depreciation, plus interest on the investment. This value is generally slightly higher than the sum of the average annual depreciation and interest values calculated from the previous equations, because capital recovery assumes that interest charges are computed on values at the beginning of each year, and are compounded annually.

$$\frac{\text{capital}}{\text{recovery}} = [\text{amortization factor} \times (\text{beginning value} - \text{salvage value})] + (\text{interest rate} \times \text{salvage value})$$

---

and other factors. Table 22-2 shows some average repair costs per 100 hours of use as a percent of the new list price for various types of machines. These rates are based on average costs over the useful life of the machine. Machinery repair costs usually increase over time, however. If a new machine is purchased under warranty, repair costs to the operator should be very low at first. As more parts wear out, though, repair costs per hour or per year increase rapidly. Repair costs are highly variable, so any rule of thumb for estimating repair costs needs to be used with caution. The best source of information is detailed records of actual repair costs for each machine under the existing level of use, cropping pattern, and maintenance program.

Fuel and Lubrication

Gasoline, diesel fuel, oil and other lubricants, and filters are included in this category. These costs are minor for nonpowered equipment but are important for self-powered machinery, such as tractors, harvesters, and sprayers. Fuel use per hour will depend on engine size, load, speed, and field conditions. Farm records can be used to estimate average fuel use. Data from tractor tests indicate that average fuel consumption in gallons per hour can be estimated from the maximum power takeoff horsepower of a tractor, as follows:

$$\frac{\text{Gallons}}{\text{per hour}} = 0.060 \times \text{PTO hp (gasoline)}$$

$$\frac{\text{Gallons}}{\text{per hour}} = 0.044 \times \text{PTO hp (diesel)}$$

Fuel cost per hour then can be estimated by multiplying the estimated fuel use by the purchase price of the fuel.

Costs for lubricants and filters average about 10 to 15 percent of fuel costs for self-powered machines. The cost of lubricants for nonpowered machines is generally small enough that it can be ignored when estimating operating costs.

Labor

The amount of labor needed for farm machinery depends on the operation being performed, field speed and efficiency, and the size of machinery being used. Labor costs are generally estimated separately from machinery costs but need to be included in any estimate of the total cost of performing a given machine operation. Machinery-related labor costs will be underestimated if only the time spent actually operating

TABLE 22-2	Average Repair Costs per 100 Hours of Use, Percent of New List Price

Machine	Percent of list price
Two-wheel-drive tractor	0.83
Four-wheel-drive tractor	0.50
Moldboard plow	5.00
Tandem disk	3.00
Chisel plow	3.75
Rotary tiller	5.33
Field cultivator	3.50
Rotary hoe	3.00
Row crop cultivator	4.00
Planter, grain drill	5.00
Mower, bar	7.50
Mower, rotary	8.75
Mower-conditioner	3.20
Windrower, self-propelled	1.83
Rake	2.40
Square baler	4.00
Round baler	6.00
Forage harvester, pull type	2.60
Forage harvester, self-propelled	1.25
Cotton picker, self-propelled	2.67
Sugar beet harvester	6.67
Combine, self-propelled	1.33
Grain wagon	2.67
Forage wagon	2.50
Boom sprayer	4.67
Orchard sprayer	3.00
Fertilizer spreader	6.67

Source: Hunt, Donnell: *Farm Power and Machinery Management,* 9th edition, based on data from the American Society of Agricultural Engineers.

the machine in the field is considered. The total labor charge should also include time spent fueling, lubricating, repairing, adjusting, and moving machinery between fields and the farmstead. These activities add as much as 10 to 25 percent to machinery field time.

The value placed on machinery labor should reflect local wage rates. The level of skill needed for various operations should be taken into account, however. A higher hourly labor value should be used for applying pesticides, planting, harvesting, and other highly skilled operations.

Custom Hire or Rental

When a custom operator is hired to perform certain machinery operations, these costs should be included in the machinery variable or operating costs for the farm. Custom machinery rates are usually quoted by the acre, hour, bushel, or ton, and include the value of the custom operator's labor as well as all other ownership and operating costs.

In other cases, a farmer may rent a machine that will be used for only a few days or weeks. The operator supplies the labor and fuel, and pays an hourly or daily charge for using the machine. This rental expense should also be included in the farm's total machinery cost.

Other Operating Costs

Some specialized machines have additional operating costs associated with their use. Items such as twine, plastic wraps, and bags may need to be included in the variable costs for certain operations.

Machinery Costs and Use

Annual total ownership or fixed costs are usually assumed to be constant regardless of the amount of machine use during the year. Operating or variable costs, however, increase with the amount of use, generally at a constant rate per acre or per hour. The result is that annual total costs also increase at a constant rate. These relations are shown in Figure 22-1a.

For decision-making purposes, it is often useful to express machinery costs in terms of the average cost per acre, per hour, or per unit of output. Average fixed costs will decline as the acres, hours, or units of output per year increase, while average variable costs will be constant if

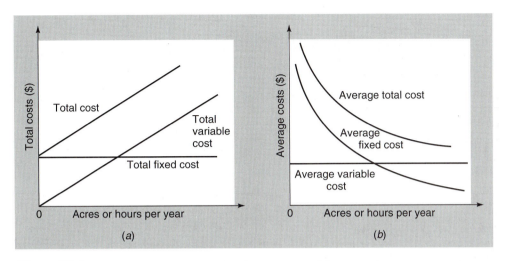

Figure 22-1 Relations between total and average machinery costs.

total variable costs increase at a constant rate. Because average total cost is the sum of average fixed and average variable costs, it also declines with added use. These relations are shown in Figure 22-1b, where the vertical distance between the average fixed cost and average total cost curves is constant and equal to average variable cost.

Machinery cost estimates are only as accurate as the estimate of annual use. As much effort should go into estimating this value as in estimating the various cost components. Use of actual field records and tractor-hour meters will improve the accuracy of annual use estimates.

EXAMPLES OF MACHINERY COST CALCULATIONS

Table 22-3 shows an example of how to calculate total annual costs and average costs per hour and per acre for a new combine. The first step is to list the basic data, such as list price, purchase cost, salvage value, ownership life, estimated annual use, and interest rate on the capital invested. Next, total ownership costs are calculated and converted into average ownership cost per hour of use. For the combine in the example, ownership costs are estimated to be $21,579 per year. Based on the assumed 300 hours of annual use, the average ownership cost per hour is estimated to be $71.93. Remember that this value will change if the actual hours of use are more or less than 300.

If the capital recovery method had been used, the estimate of fixed costs would have been slightly higher. The capital recovery factor from Appendix Table 1 for a 10-year life and 8 percent interest rate is 0.14903, so the capital recovery amount would be

$$[0.14903 \times (160,000 - 38,500)] + (0.08 \times 38,500) = \$21,187$$

Total fixed costs would be ($21,187 + $1,489) = $22,676 per year, or $75.59 per hour, using this method.

Total operating costs, including labor, are calculated in step 3 of Table 22-3. The labor requirement was increased by 20 percent over the estimated machine use, to account for time required to service, adjust, and transport the combine. Average operating cost per hour is estimated to be $44.40.

The total cost per hour is estimated at $116.33 (step 4). This figure can be converted to a cost per acre or per unit of output if the field

TABLE 22-3	**Calculating Machinery Costs for a New Combine**

Step 1: **List basic data**

New combine, 24-foot header, 240-hp diesel engine:	
List price	$175,000
Purchase cost	$160,000
Salvage value (22% of new list price, Table 22-1)	$ 38,500
Average value ($160,000 + 38,500) ÷ 2	$ 99,250
Ownership life	10 years
Estimated annual use	300 hours
Interest rate	8%

Step 2: **Calculate ownership costs**

Depreciation ($160,000 − 38,500) ÷ 10 years	$ 12,150
Interest (8% × $99,250)	$ 7,940
Taxes, insurance, and housing (1.5% × $99,250)	$ 1,489
Total annual ownership costs	$ 21,579
Ownership costs per hour ($21,579 ÷ 300 hours)	$ 71.93

Step 3: **Calculate operating costs**

Repairs (1.33% × $175,000 list price × 300 hours ÷ 100)	$ 6,983
Diesel fuel (240 horsepower × 0.044 gallon/hp-hour × $0.85 per gallon × 300 hours)	$ 2,693
Lubrication and filters (15% of fuel costs)	$ 404
Labor (300 hours × $9.00 per hour × 1.20)[*]	$ 3,240
Total annual operating costs	$ 13,320
Operating cost per hour ($13,320 ÷ 300 hours)	$ 44.40

Step 4: **Calculate total cost per hour**

Ownership cost per hour	$ 71.93
Operating cost per hour	$ 44.40
Total cost per hour	$ 116.33

Step 5: **Calculate cost per acre**

Performance rate: 8 acres per hour ($116.33 ÷ 8 acres)	$ 14.54

[*]Labor requirements are increased by 20% to account for time spent servicing, adjusting, repairing, and transporting machinery.

capacity per hour is known. In step 5 of the example, the performance rate of the combine is assumed to be 8 acres per hour, resulting in a total cost per acre of $14.54.

When the same tractor is used to pull several implements, or more than one harvesting head is used on the same self-propelled unit, ownership and operating costs must be calculated separately for both the power unit and for the attachment. They can then be added together to find the combined cost of performing the operation. In Table 22-4, ownership costs and operating costs per hour have been estimated for a 185-horsepower tractor and a 16-foot chisel plow, using the methods explained previously.

Notice that fuel and lubrication costs and the labor charge were assigned only to the tractor. The costs *per hour* for the tractor and chisel are combined to find the total cost of chiseling per hour, then are divided by the field capacity to find the cost per acre. Note that it is not correct to combine the *annual* costs of two machines when only part of the annual use of the power unit occurs with this implement. Instead, the *hourly* costs should be combined. For a self-propelled machine or an implement alone, however, the cost per acre can be found by dividing the annual cost by the annual *acres* of use, without calculating hourly costs first.

FACTORS IN MACHINERY SELECTION

One of the more difficult problems in farm management is selecting machinery with the proper capacity. This process is complicated not only by the wide range of types and sizes available, but also by capital availability, labor requirements, the particular crop and livestock enterprises in the farm plan, tillage practices, and climatic factors. The objective in selecting machinery is to purchase the machine that will satisfactorily perform the required task within the time available for the lowest possible total cost. This does not necessarily result in the purchase of the smallest machine available, however. Labor and timeliness costs must also be considered.

Machinery Size

The first step in machinery selection is to determine the field capacity for each size available.

TABLE 22-4	Combined Cost of a Tractor and Implement	
	185-hp tractor	16-ft. chisel plow
Annual ownership costs	$8,500	$ 930
Annual hours of use	500	100
Ownership cost per hour	$17.00	$ 9.30
Operating costs per hour:		
Fuel and lubrication	7.95	
Repairs	5.12	2.58
Labor	7.00	
Total cost per hour	$37.07	$11.88
Combined cost per hour	$48.95	
Field capacity, acres per hour	12.00	
Combined cost per acre	$ 4.08	

The formula for finding capacity in acres per hour is

$$\text{Field capacity} = \frac{\text{speed (mph)} \times \text{width (feet)} \times \text{field efficiency (\%)}}{8.25}$$

For example, a 12-foot-wide windrower operated at 8 miles per hour with a field efficiency of 82 percent would have an effective field capacity of

$$\frac{8 \text{ mph} \times 12 \text{ ft.} \times 82\%}{8.25} = 9.54 \text{ acres/hr.}$$

Field efficiency is included in the equation to recognize that a machine is not always used at 100 percent of its maximum capacity because of work overlap and time spent turning, adjusting, lubricating, and handling materials. Operations such as planting, which require frequent stops to refill seed, pesticide, and fertilizer containers, may have field efficiencies as low as 50 or 60 percent. On the other hand, field efficiencies as high as 85 to 90 percent are possible for some tillage operations, particularly in large fields when turning time and work overlap are minimized. Larger machines typically have higher field efficiencies due to less overlapping and less time needed for adjustments relative to the area covered.

The next step in selecting machinery is to compute the minimum field capacity (acres per hour) needed to complete the task in the time available. This value is found by dividing the number of acres the machine must cover by the number of field hours available to complete the operation. In turn, the number of field hours available depends on how many days the weather will be suitable for performing the operation, and the number of labor or field hours available each day. The formula for finding the minimum field capacity in acres per hour is

$$\text{Minimum field capacity} = \frac{\text{acres to cover}}{\text{hours per day} \times \text{days available}}$$

Suppose the operator wants to be able to windrow 180 acres in 2 days and can operate the windrower 10 hours per day. The minimum capacity needed is

$$[180 \text{ acres} \div (10 \text{ hours} \times 2 \text{ days})]$$
$$= 9.00 \text{ acres per hour.}$$

The 12-foot windrower, with a capacity of 9.54 acres per hour, should be large enough.

When several field operations must be completed within a certain number of days, it may be more convenient to calculate the days needed for each operation and add them together to test whether or not an entire set of machinery has sufficient capacity. The formula for each operation is

$$\text{Field days needed} = \frac{\text{acres to cover}}{\text{hours per day} \times \text{acres completed per hour}}$$

Using the same example, assume the operator also owns a baler with an effective field capacity of 5 acres per hour. If baling can be performed only 8 hours per day, then 180 acres requires

$$\frac{180}{8 \text{ hrs.} \times 5 \text{ acres/hr.}} = 4.5 \text{ days}$$

Thus, a total of 6.5 suitable field days are required each time the 180 acres are both windrowed and baled. The operator must decide if typical weather patterns will permit this many days to be available without a risk of significant crop losses.

The size of machinery required to complete field operations in a timely manner can be reduced by: (1) increasing the labor supply so that machinery can be operated more hours per day or two operations can be performed at the same time, (2) reducing the number of field operations performed, or (3) producing several crops with different critical planting and harvesting dates, instead of just one crop. Often these adjustments are less costly than acquiring larger machinery. A partial budget can be used to analyze any of these changes.

Machinery selection may also involve a choice between one large machine or two smaller ones. The purchase cost and annual fixed costs will be higher for two machines, because the same capacity can usually be obtained at a lower cost in one large machine. Two tractors and two operators will also be needed if tractor-drawn equipment is involved. The primary advantage of owning two machines is an increase in reliability. If one machine breaks down, work is not completely stopped but can continue at half speed utilizing the remaining machine.

Timeliness

Some field operations do not have to be completed within a fixed time period, but the later they are performed, the lower the harvested yield is likely to be. The yield reduction may be in terms of quality, such as for fruits and vegetables that become overly mature, or in quantity, such as for grain that suffers from too short a growing season or has excessive field losses during harvesting. Figure 22-2 shows a typical

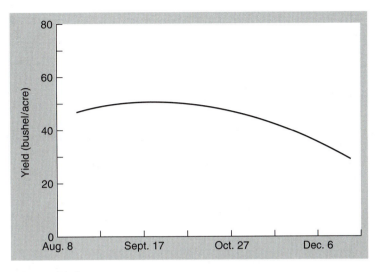

Figure 22-2   Hard red winter wheat yields as a function of planting date at Stillwater, Oklahoma.

Source: F.M. Epplin, D.E. Black, and E.G. Krenzer, *Current Farm Economics,* Vol. 64, No. 3, September 1991.

relation between planting time and potential yield for winter wheat, where planting too early or too late reduces yield.

A yield decrease reduces profit and should be included as part of the cost of utilizing a smaller machine. This cost is referred to as *timeliness cost.* The dollar cost of poor timeliness is difficult to estimate, because it will vary from year to year depending on weather conditions and prices. However, some estimate should be included when comparing the cost of owning machines of different sizes. Figure 22-3 is a hypothetical example of how timeliness and other machinery costs change as larger machinery is used to perform the same amount of work. At first, larger machinery reduces timeliness and labor costs and lowers total costs. After some point, no more gains in timeliness are available, and higher ownership costs cause total costs to rise. Timeliness concerns and labor costs must be balanced against higher ownership costs to select the least-cost machine size.

Partial budgeting is a useful tool to use when making decisions about machinery sizes where timeliness is important. Table 22-5 contains an example that emphasizes the importance of considering all costs. Changing from a 20-foot grain drill to a 40-foot drill will double annual ownership costs. However, the manager estimates that improved timeliness of planting will result in an average yield increase of 1 bushel of wheat per acre each year on 1,000 acres. The value of this added yield, combined with slightly lower labor and repair costs, results in a projected increase in net income of over $2,000 per year from purchasing the larger drill. Of course, the date of planting also depends on how long it takes to complete tillage and other pre-plant operations. Thus, the total machinery complement should be analyzed when timeliness factors are considered.

ALTERNATIVES FOR ACQUIRING MACHINERY

Efficient machinery management means having the right size and type of machine available to

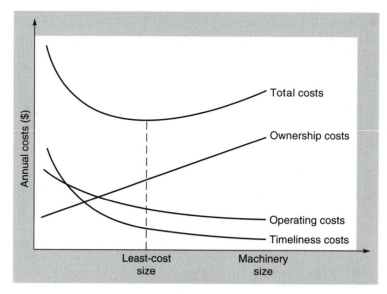

Figure 22-3   **Hypothetical effect of timeliness and machine size on cost.**

use at the right time at a minimum cost. Once the size and type have been selected, there are several common alternatives for acquiring the use of a farm machinery item.

Ownership

Most farm and ranch managers prefer to own their own machinery. Ownership gives them complete control over the use and disposal of each machine. However, machinery and vehicles represent a large investment on commercial farms and ranches, as much as $150 to $200 per acre on many cash grain farms, and over $2,000 per acre on some horticultural farms. Managers must be careful to control the size of this investment and the related operating costs. Machinery investment can be reduced by: (1) using smaller machinery, (2) increasing annual machine use to lower the average ownership cost per unit of output, (3) keeping machinery longer before trading, (4) purchasing used machinery, and (5) utilizing alternatives to ownership such as rental, leasing, and custom hiring.

Some machinery dealers offer "rollover" ownership plans. The operator purchases a new machine, then trades it for another new machine each year. The cost to trade may depend on the number of hours that the machine to be traded was used. Rollover plans allow for the use of a new machine each year, usually under warranty, but can lead to higher ownership costs per acre.

Rental

When investment capital is very limited or interest rates are high, renting a machine may be preferable to owning it. Short-term rental arrangements usually cover a few days to a whole season. The operator pays a rental fee plus the cost of insurance and daily maintenance, but not major repairs. Machinery rental is especially attractive when: (1) a specialized machine is needed for relatively low use, (2) extra capacity or a replacement machine is needed for a short time, or (3) the operator wants to experiment with a new machine or production practice without making a long-term capital investment.

TABLE 22-5	Example of a Partial Budget for Selecting the Most Profitable Machine

Adjustment: Change from a 20-foot grain drill ($15,000 purchase cost) to a 40-foot drill ($30,000 purchase cost). Assumes a 44% salvage value, 8-year ownership period, 10% interest rate. The 20-foot machine requires 137 hours to drill 1,000 acres, the 40-foot machine requires 65 hours.

Additional revenue:

1 bushel per acre increase in yield $\times$ 1,000 acres $\times$ $3.75 per bushel

Total additional revenue			$3,750

Reduced costs:	**Ownership**	**Operating**	
Depreciation	$1,050		
Interest	1,080		
Taxes and insurance	108		
Repairs		$1,028	
Labor (72 hours less at $7/hour)		504	
Subtotal	$2,238	$1,532	
Total reduced costs			$3,770
Total of additional revenue and reduced costs			$7,520

Additional costs:	**Ownership**	**Operating**	
Depreciation	$2,100		
Interest	2,160		
Taxes and insurance	216		
Repairs		$ 975	
Subtotal	$4,476	$ 975	
Total additional costs			$5,451

Reduced revenue:			**0**
Total additional costs and reduced revenue			$5,451
Net change in farm income			$2,069

Leasing

A lease is a long-term contract whereby the machine owner (the lessor), often a machinery dealer or leasing company, grants control and use of the machine to the user (the lessee) for a specified period of time for a monthly, semi-annual, or annual lease payment. Most machinery leases are for three to five years or longer, with the first payment usually due at the start. As with any formal agreement, the lease should be in writing and should cover such items as payment rates and dates; excess use penalties; payment of repairs, taxes, and insurance; responsibilities for loss or damage; and provisions for early cancellation.

Some leases allow the lessee to purchase the machine at the end of the lease period for a specified price. These are called *capital leases*. To maintain the income tax deductibility of capital lease payments, the purchase at the end of the lease must be optional and for a value approximately equal to the market value of the machine at that time. Otherwise, the lease may be considered to be a sales contract or finance lease, and depreciation and interest become tax deductible for the lessee instead of the lease payments, just as if the operator owned the machine.

Leasing machinery may help operators reduce the amount of capital they have tied up in noncurrent assets. Although lease payments represent a cash flow obligation just like a loan payment, they are generally smaller. Leasing reduces the risk of obsolescence, because the lessee is not obligated to keep the machine beyond the term of the lease agreement. Moreover, operators who prefer the reliability and performance of a newer machine often lease it for a few years and then exchange it for a new one. Some operators who have very little taxable farm income and cannot utilize depreciation deductions, such as section 179 expensing, may find leasing to have a lower after-tax cost than owning.

There are also some disadvantages to leasing farm machinery. The practice is not well established in many areas, and the desired model may not be available for lease. Lease payments are operating expenses, and a late payment may cause cancellation of the lease. Further, the lessee may not be allowed to cancel the lease early without paying a substantial penalty. A lease does not allow the operator to build equity in the machine. Unless the purchase option is exercised when the lease expires, the machine reverts to the owner and the operator has no financial interest in it.

The decision to lease or own equipment should be studied carefully, as shown in Box 22-2. Both the economic cost (net present value) and the cash flow requirements should be analyzed, and matched to the financial situation of the business.

Custom Hire

Custom hiring is an important practice in some areas for operations such as applying chemicals and harvesting grain or forages. The decision of whether to own a machine or custom hire the service depends on the costs involved, the skills needed, and the amount of work to be done. For machines that will be used very little, it is often more economical to hire the work done on a custom basis. However, the availability and dependability of custom operators must be considered. A manager may not want to rely on a custom operator for a task such as planting, where timeliness is often critical.

Total costs per acre or per unit of output should be compared when deciding between machine ownership or custom hiring. Custom charges are typically a fixed rate per acre, hour, or ton, while ownership costs will decline with increased use. These relations are shown in Figure 22-4. At low levels of use, hiring a custom operator is less expensive, while for higher usage, the cost is lower if the machine is owned. The point where the cost advantage changes, or the break-even point, is shown as output level a in Figure 22-4.

When the necessary cost data are available, the break-even point can be found from the following equation:

$$\text{Break-even units} = \frac{\text{total annual fixed costs}}{\text{custom rate} - \text{variable costs per unit}}$$

For example, the ownership or fixed costs for the combine in Table 22-3 were $21,579 per year. The variable costs for operating it were $5.55 per acre for a performance rate of 8 acres per hour ($44.40 per hour divided by 8 acres per hour). If the custom hire rate for a similar combine is $21.00 per acre, the break-even point would be

$$\frac{\$21,579}{\$21.00 - 5.55} = 1,397 \text{ acres}$$

If the machine would be used on less than 1,397 acres, it would be less costly to custom hire the work done, while above 1,397 acres, it would be less expensive to own the machine. A determination of the break-even point provides a useful guide to help managers choose between machine ownership and custom hiring.

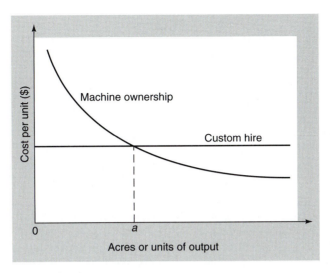

Figure 22-4   Cost per unit of output for machine ownership and custom hiring.

Labor use is another consideration in custom hiring. The custom operator typically provides the labor necessary to operate the machine, which frees the farm operator's labor for other uses. This can be an advantage if it reduces the amount of hired labor needed or if the owner's labor has a high opportunity cost at the time the custom work is being performed. It can also be an advantage for operations that take special skills to perform, such as the application of chemicals.

Many operators who own machinery find it advantageous to do custom work for other farmers. This helps spread their fixed ownership costs over more acres. It is important for custom operators to accurately estimate their costs so that they can arrive at a fair and profitable charge.

IMPROVING MACHINERY EFFICIENCY

Several values can be used to measure the efficiency of machinery use. One is *machinery investment per crop acre,* which is calculated by dividing the current value of all crop machinery by the number of crop acres on the farm. The current value of all machinery for a given year can be found by taking the average of the beginning and ending machinery inventory values for the year. Ideally, this average should be the fair market value, but current book or depreciated value (cost less accumulated depreciation) can be used if market values are not available.

A second measure of machinery efficiency is *machinery cost per crop acre.* It is found by dividing the total annual machinery costs, both ownership and operating, by the number of crop acres. Some farm record analyses also include pickup and truck expenses, machinery lease payments, and custom hiring expenses in machinery costs. When possible, the cost of machinery used for livestock purposes should be excluded, in order to make a fair comparison among farms.

These values should be used with caution. Numerous studies have shown that investment per acre and cost per acre decline with increases in farm size, and also vary by farm type. It is important to compare values only to those calculated in the same manner for farms of the same approximate size and type. Table 22-6 shows recent machinery costs per acre, excluding labor, for a group of grain producers in Iowa. Low-profit producers incurred $9 per acre higher machinery costs than high-profit producers, and had $17 per acre more invested in machinery.

Box 22-2   Buy versus Lease Comparison Example

The Struthers family farm needs to replace one of its large tillage tractors. They have located the model they want at a local dealership. The dealer offers them two alternatives:

1. Purchase the tractor for $65,000, minus $11,000 allowed for the trade-in value of their old tractor that has a tax basis of $0. They can pay off the $54,000 difference in five equal annual payments of $13,883 starting in 12 months. The annual interest rate is 9 percent.
2. Lease the tractor for 5 years. The annual lease payments would be $11,000 per year. The first one is due when the lease is signed, but can be paid by the trade-in allowance. After 5 years they can purchase the tractor for $32,000.

The Struthers have a cost of capital of 7 percent and a marginal income tax rate of 40 percent, so their after-tax discount rate is $.07 \times (1.00 - .40) = .042$, or 4.2 percent. Purchasing the tractor would allow them to take full advantage of the $25,000 section 179 depreciation expensing option discussed in Chapter 16.

By using the investment analysis techniques discussed in Chapter 17 they are able to calculate the net present value of the cost of acquiring the tractor under either option, as shown below.

Purchase with loan	Year 1	Year 2	Year 3	Year 4	Year 5	Beyond	Total
a. Loan principal	$9,023	$9,835	$10,720	$11,685	$12,737		$54,000
b. Loan interest	4,860	4,048	3,163	2,198	1,146		15,415
c. Tax depreciation	28,106	5,548	4,359	3,553	3,553	8,884	54,000
d. Tax savings (b + c) × .40	13,186	3,838	3,009	2,300	1,880	3,553	27,766
e. Net cash outflow (a + b − d)	697	10,045	10,874	11,583	12,003	−3,553	41,649
f. Discount factor	.960	.921	.884	.848	.814	varies	
g. Present value (e × f)	669	9,251	9,612	9,825	9,772	−2,687	36,441

Lease and purchase later	Year 1	Year 2	Year 3	Year 4	Year 5	Beyond	Total
a. Lease payment	$11,000	$11,000	$11,000	$11,000			$44,000
b. Purchase cost					32,000		32,000
c. Tax depreciation					$25,750	6,252	32,000
d. Tax savings (a + c) × .40	4,400	4,400	4,400	4,400	10,300	2,501	30,400
e. Net cash outflow (a + b − d)	6,600	6,600	6,600	6,600	21,700	−2,501	45,600
f. Discount factor	.960	.921	.884	.848	.814	varies	
g. Present value (e × f)	6,334	6,079	5,834	5,599	17,665	−1,768	39,742

The net present value of purchasing the tractor and paying off the loan is $36,441 after taxes, which is less than the $39,742 cost of leasing the tractor and then exercising the purchase option. This means that in this example purchasing is cheaper in the long run. However, in years 2, 3, and 4 the net cash outflow is considerably higher for purchasing than leasing. If cash flow is expected to be tight during these years, the leasing option might still be preferred.

TABLE 22-6	**Machinery Costs for Iowa Grain Producers (Per Crop Acre)**

	High-profit third	Low-profit third
Total machinery cost	$ 54	$ 63
Machinery investment	$228	$245

Source: 2001 Iowa Farm Costs and Returns, Iowa State University Extension, 2002.

Several techniques can be employed to improve machinery efficiency. Choosing the optimum size of machine and the least-cost alternative for acquiring machinery services have already been discussed. There are three other areas that have a large impact on efficiency: maintenance and operation, machinery use, and replacement decisions.

Maintenance and Operation

Repairs are a large part of machinery variable costs, but they are a cost that can be controlled by proper use and maintenance. Agricultural engineers report that excessive repair costs can generally be traced to: (1) overloading or exceeding the rated machine capacity, (2) excessive speed, (3) poor daily and periodic maintenance, and (4) improper adjustment. All these items can be corrected by constant attention and proper training of machine operators. Modern farm machines have more monitoring systems and automatic adjustment controls that help maintain efficient operating levels.

A system of scheduling and recording repairs and maintenance is essential for controlling repair costs. Adherence to the manufacturer's recommended maintenance schedule will keep warranties in effect, prevent unnecessary breakdowns, and reduce lifetime repair costs. Complete records of repairs on individual machines will help identify machines with higher than average repair costs. These machines should be considered for early replacement.

The manner in which a machine is operated affects both repair costs and field efficiency. Speed should be adjusted to load a machine to capacity without overloading it or lowering the quality of the work being done. Practices that improve field efficiency reduce costs by allowing more work to be done in a given time period, either by permitting the same work to be done in less time or by allowing a smaller machine to be used. Small and irregular-shaped fields that require frequent turns, frequent stops, and work overlap reduce field efficiency. For example, a 30-foot disk operated with a 3-foot overlap loses 10 percent of its potential efficiency from this factor alone.

Machinery Use

Using an expensive specialized machine on only a few acres also contributes to high machinery costs. Because of this, some farmers purchase low-use machines jointly with other operators. Some producers have even formed machinery cooperatives with 5 to 10 members owning all their machinery in common. This not only decreases the fixed costs per unit but also decreases the investment required from each individual. It is important that the joint owners are compatible and can agree on the details of the machine's use. Whose work will be done first and the division of expenses such as repairs, insurance, and taxes should be agreed to before the machine is purchased. Normally, operators provide their own fuel and labor, then pay a fixed rate per acre into a fund that is used to pay ownership costs and repairs.

Some owners exchange the use of specialized machinery. If the use or value of the machines is not equal, some payment may be exchanged as well. Many operations such as harvesting are accomplished more efficiently when two or more people work together, anyway.

Machinery costs per unit of output can also be reduced by performing custom work for other operators. If custom work does not interfere with timely completion of the owner's work, it will provide additional income to help pay ownership costs. The custom rate charged should reflect the opportunity cost of labor in the operator's own business as well as the costs of owning and operating the machine. Some skilled operators do custom machinery work as a part-time or full-time alternative to farming their own or rented land.

New versus Used Machines

Farm machines, particularly tractors and other self-powered machines, decline in market value most rapidly during the first few years of their useful lives. Buying used machinery is an economical way to lower machinery investment and ownership costs. Purchasing a machine is essentially purchasing the service it can provide over time. Offsetting some of the lower ownership costs may be higher repair costs and decreases in reliability and timeliness. A used machine may also become obsolete sooner than a new one. The owner's ability as a mechanic and availability of the facilities and time to do major repair work at home are often crucial factors for utilizing used machinery. Used machinery should be considered when capital is limited, interest rates are high, the machine will have a relatively small annual use, and reliability and timeliness are not critical.

Replacement

When to replace or trade a machine is one of the more difficult decisions in machinery management. There is no simple, easy decision rule that applies to all types of machines and conditions. Besides costs and reliability, replacement decisions must also consider the effects of a purchase or trade on income taxes and cash flow.

The decision to replace a machine can be made for any of the following reasons:

1. The present machine is worn out—its age and accumulated use are such that it is no longer capable of performing the required task reliably.

2. The machine is obsolete—new developments in machinery technology, or changes in farming practices, allow a newer machine to perform the job better, or with greater safety and comfort.

3. Costs are increasing with the present machine—repair, fuel, and timeliness costs are increasing rapidly, both total and per unit of output.

4. The capacity is too small—the area in production has increased, or timeliness has become so critical that the old machine cannot complete the job on time.

5. Income taxes—in a high-profit year, machines may be replaced to take advantage of the tax-reducing benefits of fast depreciation deductions, or because the owner has a higher marginal tax rate that year. However, replacement decisions should not be made on the basis of tax savings alone.

6. Cash flow—many machines are replaced in years of above-average cash income to avoid borrowing funds later. Likewise, replacement of machinery is often postponed in years when cash flow is tight.

7. Pride and prestige—there is a certain "pride of ownership" involved in the purchase of new and larger machinery. While this may be important to some individuals, it can be a costly reason, and one that is difficult to justify in an economic analysis.

These reasons can be used individually or in combination to determine the replacement age for a specific machine. Annual cost and repair records on each machine are very useful for making the replacement decision. In Figure 22-5, the annual costs for a 165-horsepower tractor purchased new are estimated for each year of a 20-year ownership period. At first, total costs decline because depreciation and interest are decreasing. Eventually, though, increasing repair costs more than offset the declining ownership costs. Total annual costs are

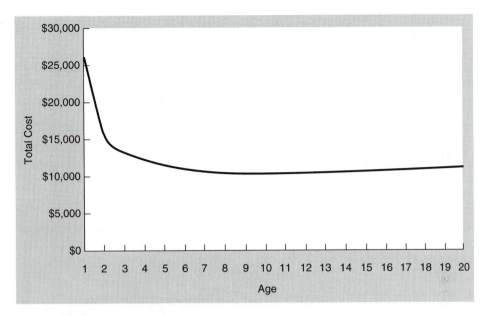

Figure 22-5   **Estimated annual cost of a 165-horsepower tractor.**

lowest around year 10, but this depends on how many hours the tractor is used annually. The higher the annual use, the faster annual repair costs will increase.

Most farm and ranch managers have an overall strategy for replacing machinery. The appropriate strategy for each operation depends on the financial resources available, the mechanical skills of the work force, and the particular priorities and objectives of the manager.

Keep and Repair

One strategy is to keep and repair equipment as long as possible. This is usually the least-cost strategy in the long run, especially if most of the repair work can be done on the farm. The risk of suffering a breakdown at a critical time is higher, though.

Trade Often

Operations that emphasize maximum utilization of machinery usually prefer newer and more reliable equipment. They may even find that leasing machinery and exchanging it for a new

model every year or two is the strategy that best meets their objectives.

Trade When Income is High

Managers who want to avoid the use of credit may wait until a year when above-average cash income occurs to trade machinery. This strategy also helps reduce taxable income in a high-income year.

Invest Each Year

Finally, some operators prefer to upgrade part of their machinery line each year, by trading or disposing of their least reliable units. This avoids having to invest large amounts of capital in any one year, and works especially well for businesses that experience only small year-to-year variations in cash net income.

Although machinery investment decisions are made infrequently, they often involve many thousands of dollars. Taking time to carefully analyze each decision can have a very large impact on the long-run profitabiliy of the farm or ranch.

SUMMARY

Machinery investment is the second largest investment on most farms and ranches after real estate. Annual machinery costs are a large part of a farm's total annual costs. Ownership costs include depreciation, interest, taxes, insurance, housing, and lease payments. Repairs, fuel, lubrication, labor, custom hire, and rental payments are included in operating costs.

Selection of the optimum machine size to own should consider both total costs and the effects of timeliness in completing operations. Rental, leasing, and custom hiring are other alternatives for acquiring the use of farm machinery, particularly for specialized machines with low annual use. Operators who are short of capital or who are skilled in machinery repair can benefit from investing in used rather than new machinery.

Machinery efficiency can be improved by proper maintenance and operation, by owning equipment jointly with other operators, or by exchanging the use of individually owned machines. The proper time to trade machinery depends on repair costs, reliability, obsolescence, cash flow, income tax considerations, and personal pride.

QUESTIONS FOR REVIEW AND FURTHER THOUGHT

1. What is the total annual fixed cost for a $45,000 tractor with a 12-year life and a $9,000 salvage value, when insurance and taxes are 2 percent of average value and there is a 9 percent opportunity cost on capital? What is the average fixed cost per hour if the tractor is used 400 hours per year? If it is used 700 hours per year?

2. What is the field capacity in acres per hour for a 28-foot-wide tandem disk operated at 5 miles per hour with 80 percent field efficiency? How much would it change if the field efficiency could be increased to 90 percent? How much time would be saved on 800 acres?

3. Assume that owning and operating a certain machine to till 850 acres has an ownership cost of $5,000 per year, plus operating costs of $5.00 per acre, including labor. Leasing a machine with a field capacity of 5.0 acres per hour to do the same work would cost $40 per hour, plus the same operating costs. Hiring the work done on a custom basis would cost $12 per acre. Which alternative has the lowest total cost?

4. Assume that a self-propelled windrower has annual fixed costs of $7,150 and variable costs of $4.25 per acre. A custom operator charges $8.60 per acre. What is the break-even point in acres per year?

5. List ways to improve the field efficiency of machinery operations such as planting, disking, and combining grain.

6. What are the advantages of owning, leasing, and custom hiring farm machinery?

7. If you decide to use your forage harvester to do custom work for others, what would happen to total ownership cost? Average ownership cost per acre harvested? Total operating cost? Average operating cost per acre harvested?

8. What factors are important in machinery replacement decisions? How would you rank them in order of importance? Would your ranking be different for different types of machines?

APPENDIX

TABLE 1 | **Amortization Factors for Equal Total Payments**

Interest Rate

Years	4%	4.5%	5%	5.5%	6%	6.5%	7%	7.5%	8%	8.5%	9%	9.5%	10%
1	1.04000	1.04500	1.05000	1.05500	1.06000	1.06500	1.07000	1.07500	1.08000	1.08500	1.09000	1.09500	1.10000
2	0.53020	0.53400	0.53780	0.54162	0.54544	0.54926	0.55309	0.55693	0.56077	0.56462	0.56847	0.57233	0.57619
3	0.36035	0.36377	0.36721	0.37065	0.37411	0.37758	0.38105	0.38454	0.38803	0.39154	0.39505	0.39858	0.40211
4	0.27549	0.27874	0.28201	0.28529	0.28859	0.29190	0.29523	0.29857	0.30192	0.30529	0.30867	0.31206	0.31547
5	0.22463	0.22779	0.23097	0.23418	0.23740	0.24063	0.24389	0.24716	0.25046	0.25377	0.25709	0.26044	0.26380
6	0.19076	0.19388	0.19702	0.20018	0.20336	0.20657	0.20980	0.21304	0.21632	0.21961	0.22292	0.22625	0.22961
7	0.16661	0.16970	0.17282	0.17596	0.17914	0.18233	0.18555	0.18880	0.19207	0.19537	0.19869	0.20204	0.20541
8	0.14853	0.15161	0.15472	0.15786	0.16104	0.16424	0.16747	0.17073	0.17401	0.17733	0.18067	0.18405	0.18744
9	0.13449	0.13757	0.14069	0.14384	0.14702	0.15024	0.15349	0.15677	0.16008	0.16342	0.16680	0.17020	0.17364
10	0.12329	0.12638	0.12950	0.13267	0.13587	0.13910	0.14238	0.14569	0.14903	0.15241	0.15582	0.15927	0.16275
11	0.11415	0.11725	0.12039	0.12357	0.12679	0.13006	0.13336	0.13670	0.14008	0.14349	0.14695	0.15044	0.15396
12	0.10655	0.10967	0.11283	0.11603	0.11928	0.12257	0.12590	0.12928	0.13270	0.13615	0.13965	0.14319	0.14676
13	0.10014	0.10328	0.10646	0.10968	0.11296	0.11628	0.11965	0.12306	0.12652	0.13002	0.13357	0.13715	0.14078
14	0.09467	0.09782	0.10102	0.10428	0.10758	0.11094	0.11434	0.11780	0.12130	0.12484	0.12843	0.13207	0.13575
15	0.08994	0.09311	0.09634	0.09963	0.10296	0.10635	0.10979	0.11329	0.11683	0.12042	0.12406	0.12774	0.13147
16	0.08582	0.08902	0.09227	0.09558	0.09895	0.10238	0.10586	0.10939	0.11298	0.11661	0.12030	0.12403	0.12782
17	0.08220	0.08542	0.08870	0.09204	0.09544	0.09891	0.10243	0.10600	0.10963	0.11331	0.11705	0.12083	0.12466
18	0.07899	0.08224	0.08555	0.08892	0.09236	0.09585	0.09941	0.10303	0.10670	0.11043	0.11421	0.11805	0.12193
19	0.07614	0.07941	0.08275	0.08615	0.08962	0.09316	0.09675	0.10041	0.10413	0.10790	0.11173	0.11561	0.11955
20	0.07358	0.07688	0.08024	0.08368	0.08718	0.09076	0.09439	0.09809	0.10185	0.10567	0.10955	0.11348	0.11746
21	0.07128	0.07460	0.07800	0.08146	0.08500	0.08861	0.09229	0.09603	0.09983	0.10370	0.10762	0.11159	0.11562
22	0.06920	0.07255	0.07597	0.07947	0.08305	0.08669	0.09041	0.09419	0.09803	0.10194	0.10590	0.10993	0.11401
23	0.06731	0.07068	0.07414	0.07767	0.08128	0.08496	0.08871	0.09254	0.09642	0.10037	0.10438	0.10845	0.11257
24	0.06559	0.06899	0.07247	0.07604	0.07968	0.08340	0.08719	0.09105	0.09498	0.09897	0.10302	0.10713	0.11130
25	0.06401	0.06744	0.07095	0.07455	0.07823	0.08198	0.08581	0.08971	0.09368	0.09771	0.10181	0.10596	0.11017
30	0.05783	0.06139	0.06505	0.06881	0.07265	0.07658	0.08059	0.08467	0.08883	0.09305	0.09734	0.10168	0.10608
35	0.05358	0.05727	0.06107	0.06497	0.06897	0.07306	0.07723	0.08148	0.08580	0.09019	0.09464	0.09914	0.10369
40	0.05052	0.05434	0.05828	0.06232	0.06646	0.07069	0.07501	0.07940	0.08386	0.08838	0.09296	0.09759	0.10226

TABLE 1 | Amortization Factors for Equal Total Payments (Continued)

Interest Rate

Years	10.5%	11%	11.5%	12%	12.5%	13%	13.5%	14%	14.5%	15%	15.5%	16%	16.5%
1	1.10500	1.11000	1.11500	1.12000	1.12500	1.13000	1.13500	1.14000	1.14500	1.15000	1.15500	1.16000	1.16500
2	0.58006	0.58393	0.58781	0.59170	0.59559	0.59948	0.60338	0.60729	0.61120	0.61512	0.61904	0.62296	0.62689
3	0.40566	0.40921	0.41278	0.41635	0.41993	0.42352	0.42712	0.43073	0.43435	0.43798	0.44161	0.44526	0.44891
4	0.31889	0.32233	0.32577	0.32923	0.33271	0.33619	0.33969	0.34320	0.34673	0.35027	0.35381	0.35738	0.36095
5	0.26718	0.27057	0.27398	0.27741	0.28085	0.28431	0.28779	0.29128	0.29479	0.29832	0.30185	0.30541	0.30898
6	0.23298	0.23638	0.23979	0.24323	0.24668	0.25015	0.25365	0.25716	0.26069	0.26424	0.26780	0.27139	0.27499
7	0.20880	0.21222	0.21566	0.21912	0.22260	0.22611	0.22964	0.23319	0.23677	0.24036	0.24398	0.24761	0.25127
8	0.19087	0.19432	0.19780	0.20130	0.20483	0.20839	0.21197	0.21557	0.21920	0.22285	0.22653	0.23022	0.23395
9	0.17711	0.18060	0.18413	0.18768	0.19126	0.19487	0.19851	0.20217	0.20586	0.20957	0.21332	0.21708	0.22087
10	0.16626	0.16980	0.17338	0.17698	0.18062	0.18429	0.18799	0.19171	0.19547	0.19925	0.20306	0.20690	0.21077
11	0.15752	0.16112	0.16475	0.16842	0.17211	0.17584	0.17960	0.18339	0.18722	0.19107	0.19495	0.19886	0.20280
12	0.15038	0.15403	0.15771	0.16144	0.16519	0.16899	0.17281	0.17667	0.18056	0.18448	0.18843	0.19241	0.19643
13	0.14445	0.14815	0.15190	0.15568	0.15950	0.16335	0.16724	0.17116	0.17512	0.17911	0.18313	0.18718	0.19127
14	0.13947	0.14323	0.14703	0.15087	0.15475	0.15867	0.16262	0.16661	0.17063	0.17469	0.17878	0.18290	0.18705
15	0.13525	0.13907	0.14292	0.14682	0.15076	0.15474	0.15876	0.16281	0.16690	0.17102	0.17517	0.17936	0.18357
16	0.13164	0.13552	0.13943	0.14339	0.14739	0.15143	0.15550	0.15962	0.16376	0.16795	0.17216	0.17641	0.18069
17	0.12854	0.13247	0.13644	0.14046	0.14451	0.14861	0.15274	0.15692	0.16112	0.16537	0.16964	0.17395	0.17829
18	0.12586	0.12984	0.13387	0.13794	0.14205	0.14620	0.15039	0.15462	0.15889	0.16319	0.16752	0.17188	0.17628
19	0.12353	0.12756	0.13164	0.13576	0.13993	0.14413	0.14838	0.15266	0.15698	0.16134	0.16572	0.17014	0.17459
20	0.12149	0.12558	0.12970	0.13388	0.13810	0.14235	0.14665	0.15099	0.15536	0.15976	0.16420	0.16867	0.17316
21	0.11971	0.12384	0.12802	0.13224	0.13651	0.14081	0.14516	0.14954	0.15396	0.15842	0.16290	0.16742	0.17196
22	0.11813	0.12231	0.12654	0.13081	0.13512	0.13948	0.14387	0.14830	0.15277	0.15727	0.16179	0.16635	0.17094
23	0.11675	0.12097	0.12524	0.12956	0.13392	0.13832	0.14276	0.14723	0.15174	0.15628	0.16085	0.16545	0.17007
24	0.11552	0.11979	0.12410	0.12846	0.13287	0.13731	0.14179	0.14630	0.15085	0.15543	0.16004	0.16467	0.16933
25	0.11443	0.11874	0.12310	0.12750	0.13194	0.13643	0.14095	0.14550	0.15008	0.15470	0.15934	0.16401	0.16871
30	0.11053	0.11502	0.11956	0.12414	0.12876	0.13341	0.13809	0.14280	0.14754	0.15230	0.15708	0.16189	0.16671
35	0.10829	0.11293	0.11760	0.12232	0.12706	0.13183	0.13662	0.14144	0.14628	0.15113	0.15601	0.16089	0.16579
40	0.10697	0.11172	0.11650	0.12130	0.12613	0.13099	0.13586	0.14075	0.14565	0.15056	0.15549	0.16042	0.16537

TABLE 2 — Future Value of a $1 Investment

Interest Rate

Years	4%	4.5%	5%	5.5%	6%	6.5%	7%	7.5%	8%	8.5%	9%	9.5%	10%
1	1.0400	1.0450	1.0500	1.0550	1.0600	1.0650	1.0700	1.0750	1.0800	1.0850	1.0900	1.0950	1.1000
2	1.0816	1.0920	1.1025	1.1130	1.1236	1.1342	1.1449	1.1556	1.1664	1.1772	1.1881	1.1990	1.2100
3	1.1249	1.1412	1.1576	1.1742	1.1910	1.2079	1.2250	1.2423	1.2597	1.2773	1.2950	1.3129	1.3310
4	1.1699	1.1925	1.2155	1.2388	1.2625	1.2865	1.3108	1.3355	1.3605	1.3859	1.4116	1.4377	1.4641
5	1.2167	1.2462	1.2763	1.3070	1.3382	1.3701	1.4026	1.4356	1.4693	1.5037	1.5386	1.5742	1.6105
6	1.2653	1.3023	1.3401	1.3788	1.4185	1.4591	1.5007	1.5433	1.5869	1.6315	1.6771	1.7238	1.7716
7	1.3159	1.3609	1.4071	1.4547	1.5036	1.5540	1.6058	1.6590	1.7138	1.7701	1.8280	1.8876	1.9487
8	1.3686	1.4221	1.4775	1.5347	1.5938	1.6550	1.7182	1.7835	1.8509	1.9206	1.9926	2.0669	2.1436
9	1.4233	1.4861	1.5513	1.6191	1.6895	1.7626	1.8385	1.9172	1.9990	2.0839	2.1719	2.2632	2.3579
10	1.4802	1.5530	1.6289	1.7081	1.7908	1.8771	1.9672	2.0610	2.1589	2.2610	2.3674	2.4782	2.5937
11	1.5395	1.6229	1.7103	1.8021	1.8983	1.9992	2.1049	2.2156	2.3316	2.4532	2.5804	2.7137	2.8531
12	1.6010	1.6959	1.7959	1.9012	2.0122	2.1291	2.2522	2.3818	2.5182	2.6617	2.8127	2.9715	3.1384
13	1.6651	1.7722	1.8856	2.0058	2.1329	2.2675	2.4098	2.5604	2.7196	2.8879	3.0658	3.2537	3.4523
14	1.7317	1.8519	1.9799	2.1161	2.2609	2.4149	2.5785	2.7524	2.9372	3.1334	3.3417	3.5629	3.7975
15	1.8009	1.9353	2.0789	2.2325	2.3966	2.5718	2.7590	2.9589	3.1722	3.3997	3.6425	3.9013	4.1772
16	1.8730	2.0224	2.1829	2.3553	2.5404	2.7390	2.9522	3.1808	3.4259	3.6887	3.9703	4.2719	4.5950
17	1.9479	2.1134	2.2920	2.4848	2.6928	2.9170	3.1588	3.4194	3.7000	4.0023	4.3276	4.6778	5.0545
18	2.0258	2.2085	2.4066	2.6215	2.8543	3.1067	3.3799	3.6758	3.9960	4.3425	4.7171	5.1222	5.5599
19	2.1068	2.3079	2.5270	2.7656	3.0256	3.3086	3.6165	3.9515	4.3157	4.7116	5.1417	5.6088	6.1159
20	2.1911	2.4117	2.6533	2.9178	3.2071	3.5236	3.8697	4.2479	4.6610	5.1120	5.6044	6.1416	6.7275
21	2.2788	2.5202	2.7860	3.0782	3.3996	3.7527	4.1406	4.5664	5.0338	5.5466	6.1088	6.7251	7.4002
22	2.3699	2.6337	2.9253	3.2475	3.6035	3.9966	4.4304	4.9089	5.4365	6.0180	6.6586	7.3639	8.1403
23	2.4647	2.7522	3.0715	3.4262	3.8197	4.2564	4.7405	5.2771	5.8715	6.5296	7.2579	8.0635	8.9543
24	2.5633	2.8760	3.2251	3.6146	4.0489	4.5331	5.0724	5.6729	6.3412	7.0846	7.9111	8.8296	9.8497
25	2.6658	3.0054	3.3864	3.8134	4.2919	4.8277	5.4274	6.0983	6.8485	7.6868	8.6231	9.6684	10.8347
30	3.2434	3.7453	4.3219	4.9840	5.7435	6.6144	7.6123	8.7550	10.0627	11.5583	13.2677	15.2203	17.4494
35	3.9461	4.6673	5.5160	6.5138	7.6861	9.0623	10.6766	12.5689	14.7853	17.3796	20.4140	23.9604	28.1024
40	4.8010	5.8164	7.0400	8.5133	10.2857	12.4161	14.9745	18.0442	21.7245	26.1330	31.4094	37.7194	45.2593

TABLE 2 | Future Value of a $1 Investment (Continued)

YEARS	Interest Rate												
	10.5%	11%	11.5%	12%	12.5%	13%	13.5%	14%	14.5%	15%	15.5%	16%	16.5%
1	1.1050	1.1100	1.1150	1.1200	1.1250	1.1300	1.1350	1.1400	1.1450	1.1500	1.1550	1.1600	1.1650
2	1.2210	1.2321	1.2432	1.2544	1.2656	1.2769	1.2882	1.2996	1.3110	1.3225	1.3340	1.3456	1.3572
3	1.3492	1.3676	1.3862	1.4049	1.4238	1.4429	1.4621	1.4815	1.5011	1.5209	1.5408	1.5609	1.5812
4	1.4909	1.5181	1.5456	1.5735	1.6018	1.6305	1.6595	1.6890	1.7188	1.7490	1.7796	1.8106	1.8421
5	1.6474	1.6851	1.7234	1.7623	1.8020	1.8424	1.8836	1.9254	1.9680	2.0114	2.0555	2.1003	2.1460
6	1.8204	1.8704	1.9215	1.9738	2.0273	2.0820	2.1378	2.1950	2.2534	2.3131	2.3741	2.4364	2.5001
7	2.0116	2.0762	2.1425	2.2107	2.2807	2.3526	2.4264	2.5023	2.5801	2.6600	2.7420	2.8262	2.9126
8	2.2228	2.3045	2.3889	2.4760	2.5658	2.6584	2.7540	2.8526	2.9542	3.0590	3.1671	3.2784	3.3932
9	2.4562	2.5580	2.6636	2.7731	2.8865	3.0040	3.1258	3.2519	3.3826	3.5179	3.6580	3.8030	3.9531
10	2.7141	2.8394	2.9699	3.1058	3.2473	3.3946	3.5478	3.7072	3.8731	4.0456	4.2249	4.4114	4.6053
11	2.9991	3.1518	3.3115	3.4785	3.6532	3.8359	4.0267	4.2262	4.4347	4.6524	4.8798	5.1173	5.3652
12	3.3140	3.4985	3.6923	3.8960	4.1099	4.3345	4.5704	4.8179	5.0777	5.3503	5.6362	5.9360	6.2504
13	3.6619	3.8833	4.1169	4.3635	4.6236	4.8980	5.1874	5.4924	5.8140	6.1528	6.5098	6.8858	7.2818
14	4.0464	4.3104	4.5904	4.8871	5.2016	5.5348	5.8877	6.2613	6.6570	7.0757	7.5188	7.9875	8.4833
15	4.4713	4.7846	5.1183	5.4736	5.8518	6.2543	6.6825	7.1379	7.6222	8.1371	8.6842	9.2655	9.8830
16	4.9408	5.3109	5.7069	6.1304	6.5833	7.0673	7.5846	8.1372	8.7275	9.3576	10.0302	10.7480	11.5137
17	5.4596	5.8951	6.3632	6.8660	7.4062	7.9861	8.6085	9.2765	9.9929	10.7613	11.5849	12.4677	13.4135
18	6.0328	6.5436	7.0949	7.6900	8.3319	9.0243	9.7707	10.5752	11.4419	12.3755	13.3806	14.4625	15.6267
19	6.6663	7.2633	7.9108	8.6128	9.3734	10.1974	11.0897	12.0557	13.1010	14.2318	15.4546	16.7765	18.2051
20	7.3662	8.0623	8.8206	9.6463	10.5451	11.5231	12.5869	13.7435	15.0006	16.3665	17.8501	19.4608	21.2089
21	8.1397	8.9492	9.8350	10.8038	11.8632	13.0211	14.2861	15.6676	17.1757	18.8215	20.6168	22.5745	24.7084
22	8.9944	9.9336	10.9660	12.1003	13.3461	14.7138	16.2147	17.8610	19.6662	21.6447	23.8124	26.1864	28.7853
23	9.9388	11.0263	12.2271	13.5523	15.0144	16.6266	18.4037	20.3616	22.5178	24.8915	27.5034	30.3762	33.5348
24	10.9823	12.2392	13.6332	15.1786	16.8912	18.7881	20.8882	23.2122	25.7829	28.6252	31.7664	35.2364	39.0681
25	12.1355	13.5855	15.2010	17.0001	19.0026	21.2305	23.7081	26.4619	29.5214	32.9190	36.6902	40.8742	45.5143
30	19.9926	22.8923	26.1967	29.9599	34.2433	39.1159	44.6556	50.9502	58.0985	66.2118	75.4153	85.8499	97.6737
35	32.9367	38.5749	45.1461	52.7996	61.7075	72.0685	84.1115	98.1002	114.3384	133.1755	155.0135	180.3141	209.6078
40	54.2614	65.0009	77.8027	93.0510	111.1990	132.7816	158.4289	188.8835	225.0191	267.8635	318.6246	378.7212	449.8182

TABLE 3 Future Value of a $1 Annuity

Interest Rate

Years	4%	4.5%	5%	5.5%	6%	6.5%	7%	7.5%	8%	8.5%	9%	9.5%	10%
1	1.0000	1.0000	1.0000	1.0000	1.0000	1.0000	1.0000	1.0000	1.0000	1.0000	1.0000	1.0000	1.0000
2	2.0400	2.0450	2.0500	2.0550	2.0600	2.0650	2.0700	2.0750	2.0800	2.0850	2.0900	2.0950	2.1000
3	3.1216	3.1370	3.1525	3.1680	3.1836	3.1992	3.2149	3.2306	3.2464	3.2622	3.2781	3.2940	3.3100
4	4.2465	4.2782	4.3101	4.3423	4.3746	4.4072	4.4399	4.4729	4.5061	4.5395	4.5731	4.6070	4.6410
5	5.4163	5.4707	5.5256	5.5811	5.6371	5.6936	5.7507	5.8084	5.8666	5.9254	5.9847	6.0446	6.1051
6	6.6330	6.7169	6.8019	6.8881	6.9753	7.0637	7.1533	7.2440	7.3359	7.4290	7.5233	7.6189	7.7156
7	7.8983	8.0192	8.1420	8.2669	8.3938	8.5229	8.6540	8.7873	8.9228	9.0605	9.2004	9.3426	9.4872
8	9.2142	9.3800	9.5491	9.7216	9.8975	10.0769	10.2598	10.4464	10.6366	10.8306	11.0285	11.2302	11.4359
9	10.5828	10.8021	11.0266	11.2563	11.4913	11.7319	11.9780	12.2298	12.4876	12.7512	13.0210	13.2971	13.5795
10	12.0061	12.2882	12.5779	12.8754	13.1808	13.4944	13.8164	14.1471	14.4866	14.8351	15.1929	15.5603	15.9374
11	13.4864	13.8412	14.2068	14.5835	14.9716	15.3716	15.7836	16.2081	16.6455	17.0961	17.5603	18.0385	18.5312
12	15.0258	15.4640	15.9171	16.3856	16.8699	17.3707	17.8885	18.4237	18.9771	19.5492	20.1407	20.7522	21.3843
13	16.6268	17.1599	17.7130	18.2868	18.8821	19.4998	20.1406	20.8055	21.4953	22.2109	22.9534	23.7236	24.5227
14	18.2919	18.9321	19.5986	20.2926	21.0151	21.7673	22.5505	23.3659	24.2149	25.0989	26.0192	26.9774	27.9750
15	20.0236	20.7841	21.5786	22.4087	23.2760	24.1822	25.1290	26.1184	27.1521	28.2323	29.3609	30.5402	31.7725
16	21.8245	22.7193	23.6575	24.6411	25.6725	26.7540	27.8881	29.0772	30.3243	31.6320	33.0034	34.4416	35.9497
17	23.6975	24.7417	25.8404	26.9964	28.2129	29.4930	30.8402	32.2580	33.7502	35.3207	36.9737	38.7135	40.5447
18	25.6454	26.8551	28.1324	29.4812	30.9057	32.4101	33.9990	35.6774	37.4502	39.3230	41.3013	43.3913	45.5992
19	27.6712	29.0636	30.5390	32.1027	33.7600	35.5167	37.3790	39.3532	41.4463	43.6654	46.0185	48.5135	51.1591
20	29.7781	31.3714	33.0660	34.8683	36.7856	38.8253	40.9955	43.3047	45.7620	48.3770	51.1601	54.1222	57.2750
21	31.9692	33.7831	35.7193	37.7861	39.9927	42.3490	44.8652	47.5525	50.4229	53.4891	56.7645	60.2638	64.0025
22	34.2480	36.3034	38.5052	40.8643	43.3923	46.1016	49.0057	52.1190	55.4568	59.0356	62.8733	66.9889	71.4027
23	36.6179	38.9370	41.4305	44.1118	46.9958	50.0982	53.4361	57.0279	60.8933	65.0537	69.5319	74.3529	79.5430
24	39.0826	41.6892	44.5020	47.5380	50.8156	54.3546	58.1767	62.3050	66.7648	71.5832	76.7898	82.4164	88.4973
25	41.6459	44.5652	47.7271	51.1526	54.8645	58.8877	63.2490	67.9779	73.1059	78.6678	84.7009	91.2459	98.3471
30	56.0849	61.0071	66.4388	72.4355	79.0582	86.3749	94.4608	103.3994	113.2832	124.2147	136.3075	149.6875	164.4940
35	73.6522	81.4966	90.3203	100.2514	111.4348	124.0347	138.2369	154.2516	172.3168	192.7017	215.7108	241.6885	271.0244
40	95.0255	107.0303	120.7998	136.6056	154.7620	175.6319	199.6351	227.2565	259.0565	295.6825	337.8824	386.5200	442.5926

TABLE 3 | Future Value of a $1 Annuity (Continued)

Years	10.5%	11%	11.5%	12%	12.5%	13%	13.5%	14%	14.5%	15%	15.5%	16%	16.5%
1	1.0000	1.0000	1.0000	1.0000	1.0000	1.0000	1.0000	1.0000	1.0000	1.0000	1.0000	1.0000	1.0000
2	2.1050	2.1100	2.1150	2.1200	2.1250	2.1300	2.1350	2.1400	2.1450	2.1500	2.1550	2.1600	2.1650
3	3.3260	3.3421	3.3582	3.3744	3.3906	3.4069	3.4232	3.4396	3.4560	3.4725	3.4890	3.5056	3.5222
4	4.6753	4.7097	4.7444	4.7793	4.8145	4.8498	4.8854	4.9211	4.9571	4.9934	5.0298	5.0665	5.1034
5	6.1662	6.2278	6.2900	6.3528	6.4163	6.4803	6.5449	6.6101	6.6759	6.7424	6.8094	6.8771	6.9455
6	7.8136	7.9129	8.0134	8.1152	8.2183	8.3227	8.4284	8.5355	8.6439	8.7537	8.8649	8.9775	9.0915
7	9.6340	9.7833	9.9349	10.0890	10.2456	10.4047	10.5663	10.7305	10.8973	11.0668	11.2390	11.4139	11.5915
8	11.6456	11.8594	12.0774	12.2997	12.5263	12.7573	12.9927	13.2328	13.4774	13.7268	13.9810	14.2401	14.5041
9	13.8684	14.1640	14.4663	14.7757	15.0921	15.4157	15.7468	16.0853	16.4317	16.7858	17.1481	17.5185	17.8973
10	16.3246	16.7220	17.1300	17.5487	17.9786	18.4197	18.8726	19.3373	19.8142	20.3037	20.8060	21.3215	21.8504
11	19.0387	19.5614	20.0999	20.6546	21.2259	21.8143	22.4204	23.0445	23.6873	24.3493	25.0310	25.7329	26.4557
12	22.0377	22.7132	23.4114	24.1331	24.8791	25.6502	26.4471	27.2707	28.1220	29.0017	29.9108	30.8502	31.8209
13	25.3517	26.2116	27.1037	28.0291	28.9890	29.9847	31.0175	32.0887	33.1997	34.3519	35.5469	36.7862	38.0713
14	29.0136	30.0949	31.2207	32.3926	33.6126	34.8827	36.2048	37.5811	39.0136	40.5047	42.0567	43.6720	45.3531
15	33.0600	34.4054	35.8110	37.2797	38.8142	40.4175	42.0925	43.8424	45.6706	47.5804	49.5755	51.6595	53.8364
16	37.5313	39.1899	40.9293	42.7533	44.6660	46.6717	48.7750	50.9804	53.2928	55.7175	58.2597	60.9250	63.7194
17	42.4721	44.5008	46.6362	48.8837	51.2493	53.7391	56.3596	59.1176	62.0203	65.0751	68.2899	71.6730	75.2331
18	47.9317	50.3959	52.9993	55.7497	58.6554	61.7251	64.9681	68.3941	72.0132	75.8364	79.8749	84.1407	88.6465
19	53.9645	56.9395	60.0942	63.4397	66.9873	70.7494	74.7388	78.9692	83.4551	88.2118	93.2555	98.6032	104.2732
20	60.6308	64.2028	68.0051	72.0524	76.3608	80.9468	85.8286	91.0249	96.5561	102.4436	108.7101	115.3797	122.4783
21	67.9970	72.2651	76.8257	81.6987	86.9058	92.4699	98.4154	104.7684	111.5568	118.8101	126.5601	134.8405	143.6872
22	76.1367	81.2143	86.6606	92.5026	98.7691	105.4910	112.7015	120.4360	128.7325	137.6316	147.1769	157.4150	168.3956
23	85.1311	91.1479	97.6266	104.6029	112.1152	120.2048	128.9162	138.2970	148.3987	159.2764	170.9894	183.6014	197.1809
24	95.0699	102.1742	109.8536	118.1552	127.1296	136.8315	147.3199	158.6586	170.9165	184.1678	198.4927	213.9776	230.7157
25	106.0522	114.4133	123.4868	133.3339	144.0208	155.6196	168.2081	181.8708	196.6994	212.7930	230.2591	249.2140	269.7838
30	180.8815	199.0209	219.1014	241.3327	265.9464	293.1992	323.3748	356.7868	393.7825	434.7451	480.0988	530.3117	585.9014
35	304.1588	341.5896	383.8792	431.6635	485.6604	546.6808	615.6404	693.5727	781.6440	881.1702	993.6353	1120.7130	1264.2895
40	507.2516	581.8261	667.8496	767.0914	881.5920	1013.7042	1166.1401	1342.0251	1544.9596	1779.0903	2049.1913	2360.7572	2720.1102

423

TABLE 4 | **Present Value of a $1 Lump Sum**

Interest Rate

Years	4%	4.5%	5%	5.5%	6%	6.5%	7%	7.5%	8%	8.5%	9%	9.5%	10%
1	0.96154	0.95694	0.95238	0.94787	0.94340	0.93897	0.93458	0.93023	0.92593	0.92166	0.91743	0.91324	0.90909
2	0.92456	0.91573	0.90703	0.89845	0.89000	0.88166	0.87344	0.86533	0.85734	0.84946	0.84168	0.83401	0.82645
3	0.88900	0.87630	0.86384	0.85161	0.83962	0.82785	0.81630	0.80496	0.79383	0.78291	0.77218	0.76165	0.75131
4	0.85480	0.83856	0.82270	0.80722	0.79209	0.77732	0.76290	0.74880	0.73503	0.72157	0.70843	0.69557	0.68301
5	0.82193	0.80245	0.78353	0.76513	0.74726	0.72988	0.71299	0.69656	0.68058	0.66505	0.64993	0.63523	0.62092
6	0.79031	0.76790	0.74622	0.72525	0.70496	0.68533	0.66634	0.64796	0.63017	0.61295	0.59627	0.58012	0.56447
7	0.75992	0.73483	0.71068	0.68744	0.66506	0.64351	0.62275	0.60275	0.58349	0.56493	0.54703	0.52979	0.51316
8	0.73069	0.70319	0.67684	0.65160	0.62741	0.60423	0.58201	0.56070	0.54027	0.52067	0.50187	0.48382	0.46651
9	0.70259	0.67290	0.64461	0.61763	0.59190	0.56735	0.54393	0.52158	0.50025	0.47988	0.46043	0.44185	0.42410
10	0.67556	0.64393	0.61391	0.58543	0.55839	0.53273	0.50835	0.48519	0.46319	0.44229	0.42241	0.40351	0.38554
11	0.64958	0.61620	0.58468	0.55491	0.52679	0.50021	0.47509	0.45134	0.42888	0.40764	0.38753	0.36851	0.35049
12	0.62460	0.58966	0.55684	0.52598	0.49697	0.46968	0.44401	0.41985	0.39711	0.37570	0.35553	0.33654	0.31863
13	0.60057	0.56427	0.53032	0.49856	0.46884	0.44102	0.41496	0.39056	0.36770	0.34627	0.32618	0.30734	0.28966
14	0.57748	0.53997	0.50507	0.47257	0.44230	0.41410	0.38782	0.36331	0.34046	0.31914	0.29925	0.28067	0.26333
15	0.55526	0.51672	0.48102	0.44793	0.41727	0.38883	0.36245	0.33797	0.31524	0.29414	0.27454	0.25632	0.23939
16	0.53391	0.49447	0.45811	0.42458	0.39365	0.36510	0.33873	0.31439	0.29189	0.27110	0.25187	0.23409	0.21763
17	0.51337	0.47318	0.43630	0.40245	0.37136	0.34281	0.31657	0.29245	0.27027	0.24986	0.23107	0.21378	0.19784
18	0.49363	0.45280	0.41552	0.38147	0.35034	0.32189	0.29586	0.27205	0.25025	0.23028	0.21199	0.19523	0.17986
19	0.47464	0.43330	0.39573	0.36158	0.33051	0.30224	0.27651	0.25307	0.23171	0.21224	0.19449	0.17829	0.16351
20	0.45639	0.41464	0.37689	0.34273	0.31180	0.28380	0.25842	0.23541	0.21455	0.19562	0.17843	0.16282	0.14864
21	0.43883	0.39679	0.35894	0.32486	0.29416	0.26648	0.24151	0.21899	0.19866	0.18029	0.16370	0.14870	0.13513
22	0.42196	0.37970	0.34185	0.30793	0.27751	0.25021	0.22571	0.20371	0.18394	0.16617	0.15018	0.13580	0.12285
23	0.40573	0.36335	0.32557	0.29187	0.26180	0.23494	0.21095	0.18950	0.17032	0.15315	0.13778	0.12402	0.11168
24	0.39012	0.34770	0.31007	0.27666	0.24698	0.22060	0.19715	0.17628	0.15770	0.14115	0.12640	0.11326	0.10153
25	0.37512	0.33273	0.29530	0.26223	0.23300	0.20714	0.18425	0.16398	0.14602	0.13009	0.11597	0.10343	0.09230
30	0.30832	0.26700	0.23138	0.20064	0.17411	0.15119	0.13137	0.11422	0.09938	0.08652	0.07537	0.06570	0.05731
35	0.25342	0.21425	0.18129	0.15352	0.13011	0.11035	0.09366	0.07956	0.06763	0.05754	0.04899	0.04174	0.03558
40	0.20829	0.17193	0.14205	0.11746	0.09722	0.08054	0.06678	0.05542	0.04603	0.03827	0.03184	0.02651	0.02209

TABLE 4 Present Value of a $1 Lump Sum   (Continued)

Interest Rate

Years	10.5%	11%	11.5%	12%	12.5%	13%	13.5%	14%	14.5%	15%	15.5%	16%	16.5%
1	0.90498	0.90090	0.89686	0.89286	0.88889	0.88496	0.88106	0.87719	0.87336	0.86957	0.86580	0.86207	0.85837
2	0.81898	0.81162	0.80436	0.79719	0.79012	0.78315	0.77626	0.76947	0.76276	0.75614	0.74961	0.74316	0.73680
3	0.74116	0.73119	0.72140	0.71178	0.70233	0.69305	0.68393	0.67497	0.66617	0.65752	0.64901	0.64066	0.63244
4	0.67073	0.65873	0.64699	0.63552	0.62430	0.61332	0.60258	0.59208	0.58181	0.57175	0.56192	0.55229	0.54287
5	0.60700	0.59345	0.58026	0.56743	0.55493	0.54276	0.53091	0.51937	0.50813	0.49718	0.48651	0.47611	0.46598
6	0.54932	0.53464	0.52042	0.50663	0.49327	0.48032	0.46776	0.45559	0.44378	0.43233	0.42122	0.41044	0.39999
7	0.49712	0.48166	0.46674	0.45235	0.43846	0.42506	0.41213	0.39964	0.38758	0.37594	0.36469	0.35383	0.34334
8	0.44989	0.43393	0.41860	0.40388	0.38974	0.37616	0.36311	0.35056	0.33850	0.32690	0.31575	0.30503	0.29471
9	0.40714	0.39092	0.37543	0.36061	0.34644	0.33288	0.31992	0.30751	0.29563	0.28426	0.27338	0.26295	0.25297
10	0.36845	0.35218	0.33671	0.32197	0.30795	0.29459	0.28187	0.26974	0.25819	0.24718	0.23669	0.22668	0.21714
11	0.33344	0.31728	0.30198	0.28748	0.27373	0.26070	0.24834	0.23662	0.22550	0.21494	0.20493	0.19542	0.18639
12	0.30175	0.28584	0.27083	0.25668	0.24332	0.23071	0.21880	0.20756	0.19694	0.18691	0.17743	0.16846	0.15999
13	0.27308	0.25751	0.24290	0.22917	0.21628	0.20416	0.19278	0.18207	0.17200	0.16253	0.15362	0.14523	0.13733
14	0.24713	0.23199	0.21785	0.20462	0.19225	0.18068	0.16985	0.15971	0.15022	0.14133	0.13300	0.12520	0.11788
15	0.22365	0.20900	0.19538	0.18270	0.17089	0.15989	0.14964	0.14010	0.13120	0.12289	0.11515	0.10793	0.10118
16	0.20240	0.18829	0.17523	0.16312	0.15190	0.14150	0.13185	0.12289	0.11458	0.10686	0.09970	0.09304	0.08685
17	0.18316	0.16963	0.15715	0.14564	0.13502	0.12522	0.11616	0.10780	0.10007	0.09293	0.08632	0.08021	0.07455
18	0.16576	0.15282	0.14095	0.13004	0.12002	0.11081	0.10235	0.09456	0.08740	0.08081	0.07474	0.06914	0.06399
19	0.15001	0.13768	0.12641	0.11611	0.10668	0.09806	0.09017	0.08295	0.07633	0.07027	0.06471	0.05961	0.05493
20	0.13575	0.12403	0.11337	0.10367	0.09483	0.08678	0.07945	0.07276	0.06666	0.06110	0.05602	0.05139	0.04715
21	0.12285	0.11174	0.10168	0.09256	0.08429	0.07680	0.07000	0.06383	0.05822	0.05313	0.04850	0.04430	0.04047
22	0.11118	0.10067	0.09119	0.08264	0.07493	0.06796	0.06167	0.05599	0.05085	0.04620	0.04199	0.03819	0.03474
23	0.10062	0.09069	0.08179	0.07379	0.06660	0.06014	0.05434	0.04911	0.04441	0.04017	0.03636	0.03292	0.02982
24	0.09106	0.08170	0.07335	0.06588	0.05920	0.05323	0.04787	0.04308	0.03879	0.03493	0.03148	0.02838	0.02560
25	0.08240	0.07361	0.06579	0.05882	0.05262	0.04710	0.04218	0.03779	0.03387	0.03038	0.02726	0.02447	0.02197
30	0.05002	0.04368	0.03817	0.03338	0.02920	0.02557	0.02239	0.01963	0.01721	0.01510	0.01326	0.01165	0.01024
35	0.03036	0.02592	0.02215	0.01894	0.01621	0.01388	0.01189	0.01019	0.00875	0.00751	0.00645	0.00555	0.00477
40	0.01843	0.01538	0.01285	0.01075	0.00899	0.00753	0.00631	0.00529	0.00444	0.00373	0.00314	0.00264	0.00222

TABLE 5 | Present Value of a $1 Annuity

Interest Rate

Years	4%	4.5%	5%	5.5%	6%	6.5%	7%	7.5%	8%	8.5%	9%	9.5%	10%
1	0.9615	0.9569	0.9524	0.9479	0.9434	0.9390	0.9346	0.9302	0.9259	0.9217	0.9174	0.9132	0.9091
2	1.8861	1.8727	1.8594	1.8463	1.8334	1.8206	1.8080	1.7956	1.7833	1.7711	1.7591	1.7473	1.7355
3	2.7751	2.7490	2.7232	2.6979	2.6730	2.6485	2.6243	2.6005	2.5771	2.5540	2.5313	2.5089	2.4869
4	3.6299	3.5875	3.5460	3.5052	3.4651	3.4258	3.3872	3.3493	3.3121	3.2756	3.2397	3.2045	3.1699
5	4.4518	4.3900	4.3295	4.2703	4.2124	4.1557	4.1002	4.0459	3.9927	3.9406	3.8897	3.8397	3.7908
6	5.2421	5.1579	5.0757	4.9955	4.9173	4.8410	4.7665	4.6938	4.6229	4.5536	4.4859	4.4198	4.3553
7	6.0021	5.8927	5.7864	5.6830	5.5824	5.4845	5.3893	5.2966	5.2064	5.1185	5.0330	4.9496	4.8684
8	6.7327	6.5959	6.4632	6.3346	6.2098	6.0888	5.9713	5.8573	5.7466	5.6392	5.5348	5.4334	5.3349
9	7.4353	7.2688	7.1078	6.9522	6.8017	6.6561	6.5152	6.3789	6.2469	6.1191	5.9952	5.8753	5.7590
10	8.1109	7.9127	7.7217	7.5376	7.3601	7.1888	7.0236	6.8641	6.7101	6.5613	6.4177	6.2788	6.1446
11	8.7605	8.5289	8.3064	8.0925	7.8869	7.6890	7.4987	7.3154	7.1390	6.9690	6.8052	6.6473	6.4951
12	9.3851	9.1186	8.8633	8.6185	8.3838	8.1587	7.9427	7.7353	7.5361	7.3447	7.1607	6.9838	6.8137
13	9.9856	9.6829	9.3936	9.1171	8.8527	8.5997	8.3577	8.1258	7.9038	7.6910	7.4869	7.2912	7.1034
14	10.5631	10.2228	9.8986	9.5896	9.2950	9.0138	8.7455	8.4892	8.2442	8.0101	7.7862	7.5719	7.3667
15	11.1184	10.7395	10.3797	10.0376	9.7122	9.4027	9.1079	8.8271	8.5595	8.3042	8.0607	7.8282	7.6061
16	11.6523	11.2340	10.8378	10.4622	10.1059	9.7678	9.4466	9.1415	8.8514	8.5753	8.3126	8.0623	7.8237
17	12.1657	11.7072	11.2741	10.8646	10.4773	10.1106	9.7632	9.4340	9.1216	8.8252	8.5436	8.2760	8.0216
18	12.6593	12.1600	11.6896	11.2461	10.8276	10.4325	10.0591	9.7060	9.3719	9.0555	8.7556	8.4713	8.2014
19	13.1339	12.5933	12.0853	11.6077	11.1581	10.7347	10.3356	9.9591	9.6036	9.2677	8.9501	8.6496	8.3649
20	13.5903	13.0079	12.4622	11.9504	11.4699	11.0185	10.5940	10.1945	9.8181	9.4633	9.1285	8.8124	8.5136
21	14.0292	13.4047	12.8212	12.2752	11.7641	11.2850	10.8355	10.4135	10.0168	9.6436	9.2922	8.9611	8.6487
22	14.4511	13.7844	13.1630	12.5832	12.0416	11.5352	11.0612	10.6172	10.2007	9.8098	9.4424	9.0969	8.7715
23	14.8568	14.1478	13.4886	12.8750	12.3034	11.7701	11.2722	10.8067	10.3711	9.9629	9.5802	9.2209	8.8832
24	15.2470	14.4955	13.7986	13.1517	12.5504	11.9907	11.4693	10.9830	10.5288	10.1041	9.7066	9.3341	8.9847
25	15.6221	14.8282	14.0939	13.4139	12.7834	12.1979	11.6536	11.1469	10.6748	10.2342	9.8226	9.4376	9.0770
30	17.2920	16.2889	15.3725	14.5337	13.7648	13.0587	12.4090	11.8104	11.2578	10.7468	10.2737	9.8347	9.4269
35	18.6646	17.4610	16.3742	15.3906	14.4982	13.6870	12.9477	12.2725	11.6546	11.0878	10.5668	10.0870	9.6442
40	19.7928	18.4016	17.1591	16.0461	15.0463	14.1455	13.3317	12.5944	11.9246	11.3145	10.7574	10.2472	9.7791

TABLE 5 | **Present Value of a $1 Annuity** (Continued)

Interest Rate

Years	10.5%	11%	11.5%	12%	12.5%	13%	13.5%	14%	14.5%	15%	15.5%	16%	16.5%
1	0.9050	0.9009	0.8969	0.8929	0.8889	0.8850	0.8811	0.8772	0.8734	0.8696	0.8658	0.8621	0.8584
2	1.7240	1.7125	1.7012	1.6901	1.6790	1.6681	1.6573	1.6467	1.6361	1.6257	1.6154	1.6052	1.5952
3	2.4651	2.4437	2.4226	2.4018	2.3813	2.3612	2.3413	2.3216	2.3023	2.2832	2.2644	2.2459	2.2276
4	3.1359	3.1024	3.0696	3.0373	3.0056	2.9745	2.9438	2.9137	2.8841	2.8550	2.8263	2.7982	2.7705
5	3.7429	3.6959	3.6499	3.6048	3.5606	3.5172	3.4747	3.4331	3.3922	3.3522	3.3129	3.2743	3.2365
6	4.2922	4.2305	4.1703	4.1114	4.0538	3.9975	3.9425	3.8887	3.8360	3.7845	3.7341	3.6847	3.6365
7	4.7893	4.7122	4.6370	4.5638	4.4923	4.4226	4.3546	4.2883	4.2236	4.1604	4.0988	4.0386	3.9798
8	5.2392	5.1461	5.0556	4.9676	4.8820	4.7988	4.7177	4.6389	4.5621	4.4873	4.4145	4.3436	4.2745
9	5.6463	5.5370	5.4311	5.3282	5.2285	5.1317	5.0377	4.9464	4.8577	4.7716	4.6879	4.6065	4.5275
10	6.0148	5.8892	5.7678	5.6502	5.5364	5.4262	5.3195	5.2161	5.1159	5.0188	4.9246	4.8332	4.7446
11	6.3482	6.2065	6.0697	5.9377	5.8102	5.6869	5.5679	5.4527	5.3414	5.2337	5.1295	5.0286	4.9310
12	6.6500	6.4924	6.3406	6.1944	6.0535	5.9176	5.7867	5.6603	5.5383	5.4206	5.3069	5.1971	5.0910
13	6.9230	6.7499	6.5835	6.4235	6.2698	6.1218	5.9794	5.8424	5.7103	5.5831	5.4605	5.3423	5.2283
14	7.1702	6.9819	6.8013	6.6282	6.4620	6.3025	6.1493	6.0021	5.8606	5.7245	5.5935	5.4675	5.3462
15	7.3938	7.1909	6.9967	6.8109	6.6329	6.4624	6.2989	6.1422	5.9918	5.8474	5.7087	5.5755	5.4474
16	7.5962	7.3792	7.1719	6.9740	6.7848	6.6039	6.4308	6.2651	6.1063	5.9542	5.8084	5.6685	5.5342
17	7.7794	7.5488	7.3291	7.1196	6.9198	6.7291	6.5469	6.3729	6.2064	6.0472	5.8947	5.7487	5.6088
18	7.9451	7.7016	7.4700	7.2497	7.0398	6.8399	6.6493	6.4674	6.2938	6.1280	5.9695	5.8178	5.6728
19	8.0952	7.8393	7.5964	7.3658	7.1465	6.9380	6.7395	6.5504	6.3701	6.1982	6.0342	5.8775	5.7277
20	8.2309	7.9633	7.7098	7.4694	7.2414	7.0248	6.8189	6.6231	6.4368	6.2593	6.0902	5.9288	5.7748
21	8.3538	8.0751	7.8115	7.5620	7.3256	7.1016	6.8889	6.6870	6.4950	6.3125	6.1387	5.9731	5.8153
22	8.4649	8.1757	7.9027	7.6446	7.4006	7.1695	6.9506	6.7429	6.5459	6.3587	6.1807	6.0113	5.8501
23	8.5656	8.2664	7.9845	7.7184	7.4672	7.2297	7.0049	6.7921	6.5903	6.3988	6.2170	6.0442	5.8799
24	8.6566	8.3481	8.0578	7.7843	7.5264	7.2829	7.0528	6.8351	6.6291	6.4338	6.2485	6.0726	5.9055
25	8.7390	8.4217	8.1236	7.8431	7.5790	7.3300	7.0950	6.8729	6.6629	6.4641	6.2758	6.0971	5.9274
30	9.0474	8.6938	8.3637	8.0552	7.7664	7.4957	7.2415	7.0027	6.7778	6.5660	6.3661	6.1772	5.9986
35	9.2347	8.8552	8.5030	8.1755	7.8704	7.5856	7.3193	7.0700	6.8362	6.6166	6.4100	6.2153	6.0317
40	9.3483	8.9511	8.5839	8.2438	7.9281	7.6344	7.3607	7.1050	6.8659	6.6418	6.4314	6.2335	6.0471

GLOSSARY

A

Account payable   An expense that has been incurred but not yet paid.

Account receivable   Income that has been earned but for which no cash payment has been received.

Accounting   A comprehensive system for recording and summarizing business transactions.

Accounting period   The period of time over which accounting transactions are summarized.

Accounting profit   (See profit, accounting.)

Accrual accounting   An accounting system that recognizes income when it is earned and expenses when they are incurred.

Accrued expense   An expense that has been incurred, sometimes accumulating over time, but has not been paid.

Accrued liability   A liability that has been incurred but not yet paid, such as accrued interest.

Accumulated depreciation   The sum of all depreciation taken on an asset from time of purchase to the present.

Adjusted basis   The income tax basis of an asset, equal to the original basis reduced by the amount of depreciation expense claimed and/or increased by the cost of any improvements made.

Amortized loan   A loan that is scheduled to be repaid in a series of periodic payments.

Annual percentage rate (APR)   The true annual rate at which interest is charged on a loan.

Annuity   A series of equal periodic payments.

Appraisal   The process of estimating the market value of an asset.

Appreciation   An increase in the market value of an asset.

Asset   Physical or financial property that has value and is owned by a business or individual.

Average fixed cost (AFC)   Total fixed cost divided by total output; average fixed cost per unit of output.

Average physical product (APP)   The average amount of physical output produced for each unit of input used; total output divided by total input.

Average total cost (ATC)   Total cost divided by total output; average cost per unit of output.

Average variable cost (AVC)   Total variable cost divided by total output; average variable cost per unit of output.

B

Balance sheet   A financial report summarizing the assets, liabilities, and equity of a business at a point in time.

Balloon payment loan   A loan amortization method in which a large portion of the principal is due with the final payment.

Bankruptcy   A legal action that a business can take when it no longer has the financial resources to pay its debts and must reorganize or go out of business.

Basis (marketing)   The difference between the local cash price and the futures contract price of the same commodity at a point in time.

Basis (tax)   The value of an asset for income tax purposes.

Bonus (wage)   A payment made to an employee, in addition to the normal salary, based on superior performance or other criteria.

Book value   The original cost of an asset minus the total accumulated depreciation expense taken to date.

Boot   When trading a used asset for a new one, the cash paid to make up for the difference in value.

Borrowing capacity   The maximum amount an individual or business can borrow based on ability to repay and other factors.

Break-even price   The selling price for which total income will just equal total expenses for a given level of production.

Break-even yield   The yield level at which total income will just equal total expenses at a given selling price.

Breeding livestock   Livestock owned for the primary purpose of producing offspring.

Budget   An estimate of future income, expenses, or cash flows.

Bushel lease   A leasing arrangement in which the rent is paid as a specified number of bushels of grain delivered to the owner.

C

C corporation   A "regular" corporation that files its own income tax return. (See also S corporation.)

Capital   A collection of physical and financial assets that have a market value.

Capital asset   An asset that is expected to last through more than one production cycle and can be used to produce other saleable assets or services.

Capital budgeting   A process for determining the profitability of a capital investment.

Capital gain   The amount by which the sale value of an asset exceeds its cost or adjusted tax basis.

Capital lease   A lease contract that allows the lessee to purchase the leased equipment at the end of the lease period.

Capital recovery   The annualized equivalent value of the initial investment cost of a capital asset.

Capitalization method   A procedure for estimating the value of an asset by dividing the expected annual net returns by an annual discount rate.

Cash accounting   An accounting system that recognizes income when it is actually received and expenses when they are actually paid.

Cash expenses   Expenses that require the expenditure of cash.

Cash flow   The movement of cash funds into and out of a business.

Cash flow budget   A projection of the expected cash inflows and cash outflows for a business over a period of time.

Cash rent lease   A rental arrangement in which the operator makes a cash payment to the owner for the use of certain property, pays all production costs, and keeps all the income generated.

Closed cooperative   A farmer cooperative to which members agree to sell a fixed amount of production on a regular schedule.

Coefficient of variation   A measure of the variability of the outcomes of a particular event; equal to the standard deviation divided by the mean.

Collateral   Assets pledged as security for a loan.

Commodity Credit Corporation (CCC)   A corporation owned by the U.S. Department of Agriculture. Its primary purpose is to support agricultural prices through the use of commodity loans.

Comparable sale   An actual land sale used in an appraisal to help estimate the market value of a similar piece of land.

Comparative analysis   The comparison of the performance level of a farm business to the performance level of other similar farms in the same area, or to other established standards.

Competitive enterprises   Enterprises for which the output level of one can be increased only by decreasing the output level of the other.

Complementary enterprises   Enterprises for which increasing the output level of one also increases the output level of the other.

Compounding   The process of determining the future value of an investment or loan, in which interest is charged on the accumulated interest as well as the original capital.

Compound interest   The reinvestment of each interest payment so that it becomes part of the principal which earns interest in future time periods.

Contingent liabilities   Liabilities which will come into existence only if some specific event should occur. An example would be income taxes due should an asset such as land be sold.

Contributed capital   Capital invested in a business by its owner(s), other than earnings produced by and retained in that business.

Control   The process of monitoring the progress of a farm business and taking corrective action when desired performance levels are not being met.

Cooperative   A form of business organization in which profits are distributed as patronage refunds and all members have a single vote.

Corporation   A form of business organization in which the owners have shares in a separate legal entity that itself can own assets and borrow money.

Cost recovery   A term used to describe the system or method used to compute depreciation for income tax purposes.

Credit (financial)   The capacity or ability to borrow money.

Credit (accounting)   An entry in the right hand side of an account ledger which causes a decrease in assets or an increase in liabilities and equity.

Creditor   Someone to whom a debt is owed, a lender.

Crop share lease   A lease agreement in which crop production and certain input costs are divided between the operator and the land owner.

Cumulative distribution function (CDF)   A graph of all the possible outcomes for a certain event, and the probability that each outcome, or one with a lower value, will occur.

Current assets   Assets that are normally used up or sold within a year.

Current liabilities   Liabilities that are normally paid within a year.

Current ratio   The ratio of current assets to current liabilities; a measure of liquidity.

Custom farming   An arrangement in which the land owner pays the operator a fixed cash amount to perform all the labor and machinery operations needed to produce and harvest a crop.

Custom work   An arrangement in which an operator performs one or more machinery operations for someone else for a fixed charge.

Cwt.   An abbreviation for hundredweight, equal to 100 pounds. Many livestock products and some crops are priced by this unit.

D

Debit   In accounting, an entry on the left hand side of a ledger which will increase assets or decrease liabilities and equity.

Debt   An obligation to pay, such as a loan or account payable.

Debt/asset ratio   The ratio of total liabilities to total assets; a measure of solvency.

Debt/equity ratio   The ratio of total liabilities to owner's equity; a measure of solvency.

Debt service   The payment of debts according to a specified schedule.

Decision tree   A diagram that traces out all the possible strategies and outcomes for a particular decision or sequence of related decisions.

Declining balance method   A depreciation method which results in high depreciation in the early years of life and smaller amounts in the later years.

Deferred taxes   The amount by which income taxes will increase or decrease at some future time when assets and liabilities shown on a current balance sheet are sold or paid.

Deflation   A general decrease in the level of all prices.

Depreciation   An annual, noncash expense to recognize the amount by which an asset loses value due to use, age, and obsolescence. It also spreads the original cost over the asset's useful life.

Depreciation recapture   Taxable income that results from selling a depreciable asset for more than its adjusted tax basis.

Diminishing returns   A decline in the rate at which total output increases as more and more inputs are used; a declining marginal physical product.

Discount rate   The interest rate used to find the present value of an amount to be paid or received in the future.

Discounted cash flow   The present value of a series of net cash flows to be received over time. Often used in investment analysis.

Discounting   The process of reducing the value of a sum to be paid or received in the future by the amount of interest that would be accumulated on it to that point in time.

Diseconomies of size   A production relation in which the average total cost per unit of output increases as more output is produced.

Diversification   The production of two or more commodities for which production levels and/or prices are not closely correlated.

Double-entry accounting   An accounting system in which changes in assets and liabilities, as well as income and expenses, are recorded for each transaction.

Down payment   The portion of the cost of purchasing a capital asset that is financed from owner's equity usually in the form of cash.

E

Economic efficiency   The ratio of the value of output per physical unit of input or per unit cost of the input.

Economic profit   (See profit, economic.)

Economies of size   A production relation in which average total cost per unit of output decreases as output increases.

Efficiency   A ratio showing the number of units of production generated per unit of resource.

Enterprise   An individual crop or type of livestock, such as wheat, dairy, or lettuce. A farm's production plan will often consist of several enterprises.

Enterprise analysis   An analysis of one individual enterprise, in which a portion of the whole-farm income and expenses are allocated to each enterprise.

Enterprise budget   A projection of all the costs and returns for a single enterprise.

Environmental audit   A thorough inspection of a tract of land to determine whether any environmental hazards exist.

Equal marginal principle   The principle that a limited resource should be allocated among competing uses in such a way that the marginal value products from the last unit in each use are equal.

Equity   The amount by which the value of total assets exceeds total liabilities; the amount of the owner's own capital invested in the business.

Equity/asset ratio   The ratio of owner's equity to total assets; a measure of solvency.

Expected value   The weighted average outcome from an uncertain event, based on its possible outcomes and their respective probabilities.

Expenditure   An outlay of cash for operating or investment purposes. May also be an expense.

Expense   Costs incurred in the process of producing revenue. May be cash or noncash.

Extension Service   An educational service for farmers and others provided jointly by the U.S. Department of Agriculture, state land grant universities, and county governments.

F

Fair market value   (See market value.)

Farm (includes ranch)   The definition used by the U. S. Department of Agriculture is any business entity which sold, or would have sold in a normal year, $1,000 or more of agricultural products.

Farm Credit System (FCS)   A borrower-owned cooperative established by the authority of the U. S. Congress which makes loans to farmers and ranchers.

Farm Financial Standards Council (FFSC)   A committee of agricultural financial experts that developed a set of guidelines for uniform financial reporting and analysis of farm businesses.

Farm management   The process of making decisions about the allocation of scarce resources in agricultural production for the purpose of meeting certain management goals.

Farm Service Agency (FSA)   An agency of the U.S. Department of Agriculture that administers farm commodity and conservation programs and provides direct and guaranteed loans to farmers and ranchers.

Farming systems analysis   The study of how various agricultural enterprises and activities interact with each other.

Feasibility analysis   An analysis of the cash inflows generated by an investment compared to the cash outflows required.

Federal Insurance Contributions Act (FICA)   A federal law which created a retirement and disability program commonly called Social Security.

Feeder livestock   Young livestock that are purchased for the purpose of being fed until they reach slaughter weight.

Field efficiency   The actual accomplishment rate for a field implement as a percent of the theoretical accomplishment rate if no time were lost due to overlapping, turning, and adjusting the machine.

Financial statements   Often used as another term for balance sheet but is a general term for other documents relating to the financial condition of a business such as an income statement, statement of cash flows, etc.

Financing   The acquisition of funds to meet the cash flow requirements of an investment or production activity.

Fiscal year   An annual accounting period that does not correspond to the calendar year.

Fixed assets   Assets that are expected to have a long or indefinite productive life.

Fixed costs   Costs that will not change in the short run even if no production takes place.

Foreclosure   Legal action taken by a creditor to obtain possession of collateral whenever a borrower is unable to make loan payments.

Forward price contract   A contract between a buyer and seller that fixes the price of a commodity before it is delivered, possibly many months before delivery.

Fringe benefits   Compensation provided to employees in addition to cash wages and salary.

Future value (FV)   The value that a payment or set of payments will have at some time in the future, when interest is compounded.

Futures market   A central market where contracts for future sales of agricultural commodities are bought and sold.

G

General partnership   A partnership in which all partners are general partners; they all participate in management and have unlimited financial liability for partnership actions.

Gross income   The total income, both cash and noncash, received from an enterprise or business, before any expenses are paid.

Gross margin   The difference between gross income and variable costs; also called income above variable costs.

Gross revenue   The total of all the revenue received by a business over a period of time; same as gross income.

H

Half-year convention   A provision of the income tax depreciation system (MACRS) which allows one-half year of depreciation in the year an asset is purchased regardless of the date of purchase. May not apply if too many assets are purchased in the last quarter of the year.

Hedging   A strategy for reducing the risk of a decline in prices by selling a commodity futures contract in advance of when the actual commodity is sold.

I

Implementation   The process of carrying out management decisions.

Improvements   Renovations or additions to capital assets that improve their productivity and/or extend their useful lives.

Incentive program   Provisions in an employment contract that pays the employee a bonus for achieving certain performance levels.

Income   Economic gain resulting from the production of goods and services, including receipts from the sale of commodities, other cash payments, increases in inventories, and accounts receivable.

Income statement   A report that summarizes the income and expenses and computes the resulting profit of a business over a period of time.

Indirect costs   Costs that arise from owning an asset and are affected little by use or production. (See fixed costs and ownership costs.)

Inflation   A general increase in the level of all prices over time.

Input   A resource used in the production of an output.

Interest   The amount paid to a lender for the use of borrowed money, or the opportunity cost of investing equity capital in an alternative use.

Intermediate asset   An asset with a useful life greater than one year but less than 7–10 years. Included as a noncurrent asset when following FFSC recommendations.

Intermediate liability   A liability with an intermediate asset as collateral and payments spread over 2–10 years. Included as a noncurrent liability when following FFSC recommendations.

Internal rate of return (IRR)   The discount or interest rate at which the net present value of an investment is just equal to zero.

Internal transaction   A noncash accounting transaction carried out between two enterprises within the same business.

Inventory   A complete listing of the number, type, and value of assets owned at a point in time.

Isoquant   A line on a graph connecting points that represent all the possible combinations of inputs that can produce the same output.

J

Joint venture   Any of several forms of business operation in which more than one person is involved in ownership and management.

L

Labor share lease   A leasing agreement in which the operator receives a share of the production in exchange for contributing only labor.

Land contract   An agreement by which a land buyer makes principal and interest payments to the seller on a regular schedule.

Law of diminishing returns   A relation observed in many physical and biological production processes, in which the marginal physical product declines as more and more units of a variable input are used in combination with one or more fixed inputs.

Lease   An agreement that allows a person to use and/or possess someone else's property in exchange for a rental payment.

Lessee   A person who leases property from the owner; a tenant.

Lessor   A person who leases owned property to a lessee; a landlord.

Leverage   The practice of using credit to increase the total capital managed beyond the amount of owner equity.

Liabilities   Financial obligations (debt) that must be paid at some future time.

Lien   A creditor's legal claim against property (the collateral) to assure the payment of a debt.

Limited liability company   A form of business organization that is similar to a partnership but offers its owners the advantage of limited financial liability.

Limited partnership   A form of business in which more than one person has ownership, but some (the limited partners) do not participate in management and have liability limited to the amount of their investment.

Line of credit   An arrangement by which a lender transfers loan funds to a borrower as they are needed, up to a maximum amount.

Linear programming   A mathematical technique used to find a set of economic activities that maximizes or minimizes a certain objective, given a set of limited resources and/or other constraints.

Liquidate   To convert an asset into cash.

Liquidity   The ability of a business to meet its cash financial obligations as they come due.

Livestock share lease   A lease agreement in which both the owner and operator contribute capital and share the production of crops and livestock.

Loan repayment capacity The ability to repay loans based on collateral and revenues. Often used to describe the maximum principal and interest that can be paid in a year.

Long run That period of time which is long enough for the manager to change the amounts of all inputs or resources available for use.

Long-term liabilities Liabilities that are scheduled to be repaid over a period of ten years or longer.

Loss Occurs when expenses exceed revenue, which causes a decrease in equity.

Lumpy input A resource that can be obtained only in certain indivisible sizes, such as a tractor or a full-time employee.

M

Marginal cost (MC) The additional cost incurred from producing an additional unit of output.

Marginal input cost (MIC) The additional cost incurred by using an additional unit of input.

Marginal physical product (MPP) The additional physical product resulting from the use of an additional unit of input.

Marginal revenue (MR) The additional income received from selling one additional unit of output.

Marginal tax rate The additional tax that results from an additional dollar of taxable income at a given income level.

Marginal value product (MVP) The additional income received from using an additional unit of input.

Market livestock Animals that are fed for eventual slaughter, not for the production of offspring.

Market value The value for which an asset would be sold in an open-market transaction.

Marketable securities Stocks, bonds, and other financial instruments that can be readily and easily converted into cash.

Minimum price contract A forward price contract that guarantees the seller a minimum price but allows a higher price if the market is above the minimum when the commodity is delivered.

Mission statement A short, descriptive statement of why the farm or ranch business exists and its goals.

Modified Accelerated Cost Recovery System (MACRS) A system for calculating tax depreciation, as specified by the income tax regulations.

Mortgage A legal agreement by which a lender receives the right to acquire a borrower's property to satisfy a debt if the repayment schedule is not met.

N

Natural Resources Conservation Service (NRCS) An agency of the U.S. Department of Agriculture that provides technical and financial assistance for carrying out soil and water conservation practices.

Net farm income The difference between total revenue and total expenses, including gain or loss on the sale of all capital assets; also the return to owner equity, unpaid labor, and management.

Net farm income from operations The difference between total revenue and total expenses, not including gain or loss on the sale of certain capital assets.

Net operating loss (NOL) A negative net farm profit for income tax purposes, which can be used to offset past and/or future taxable income.

Net present value (NPV) The present value of the net cash flows that will result from an investment, minus the amount of the original investment.

Net worth The difference between the value of the assets owned by a business and the value of its liabilities. Also called owner equity.

Noncash expense An expense that does not involve the expenditure of cash, such as depreciation.

Noncurrent asset An asset that will normally be owned or used up over a period longer than a year.

Noncurrent liability A liability that will normally be paid over a period longer than a year.

Non-real estate   All assets other than land and items attached to land, such as buildings and fences.

Note payable   A liability resulting from signing a promissory note, which is a legal, written promise to repay a loan.

Note receivable   An asset resulting from lending money to a person and receiving a promissory note. (See note payable.)

O

Operating agreement   An arrangement between two or more individuals whereby they perform some of their business activities jointly while maintaining individual ownership of the resources being used.

Operating costs   Costs for the purchase of inputs and services that are used up relatively quickly, usually in one production cycle.

Operating profit margin ratio   The value represented by net farm income from operations, plus interest expense, minus opportunity cost of operator labor and management, expressed as a percentage of gross revenue.

Opportunity cost   The income that could be received by employing a resource in its most profitable alternative use.

Option   A marketing transaction in which a buyer pays a seller a premium to acquire the right to sell or buy a futures contract at a specified price.

Ordinary income   For income tax purposes, any taxable income which is not capital gain income.

Organizational chart   A diagram that shows the workers involved in a business and the lines of authority and communication among them.

Output   The result or yield from a production process, such as raising crops and livestock.

Overhead costs   Costs that are not directly related to the type and quantity of products produced; a type of fixed cost.

Owner equity   The difference between the total value of the assets of a business and the total value of its liabilities; also called net worth.

Owner withdrawals   Business assets, generally cash, transferred to the owner(s) for their personal use.

Ownership costs   Costs that result simply from owning assets, regardless of how much they are used; fixed costs.

P

Partial budget   An estimate of the changes in income and expenses that would result from carrying out a proposed change in the current farm plan.

Partnership   A form of business organization in which more than one operator owns the resources and/or provides management. (See general and limited partnerships.)

Payback period   The length of time it takes for the accumulated net returns earned from an investment to equal the original investment.

Payoff matrix   A contingency table that illustrates the possible outcomes for a particular occurrence and their respective probabilities.

Person-year equivalent   A total of 12 months of labor contributed by one or more persons.

Physical efficiency   The ratio of output received per unit of input used, all in physical units.

Precision agriculture   A production system utilizing global positioning equipment to precisely apply different levels of inputs to different locations in a field according to their individual requirements.

Prepaid expense   A payment made for an input or service prior to the accounting period in which it will be used.

Present value (PV)   The current value of a set of payments to be received or paid out over a period of time.

Price ratio (input)   The ratio of the price of the input being added to the price of the input being replaced.

Price ratio (output)   The ratio of the price of the output being gained to the price of the output being lost.

Principal   The amount borrowed, or the part of the original loan that has not yet been repaid.

Probability distribution   A set of possible outcomes to a particular event, and the probability of each occurring.

Production function   A physical or biological relation showing how much output results from using certain quantities of inputs.

Production possibility curve (PPC)   A line on a graph that connects points representing all the possible combinations of outputs that can be produced from a fixed set of resources.

Profit, accounting   Gross revenue minus total expenses where both values are computed using standard accounting principles and practices.

Profit, economic   Accounting profit less opportunity costs on all unpaid resources (generally labor, management, and equity capital) used to produce that profit.

Profitability   The degree or extent to which the value of the income derived from a set of resources exceeds their cost.

Pro forma financial statement   A financial statement which projects future financial activities and results.

Promissory note   A legal agreement that obligates a borrower to repay a loan.

Put option   A contract that gives the buyer the right to sell a futures contract for an agricultural commodity at a specified price. It is used to set a minimum selling price in advance.

R

Ranch   (See farm.)

Real estate   Land, or assets permanently attached to land.

Reduced cost   A value from the solution to a linear programming problem, which shows how much the gross margin would be reduced by forcing into the solution one unit of an enterprise not included in the optimal farm plan.

Repayment capacity   A measurement of the ability of a borrower to repay loans.

Retained farm earnings   Net income generated by a farm business that is used to increase owner equity, rather than being withdrawn to pay for living expenses, taxes, or dividends.

Return on assets (ROA)   The value represented by net farm income from operations, plus interest expense, minus the opportunity cost of operator labor and management. It is usually expressed as a percentage of the average value of total assets.

Return on equity (ROE)   The net return generated by the business before gains or losses on capital assets are realized, but after the value of unpaid labor and management is subtracted. Usually expressed as a percent of the average value of owner's equity.

Return to management   The net return generated by a business after all expenses have been paid and the opportunity costs for owner's equity and unpaid labor have been subtracted.

Revenue   (See income.)

Revenue insurance   An insurance policy that guarantees crop producers a minimum level of gross income per acre. It protects against combinations of low prices and yields.

Risk   A situation in which more than one possible outcome exists, some of which may be unfavorable.

Rule of 72   A relation used to estimate the time it will take for an investment to double in value; found by dividing 72 by the percent rate of return earned on the investment.

S

S corporation   A corporation that is taxed like a partnership; that is, all income, expenses, and capital gains are passed pro rata to the stockholders to include with their other taxable income.

Salvage value   The market value of a depreciable asset at the time it will be sold or removed from service.

Secured loan   A loan for which the borrower agrees to let the lender take possession of and sell certain assets if the repayment terms are not met.

Self-liquidating loan   A loan that will be repaid from the sale of the assets originally purchased with the loan funds.

Sensitivity analysis   A procedure for assessing the riskiness of a decision by using several possible price and/or production outcomes to budget the results, and then comparing them.

Shadow price A value obtained from a linear programming solution that shows the amount by which total gross margin would be increased if one more unit of a limiting input were available.

Short run That period of time for which at least one production input is available only in a fixed quantity.

Short-term loan A loan scheduled to be repaid in less than a year.

Signature loan A loan for which no collateral is pledged.

Single-entry accounting An accounting system in which income and expenses are recorded but changes in assets and liabilities are not.

Skill-based pay An approach to setting worker compensation based on levels of responsibility rather than specific duties.

Social security A tax on wages and self-employment income to provide retirement and disability income for individuals. (See Federal Insurance Contributions Act.)

Sole proprietorship A form of business organization in which one operator or family owns the resources and provides the management.

Solvency The degree to which the liabilities of a business are backed up by assets; the relationship between debt and equity capital.

Standard deviation A measure of the variability of possible outcomes for a particular event; equal to the square root of the variance.

Statement of cash flows A summary of the actual cash inflows and cash outflows experienced by a business during some past time period.

Statement of owner equity A financial statement showing the causes and amounts of change in owner equity for an accounting period.

Straight line depreciation A depreciation method which results in an equal amount of depreciation for each year of an asset's useful life.

Strategic management The process of charting the overall long-term course of the farm or ranch.

Subjective probability A probability based only on individual judgment and past experiences.

Substitution ratio The ratio of the amount of one input replaced to the amount of another input added, or the amount of one output lost to the amount of another output gained.

Sum-of-years digits A depreciation method which has larger depreciation in the early years of life than straight line and declines by a constant amount each year.

Sunk cost A cost that can no longer be reversed, changed, or avoided; a fixed cost.

Supplementary enterprises Enterprises for which the level of production of one can be increased without affecting the level of production of the other.

Sustainable agriculture Agricultural production practices that maximize the long-run social and economic benefits from the use of land and other agricultural resources.

Systems analysis An evaluation of individual enterprises and technologies that takes into account their interactions with other enterprises and technologies.

T

Tactical management The process of making and implementing short-term decisions that keep the farm or ranch moving toward its long-term goals.

Tangible asset Any asset that has a physical presence such as land, buildings, machinery, etc. Intangible assets include copyrights, patents, and lease rights.

Tax-free exchange A trade of one piece of farm property for another similar piece of property, such that any taxable gain is reduced or eliminated.

Technical coefficient The rate at which units of input are transformed into output.

Technology A particular system of inputs and production practices.

Tenant A farm operator who rents land, buildings, or other assets from their owner; a lessee.

Tenure The manner by which an operator gains control and use of real estate assets, such as renting or owning them.

Tillable acres Land that is or could be cultivated.

Timeliness cost Loss of revenue resulting from a lower quality or quantity of crop harvested due to late planting, harvesting, or other field operation.

Total cost (TC)   The sum of total fixed cost and total variable cost.

Total fixed cost (TFC)   The sum of all fixed costs.

Total physical product (TPP)   The quantity of output produced by a given quantity of inputs.

Total revenue (TR)   The income received from the total physical product; same as total value product.

Total value product (TVP)   Total physical product multiplied by the selling price of the product.

Total variable cost (TVC)   The sum of all variable costs.

Trend analysis   Comparison of the performance level of a farm business to the past performance of the same business.

U

Unsecured loan   A loan for which the borrower does not give the lender the right to possess certain assets if the repayment terms are not met; there is no collateral.

USDA   The United States Department of Agriculture which oversees many federal programs and policies related to agriculture including farm programs, extension, research, and food distribution programs.

Useful life   The number of years used to fully depreciate a depreciable asset. May be different than the asset's productive life.

V

Value of farm production   The market value of all crops, livestock, and other income generated by a farm business, as measured by accrual accounting, after subtracting the value of purchased livestock and feed.

Variable cash lease   A leasing arrangement in which a cash payment is made in return for the use of the owner's property, but the amount of the payment depends on the actual production and/or price received by the tenant.

Variable cost   Costs that will occur only if production actually takes place, and which tend to vary with the level of production.

Variable interest rate   An interest rate that can change during the repayment period of a loan.

Variance   A measure of the variability in the possible outcomes of a particular event.

W

Weighted average   A long-run expected outcome from an event, found by multiplying each possible outcome by its respective probability and summing the results. Also called "expected value."

Whole-farm budget   A projection of the total production, income, and expenses of a farm business for a given whole-farm plan.

Whole-farm plan   A summary of the intended kinds and size of enterprises to be carried on by a farm business.

Withdrawals   (See owner withdrawals.)

Workers' compensation insurance   An insurance plan required by law in most states, which protects employees from job-related accidents or illnesses, and sets maximum compensation limits for such occurrences.

Working capital   The difference in value between current assets and current liabilities; a measure of liquidity.

INDEX